problems
in chemistry

revised and expanded

UNDERGRADUATE CHEMISTRY

A Series of Texbooks

edited by
J. J. LAGOWSKI
Department of Chemistry
The University of Texas at Austin

Volume 1: Modern Inorganic Chemistry, *J. J. Lagowski*
Volume 2: Modern Chemical Analysis and Instrumentation, *Harold F. Walton* and *Jorge Reyes*
Volume 3: Problems in Chemistry, Revised and Expanded, *Henry O. Daley, Jr.* and *Robert F. O'Malley*

Charles Seale-Hayne Library
University of Plymouth
(01752) 588 588
LibraryandITenquiries@plymouth.ac.uk

problems
in chemistry

revised and expanded

henry o. daley, jr.
bridgewater state college
bridgewater, massachusetts

and

robert f. o'malley
boston college
chestnut hill, massachusetts

marcel dekker, inc. new york   1974

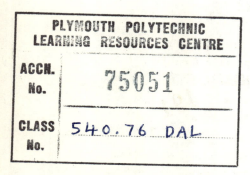
MARCEL DEKKER, INC.
305 East 45th Street, New York, New York 10017

LIBRARY OF CONGRESS CATALOG CARD NUMBER: 73-86822

ISBN: 0-8247-6107-3

Current printing (last digit):
10 9 8 7 6 5 4 3 2 1

PRINTED IN THE UNITED STATES OF AMERICA

CONTENTS

Contents

TO THE INSTRUCTOR

The purpose of a general or introductory chemistry course is multifold and varies considerably from college to college. In general it must introduce the student to the basic principles of chemistry as well as to develop a facility for working in a laboratory. The relationship between what a student does in the laboratory and what he does in the lecture part of the course should be clearly developed. Whether the course is intended for chemistry, biology, or non-science majors, students should learn that the knowledge they are acquiring in the lecture part of the course is developed or verified by laboratory experiments.

The interpretation of the results of laboratory experiments involves a chemist in one of the most basic everyday activities associated with chemistry, that is, problem solving. Chemistry is by its very nature a problem solving oriented course. All too often this aspect of chemistry is overlooked by students. Most students require practice in solving problems. The more problems a student solves, the more proficient he becomes in this aspect of chemistry.

It is the purpose of this book not only to provide the beginning chemistry student with a relatively large number of problems which illustrate chemical principles, but also to facilitate his acquisition of problem-solving skills- by helping him to see that the skills he acquires even at the introductory level are relevant to the activities of practicing chemists. The hope is that the student will come to see that he is doing what is being done, not that he is merely repeating what has been done. To this end, the problems in the present text are based, wherever possible, on experimental results obtained by chemists, and the references are cited. The problems and the techniques for solving them are intended to reflect the real features of one important aspect of chemistry.

Besides showing that problem-solving even at the introductory level is directly relevant to the practice of chemistry, there are other aims implicit in the general approach taken in this book. One of the most important is to illustrate the use of significant figures in a practical way. It is one thing to learn that significant figures are important because one must learn them in order to pass a test. It is another to learn from example that they are constantly used by chemists and therefore are important for learning chemistry. Consistent citation of references should also provide an indirect introduction to the chemical literature (the instructor, of course, also has the option of using these references to introduce students directly to the literature). Finally, it will be no small gain if the student comes to appreciate that chemical knowledge represents the accumulation of results from the activities of a great many people, each one making a contribution, however small, to the enterprise called chemistry.

A brief word concerning the use of this book. Although it is intended primarily as a self-study supplement and answers to problems are given in an appendix, it can also be used by the instructor as a source of test problems and as a supplementary text. The first 18 chapters contain about 830 problems (not counting 225 or so example problems with solutions worked out.) The problems at the end of these chapters are divided into two sections to facilitate the choosing of representative problems on the material covered in these chapters. Each section contains a complete selection of problems arranged in the same order as the material is covered in the chapter. An additional 200 problems in Chapter 19 are grouped to correspond to a standard organization of the descriptive portion of the introductory course.

In this edition of the book two new chapters have been added, one on entropy and free energy, the other on coordination chemistry.

New problems have been added to almost every chapter. In addition,
material has been added to a few chapters to reflect topics which
have been recently introduced on the freshman level such as
"parts per million" in the chapter on solutions and the Henderson-
Hasselbach equation in the chapter on acids and bases. The S. I.
system is introduced and each term is defined in the appropriate
chapter. The answers to all problems have been checked independently
by each author. Many of the problems in the text have been tested
by classroom use.

    Though there are many people who have been helpful in developing
this book I would, in particular, like to thank Dr. Vahe Marganian
of Bridgewater State College, for the many discussions we have had.
I would also like to thank my wife for the patience she has had to
type this manuscript.

H. O. D.

TO THE STUDENT

Solving problems by mathematical manipulation is an essential, daily experience of chemists. Students who seek even an elementary knowledge of chemistry must learn to solve a variety of problems. If for no other reason, then, a knowledge of the chemical principles and the mathematical manipulations necessary to solve typical chemical problems is important.

Students almost universally believe that they must understand the principles of chemistry before they can apply them to the solution of problems. Many students fail to realize that it is also necessary to remember statements of principles, definitions of terms, and many facts of a descriptive nature. The importance of a thorough understanding of the laws and theories of chemistry is not to be underestimated, but the need to commit to memory a certain body of information is equally important. A student may understand the concepts of atom, molecule, mole, equivalent weight, and so on, as he reads about them in the text, but this understanding is not enough if he cannot recall the precise definition of the terms when he attempts to apply them to the solution of a problem.

Before trying to solve problems, study the relevant sections of the text and class notes. Memorize the precise meaning of important terms and statements of principles. The first step toward the successful solution of a problem is to read the statement of the problem very carefully. Next, reflect on the meaning of the terms and determine exactly what is asked for by the problem. (This point cannot be overemphasized. Probably the greatest source of confusion in any kind of problem solving stems from the failure to determine

what is asked for. Obviously we need to "understand" the problem.
But we often forget that understanding a problem is above all a
matter of knowing what kind of answer we are looking for.) Once
you have determined both what the problem asks for and what principles
are required to provide the appropriate kind of answer, you can
proceed in an orderly fashion to perform the necessary mathematical
manipulations.

At this point the most important advice is: <u>Be neat</u>. Fewer
errors and time saved are the rewards of a neat, orderly approach
to the solution of a problem. Write as legibly as you can and leave
space between the steps.

Example
A compound containing 5.28% silicon and 94.66% iodine was prepared
at M.I.T. Calculate the empirical formula of the compound.
     [W. C. Schumb, and D. W. Breck, <u>J. Am. Chem. Soc.</u>, <u>74</u>,
     1758 (1952)]

$$\frac{5.28}{28.1} = 0.188 \text{ g atom Si}/100 \text{ g}$$

$$\frac{g}{g/\text{g atom}} = g \times \frac{g \text{ atom}}{g} = g \text{ atom}$$

$$\frac{94.7}{127} = 0.745 \text{ g atom I}/100 \text{ g}$$

$$\frac{0.745}{0.188} = 3.97$$

$$\frac{\text{g atom I}}{\text{g atom Si}}$$

$$\therefore \text{ Si I}_4 \text{ is the empirical formula}$$

If the proper procedure for solving the problem does not occur
to you, read the pertinent sections in your text and notes again.
Reconsider very carefully the definition of terms and the statements
of principles. Very often the difficulty is due to an inexact
definition or a failure to consider all that is implied in a
definition or a principle.

Although students study to acquire knowledge of a subject for
many different reasons, they must all face the practical requirements
of passing examinations. To prepare for examinations they must
learn or memorize definitions and principles, and they must practice
the application of what they have learned by answering questions
and solving problems similar to those which appear on examinations.
This book is intended to assist the student in his active preparation
for examinations in chemistry.

problems
in chemistry

revised and expanded

Chapter 1

PROPERTIES AND THEIR MEASUREMENTS

1-1 PROPERTIES

The 105 elements and the immense number of known compounds are
distinguished from one another by their properties, which are the
characteristics by which chemists identify the various elements and
compounds. Chemical properties are those which are observed when
substances are changed into totally different substances. Physical
properties can be observed without the occurrence of chemical
reactions.

1-2 EXTENSIVE AND INTENSIVE PROPERTIES

Extensive properties depend upon the particular sample present;
examples are volume, weight, and area. Intensive properties are
properties which are independent of the size and shape of a particular
sample of a substance; examples are color, odor, melting point, and
vapor pressure. One important intensive property of any substance
is its density.

1-3 DENSITY AND SPECIFIC GRAVITY

Density is defined as the mass per unit volume. It may be
expressed as

$$d = \frac{m}{V}$$

where d is density, m is mass, and V is volume.

TABLE 1-1   Densities of Selected Materials

Material	Density, g/cc
Al	2.70
Au	19.3
B	2.50
Br	3.12
Ga	6.09
Hg	13.5
Na	0.971
Pb	11.3
Pt	21.4
U	18.9
CCl_4	1.60
$SiCl_4$	1.48
$SnCl_4$	2.23
glass	2.23

A property of a system closely related to density is specific gravity. The specific gravity is defined as the ratio of the density of a substance to the density of water. In many cases the density of water is taken at $4^{\circ}C$. Since specific gravity is a ratio, it is a dimensionless quantity. For routine work the density of water at $4^{\circ}C$ is close enough to 1 g/cc so that the specific gravity and the density of a substance are the same numerical figure.

1-4   UNITS OF MEASUREMENT

Scientific knowledge is incomplete as long as it is merely descriptive or qualitative. Quantitative knowledge, expressed as

measurements, is the immediate goal of scientific investigations.
Among the fundamental units of any system of measurement, the units
of length, mass, and time are the most basic. In scientific work,
these quantities are expressed as units of the metric system.

1-5 THE METRIC SYSTEM

The fundamental units of length, mass, and time are the meter,
the kilogram, and the second, respectively. They are easy to use
because the smaller and larger multiples of each fundamental unit are
related by powers of 10.

The modern version of the metric system, developed in 1960, is
known as the International System of Units (SI). For a brief
history and use of the English and metric systems, see the National
Bureau of Standards special publication number 304A, 1970. The
defined values of the fundamental units are those adopted by the
National Bureau of Standards [Nat'l Bur. Std. (U. S.) Tech. News
Bull., 47(10)(1963) or J. Chem. Educ., 48, 569 (1971)]. This
system was established by international agreement to allow science,
industry, and commerce to work with one another.

The meter is defined as 1,650,763.73 wavelengths of the
orange-red line of the spectrum of krypton 86. The accepted
abbreviation is m. Other units of length are defined in terms of
the meter.

Kilometer (km)	= 1000 m	= 10^3 m
Centimeter (cm)	= 0.01 m	= 10^{-2} m
Millimeter (mm)	= 0.001 m	= 10^{-3} m
Micrometer (μm)*	= 0.000001 m	= 10^{-6} m
Nanometer (nm)*	= 0.000000001 m	= 10^{-9} m

---

* The terms micron and millimicron were previously used for
micrometer and nanometer, respectively. The terms are not used as
part of the SI units

Angstrom (Å)   = 0.0000000001 m   = 10^{-10} m

The unit of mass is the <u>kilogram</u> (kg), the mass of the international kilogram at Sevres, France, which is a platinum-iridium
cylinder. Smaller units are listed with their symbols.

Gram (g)   = 10^{-3} kg
Milligram (mg) = 10^{-6} kg
Microgram (μg) = 10^{-9} kg

The microgram is also called gamma (γ). Chemists refer to
solutions containing one gamma of solute per liter of solution,
for example.

The fundamental unit of time is the <u>second</u>. It is defined as
9,192,631,770 cycles of the radiation associated with a specific
transition of the cesium atom.

The fundamental unit of volume is the cubic meter (m^3).
Volumes of liquids are measured in a unit derived from this, called
the liter. The liter is defined as 0.001 m^3 or 1000 cm^3
(commonly, cc). The milliliter (ml) is of constant use in the
laboratory for the measurement of liquids. One liter contains 1000 ml

1-6 ACCURACY AND PRECISION

The accuracy of a measurement may be defined as <u>the degree of
conformity</u> <u>of</u> <u>a measure</u> <u>to a true</u> value. The <u>precision</u> of a measurement refers to its reproducibility.

The numerical value of a measurement is based in part on the
reliability of the measuring instrument and in part on the judgment
of the person making the measurement. When a length is measured by
placing a ruler adjacent to it, the reliability depends first on
the care with which the marks have been placed on the ruler. The

judgment of the user enters when he decides which of the marks lies
nearest to the length to be measured. If the length lies between
two marks, he must estimate the last digit of the numerical value of
the length. If a length is recorded as 28.6 mm, the final digit
is an estimate and is uncertain. This number has three significant
figures.

A measurement recorded as 28.62 implies that the uncertainty is
in the last digit and that a more refined technique of measurement
has been used. This number has four significant figures.

The judgment of the chemist may be involved even when more
sophisticated instruments are used. Measurements made with a
balance, a buret, a pH meter, or an infrared spectrophotometer as
well as many other instruments, may include estimates made by
the chemist.

1-7 SIGNIFICANT FIGURES

The number of significant figures in a measured quantity is
equal to the number of digits whose values are known plus the first
digit of uncertain value. The position of the decimal point is
irrelevant. A zero is significant unless it is used to indicate the
position of the decimal point (represent the order of magnitude).
A zero is never significant if there are digits only on its right,
for there it is used to indicate the position of the decimal point.
However, a zero is significant if it lies between two digits.

A problem arises when zeros have digits only on the left. They
may or may not be significant. A mass of 2000 g may have 1, 2, 3,
or 4 significant figures. The zeros may represent certain values,
or only indicate the position of the decimal point. It is good
practice to use exponential notation in such cases.

Example 1-1

Quantity	Number of Significant Figures
217	3
2.17	3
0.217	3
0.000217	3
7,043	4
70.43	4
0.07043	4
1.040	4

Example 1-2

Quantity	Number of Significant Figures
2×10^3	1
2.0×10^3	2
2.00×10^3	3

In exponential numbers the zeros to the right of the decimal point are used to represent significant figures.

1-8 SIGNIFICANT FIGURES IN CALCULATIONS

Chemists often use measured quantities from experiments in calculations and record the final results in reports and in papers submitted to journals. The final results of such calculations must contain no more significant figures than the least precise measurement. In solving a problem the student should never include more significant figures in the answer than there are in the least precise quantity in the problem. If several measured quantities are involved, they should be rounded off, if necessary, to agree with the least precise quantity.

Rounding off a number means that figures that are not significant for the purpose at hand are dropped. There are three useful rules to follow in deciding whether to drop a digit:

1. If the digit dropped is greater than 5, add 1 to the last remaining digit.

Example 1-3
         62.138 will become 62.14

2. If the digit dropped is less than 5, the last remaining digit is not changed.

Example 1-4
         28.133 will become 28.13

3. If the digit dropped is 5, the last remaining digit is left unchanged if it is even; 1 is added if it is odd.

Example 1-5
         1.8245 will become 1.824;   1.8235 will become 1.824

1-9 ADDITION AND SUBTRACTION

Before adding or subtracting, round off the numbers so that all have the same number of digits to the right of the decimal point.

Example 1-6

28.521	becomes	28.52
6.38		6.38
0.216		0.22
111.535		111.54

Example 1-7

18.406	becomes	18.406
2.12456		2.124
8.0745		8.074

Note that the numbers are not rounded off in succession, that is, 2.12456 became 2.124 not 2.125.

1-10 MULTIPLICATION AND DIVISION

Quantities should be rounded off to agree with the measurement containing the least number of significant figures.

Example 1-8

$$\frac{0.2536 \times 0.08205 \times 298.16}{0.974 \times 255.7} =$$

becomes

$$\frac{0.254 \times 0.0820 \times 298}{0.974 \times 256} =$$

1-11 UNITS IN CALCULATIONS

Solving problems in chemistry requires the use of numerical values of measurements, and the numerical values are expressed in either the fundamental units or units derived from them. The "answer" to a problem based on measurements must be expressed in the appropriate unit. To assure that the final unit is correct, it is wise to include the units in the steps of the calculations.

Example 1-9

Calculate the volume of a regular rectangular polyhedron which is 25.2 cm long, 15.3 cm wide, and 6.82 cm high.

The volume V of a rectangular polyhedron is given by the formula V = A x h, where A is the area of the base and h is the height. The area A of a rectangle is given by A = ℓ x w, where ℓ is the length and w is the width. To perform this calculation the numerical values of the dimensions are substituted for the symbols in the equations:

   V = ℓ x w x h
   V = 25.2 x 15.3 x 6.82

The units may be placed directly in the steps of the calculation along with the number:

   V = 25.2 cm x 15.3 cm x 6.82 cm
   V = 2630 cm^3 (cc)

Another approach is to multiply the units separately, below the numerical step:

   V = 25.2 x 15.3 x 6.82 = 2630
       cm x cm x cm = cm^3 (cc)

The second approach is used most frequently in this book. If the algebraic manipulation of the units leads to the unit expected in the answer, the student may be reasonably confident that his method of solution is correct. This approach provides the student with a valuable self-check. In complex problems, the inclusion of the units in the actual numerical steps may confuse rather than simplify the solution. The student is advised to use the first method for simple calculations only.

Example 1-10

Calculate the volume occupied by 261 g of copper. The mass and volume of a substance are related by its density, and the density of copper is 8.93 g/cm^3.

   d = $\frac{m}{V}$

$$V = \frac{m}{d}$$

$$V = \frac{261 \text{ g}}{8.93 \text{ g/cm}^3} = 29.2 \text{ cm}^3 \text{ (cc)}$$

or

$$V = \frac{261}{8.93} = 29.2 \text{ cm}^3$$

$$\frac{\text{g}}{\text{g/cm}^3} = \text{g} \times \frac{\text{cm}^3}{\text{g}} = \text{cm}^3 \text{ (cc)}$$

Example 1-11

What is the mass of a cylinder of silver of radius 1.63 cm and length 8.71 cm? The density of silver is 10.5 g/cm^3.

$$d = \frac{m}{V}$$

$$m = dV$$

but

$$V = \pi r^2 h$$

$$m = d\pi r^2 h$$

$$m = 10.5 \frac{\text{g}}{\text{cm}^3} \times 3.14 \times (1.63 \text{ cm})^2 \times 8.71 \text{ cm}$$

$$= 10.5 \frac{\text{g}}{\text{cm}^3} \times 3.14 \times 2.66 \text{ cm}^2 \times 8.71 \text{ cm} = 766 \text{ g}$$

or

$$m = 10.5 \times 3.14 \times 1.63^2 \times 8.71$$

$$m = 10.5 \times 3.14 \times 2.66 \times 8.71 = 766 \text{ g}$$

$$\frac{\text{g}}{\text{cm}^3} \times \text{cm}^2 \times \text{cm} = \frac{\text{g}}{\text{cm}^3} \times \text{cm}^3 = \text{g}$$

1-12 CONVERSION OF UNITS

The metric system of units is used in scientific work throughout the world. The United States is the only major country in the world

which still does not use the metric system in everyday life. However
it is not necessary to convert measurements from one system to another
frequently. When it is necessary, the most useful conversion
factors are.

    1 inch (in.) = 2.54 cm

and

    1 pound (lb) = 453.6 g

    Chemical manufacturers and distributors often list their prices
by the pound, although the use of metric units is increasing.
Dimensions of pieces of equipment may be cited in inches or feet in
catalogs and descriptive literature.

Example 1-12

What is the length in centimeters of a glass rod 7.82 in. long?

    7.82 in. x 2.54 cm/in. = 19.8 cm

or

    7.82 x 2.54 = 19.8 cm

    in. x cm/in. = cm

Example 1-13

How many grams of bromine are there in a vial that contains $\frac{1}{4}$ lb
of this substance?

    $\frac{1}{4}$ lb x 454 g/lb = 114 g

or

    $\frac{1}{4}$ x 454 = 114 g

    lb x g/lb = g

Example 1-14

A bottle containing a solution of ammonia which is 30% ammonia by
weight contains 1 lb of solution. How many grams of ammonia
are present?

    1 lb solution x 454 g/lb = 454 g solution

    454 g solution x 0.300 g ammonia/g solution = 136 g ammonia

The symbol % means "parts per 100 parts." In this instance there are 30 parts by weight of ammonia in 100 parts by weight of solution or, specifically, 30.0 g ammonia per 100 g solution. In chemistry, whenever % is used to describe the composition of a solution, it is to be assumed that percent by weight is meant unless you are told otherwise. When a % is changed to a decimal by dividing by 100, the basis is changed from 100 to 1. In this example, 30.0% becomes 0.300 g of NH_3 per 1.00 g of solution.

1-13 TEMPERATURE

It is not possible to define temperature in relation to the fundamental units of length, mass, and time; temperature is a fundamental concept by which the relative "hotness" or "coldness" of objects or environments is compared.

The temperature scale in SI is the thermodynamic or Kelvin temperature scale. This has an origin, or zero point, at absolute zero, and a fixed point at the triple point of water, defined as 273.16 kelvins. The triple point is that temperature at which the solid, liquid, and gaseous states all exist in equilibrium with one another at the same temperature.

The temperature scale most commonly used in science is the Celsius, or centigrade scale. The designation Celsius has been officially adopted, but the term centigrade is still frequently used. Temperatures on the Celsius (centigrade) scale are recorded as ^{o}C. The triple point of water is defined as 0.01 ^{o}C on the Celsius scale. To convert from Celsius degrees to Kelvin degrees you add 273.15 to the Celsius temperature.

$$K = {}^{o}C + 273.15$$

Several devices, including gas thermometers, thermocouples, and the familiar mercury thermometer, are used for the measurement of temperature. The mercury-in-glass type is the most common .

The freezing point and the boiling point of water under one

atmosphere of pressure are two convenient points for checking a thermometer. The freezing point of water is $0^{\circ}C$, while the boiling point is $100^{\circ}C$. The interval between these two points has 100 equal parts. Unless otherwise noted, all temperatures given in the text are Celsius.

On the familiar Fahrenheit scale, the freezing point of water is 32° and the boiling point of water is 212°. The interval between these two points is divided into 180 equal parts. Each part corresponds to 1° on the Fahrenheit scale. Temperatures on this scale are recorded as $^{\circ}F$. Examples are $67^{\circ}F$, $81.6^{\circ}F$, and $1600^{\circ}F$.

Since the interval from the freezing point to the boiling point of water is an interval of 100 Celsius degrees or 180 Fahrenheit degrees, a change of 100 Celsius degrees measures the same temperature change as does a change of 180 Fahrenheit degrees. Therefore a temperature change of 5 Celsius degrees is the same as a temperature change of 9 Fahrenheit degrees.

1-14 CONVERSION OF TEMPERATURES

It is occasionally necessary to convert a temperature from one scale to the other. Useful equations for these conversions are

$$^{\circ}C \ = \ \frac{5}{9} \ (^{\circ}F \ - \ 32)$$

$$^{\circ}F \ = \ \frac{9}{5} \ ^{\circ}C \ + \ 32$$

Example 1-15
In earlier measurements of properties, such as solubility, the standard temperature was established as $18.0^{\circ}C$; later it was changed to $25.0^{\circ}C$; and now some data are measured at $30.0^{\circ}C$. Convert these temperatures to the Fahrenheit scale.

$$^{\circ}F \ = \ \frac{9}{5} \ ^{\circ}C \ + \ 32$$

$$^{o}F = \frac{9}{5} \times 18.0 + 32.0 = 64.4 \ ^{o}F$$

$$^{o}F = \frac{9}{5} \times 25.0 + 32.0 = 77.0 \ ^{o}F$$

$$^{o}F = \frac{9}{5} \times 30.0 + 32.0 = 86.0 \ ^{o}F$$

Example 1-16

A temperature of 68 ^{o}F has often been described as a normal, comfortable room temperature. Convert this temperature to the Celsius scale.

$$^{o}C = \frac{5}{9} (^{o}F - 32)$$

$$^{o}C = \frac{5}{9} \times (68-32) = \frac{5}{9} \times 36 = 20 \ ^{o}C$$

1-15 SPECIFIC HEAT

The quantity of heat required to raise the temperature of a given sample of a substance depends upon the nature of the substance and the size of the particular sample. Thus it is an extensive property. In order to consider the quantity of heat absorbed by various substances, this quantity must be expressed as an intensive property, that is, one which does not depend on the particular sample. This is done by using the same mass of the substance. The property so measured is called the specific heat or the heat capacity per gram, the quantity of heat required to raise the temperature of one gram of the substance one degree Celsius. The heat absorbed or given off by a substance is given as

Q = mass x specific heat x ΔT

where ΔT is the temperature difference of the substance. It is always taken as T(final) - T(initial).

The common unit of measure of heat is the calorie. This has been defined as 4.184 joules (J) or 4.184 x 10^{7} ergs. It is

approximately equal to the heat required to raise the temperature
of 1 g of water $1^{\circ}C$.

Example 1-17

Calculate the quantity of heat required to raise the temperature
of 50 g of water from 25 to 50 $^{\circ}C$.

$$50 \text{ g} \times (50-25)^{\circ}C \times 1 \text{ cal/g}^{\circ}C = 50 \text{ g} \times 25^{\circ}C \times 1 \text{ cal/g}^{\circ}C = 1250 \text{ cal}$$

or

$$50 \times 25 \times 1 = 1250 \text{ cal}$$
$$\text{g} \times {}^{\circ}C \times \text{cal/g}^{\circ}C = \text{cal}$$

Example 1-18

How much heat is required to raise the temperature of 5.6 g of
mercury from 20 to $80^{\circ}C$? The specific heat of mercury is 0.33 cal/g$^{\circ}C$.

$$5.6 \times 60 \times 0.33 = 112 \text{ cal}$$
$$\text{g} \times {}^{\circ}C \times \text{cal/g}^{\circ}C = \text{cal}$$

For a system in which there is no heat lost or gained during
a process, the sum of the heat gained by one part of the system
plus the heat lost by another part of the system must be zero.

$$Q_{gained} + Q_{lost} = 0$$

Example 1-19

One hundred grams of water at $80^{\circ}C$ is added to 50 grams of water
originally at $20^{\circ}C$. Assuming no heat is lost, what is the final
temperature of the system?

$$Q_{gained} = 50 \text{ g} \times 1 \text{ cal/g}^{\circ}C \times (T_f - 20)^{\circ}C$$

$$Q_{lost} = 100 \text{ g} \times 1 \text{ cal/g}^{\circ}C \times (T_f - 80)^{\circ}C$$

$$50 \times 1 \times (T_f - 20) + 100 \times 1 \times (T_f - 80) = 0$$

$$\text{g} \times \text{cal/g}^{\circ}C \times {}^{\circ}C + \text{g} \times \text{cal/g}^{\circ}C \times {}^{\circ}C = 0$$

$$50T_f - 1000 + 100T_f - 8000 = 0$$

$$^{o}C + {^{o}C} = 0$$

$$150T_f = 9000$$

$$T_f = 60^{o}C$$

PROBLEMS

<u>1-1</u> State the number of significant figures in each of the following
quantities.

    a. 51.8   b. 0.0605 g

    c. 9.89577   d. 50.0 ml

    e. 5.0 ml   f. 21,000 lb

    g. 16 oz   h. 6.022×10^{23} atoms

    i. 3.6×10^{9} years   j. 23.40×10^{5} g

<u>1-2</u> Add:

    a. 94.5 g   b. 79.96%   c. 2.85×10^{-3} mg

        4.03 g   14.2% 4.19×10^{-2} mg

        ———   5.7%   ———

                 ———

    d. 35.5%   e. 0.0873 g

        33.7%   11.61 g

        8.63%   0.0048 g

        ———   2.631 g

                ———

<u>1-3</u> Subtract:

    a. 6.17 ml   b. 23.8 liters

       0.016 ml   0.0141 liters

       ———   ———

    c. 238.5 g   d. 0.07638 g

       14.92 g   0.0605 g

       ———   ———

<u>1-4</u> Carry out the following manipulations, and observe the proper
treatment of significant figures.

    a. Add: 2.619 g + 8.96 mg + 0.0121 g + 168.6 mg

    b. Subtract: 261.4 ml from 2.876 liter

1-5 Multiply:

    a. 740 x 250 =   b. 0.08205 x 298 =

    c. 0.08205 x 298.2 =   d. 0.08205 x 298.15 =

    e. $6.022 \times 10^{23} \times 1.00796 =$

1-6 Divide:

    a. 18.96 ÷ 6.939 =   b. 48.42 ÷ 44.956 =

    c. 26.3 ÷ 207.19 =   d. 760 ÷ 298.2 =

    e. 0.08205 ÷ 250 =

1-7 The interatomic distance in the hydrogen molecule is given as 0.74 Å. What is this distance in nanometers, micrometers, millimeters, and centimeters?

1-8 A quantity of lithium metal weighs 6.939 g. Express this weight in milligrams and kilograms.

1-9 A solution contains 1.00 mg of a substance in 1 liter of the solution. How many γ of the substance are present?

1-10 Calculate, in cubic centimeters and in milliliters, the volume enclosed by a hollow cylinder whose length is 13 cm and inside diameter 22 mm.

1-11 A sample of a vapor in a 223.4 cc cell weighed 0.5763 g. Calculate the density of the vapor in grams per milliliter and grams per liter.

        [A. W. Laubengayer and F. B. Schirmer, J. Am. Chem. Soc.,
        62, 1578 (1940)]

1-12 The density of fluoramine was measured in a pycnometer, or specific-gravity bottle, having a volume of 0.0343 ml which was calibrated with mercury. What weight of mercury was required to fill the pycnometer? ·

   [E. A. Lawton and J. Q. Weber, J. Am. Chem. Soc., 85, 3596 (1963)]

1-13 An experiment requires that a solution be made by using 500 g of sulfuric acid. The only container available is a 400-ml beaker. Is it large enough to contain the acid? The density of the acid is 1.84 g/ml.

1-14 An experiment requires 250 g of a compound which is listed in two manufacturers' catalogs. The first manufacturer offers the compound at \$12.30 per 100 g; the second offers the same compound at \$60.00 per lb. From which of the two manufacturers should you purchase the compound if both supply the same quality?

1-15 What is the weight of a package of 10-mm glass rod containing four 1-ft pieces?

1-16 The half-life of a first order reaction was 10 sec. What is the half-life expressed in milliseconds? What is it in microseconds?

1-17 If 108.2 g of silver oxide is required for an experiment, will 4 oz be sufficient?

1-18 In the physical chemistry laboratory, a common experiment is to determine the surface tension of a liquid by the capillary rise method. A section of a broken thermometer is often used as the capillary. A student carrying out such an experiment used a section of a thermometer 200 mm long. The thermometer section weighed 18.8480 g. When the capillary was filled with mercury, the

thermometer section weighed 18.9249 g. Assuming the capillary is a
uniform cylinder, what is the radius of the capillary in millimeters?

1-19 What quantity of nitric acid, in grams, is present in a bottle
containing 1½ lb of 70.0% HNO_3?

1-20 How many grams of ammonia must be present in 1.00 pint of a
30.0% aqueous solution? The specific gravity of the solution is
0.9016.

1-21 An experiment requires 383 g of toluene. A bottle containing
0.89 lb is available. Is it sufficient?

1-22 What is the volume occupied by 1.60 lb of concentrated
hydrobromic acid (that is, 48% HBr) if the specific gravity of the
solution is 1.50?

1-23 How many grams of hydrogen bromide are present in the solution
described in Prob. 1-22?

1-24 A bottle of concentrated H_2SO_4 is found to contain 96.5% H_2SO_4
by weight. The density of the acid is 1.84 g/cc.
        a. How many grams of H_2SO_4 are present in one liter of the
acid solution?
        b. How many grams of H_2SO_4 are present in one kilogram of the
acid solution?

1-25 What is the freezing point of benzene ($5.53^{\circ}C$) on the
Fahrenheit scale?

1-26 A considerable amount of research, especially in biochemistry,
is now being carried on at $37.0^{\circ}C$. Convert this temperature to
degrees Fahrenheit. Why is this particular temperature used?

1-27 Calculate the heat required to raise the temperature of
1.61 g of Al from 23 to $42^{\circ}C$. The specific heat of aluminum is
0.217 cal/g-deg.

1-28 A quantity of water weighing 5.0 g at a temperature of $50^{\circ}C$
was mixed with an 8.0-g sample at a temperature of $10^{\circ}C$. What was
the final temperature of the water if no heat was lost to the
surroundings?

1-29 In a freshman laboratory experiment, 50.8 g of Al (specific
heat, 0.217 cal/g-deg) at $100^{\circ}C$ was added to 50 g of water in a
calorimeter. The original temperature of the water was $22^{\circ}C$.
What is the final temperature of the mixture, assuming no heat is
lost to the surroundings?

1-30 A company plans to build a power plant next to a river which
has a flow rate of 1×10^{10} g/min. One-tenth of this water will be
diverted through the plant to cool the condenser. If the condenser
discards heat at the rate of 1×10^{10} calories per minute to the
water:

     a. What is the temperature change of the cooling water after
it leaves the condenser?

     b. What will be the temperature change of the river water when
the water from the condenser is returned to the river (assume no
heat loss)?

     c. Give the equivalent temperature changes from part (a) and
(b) in degrees Fahrenheit.

SUPPLEMENTARY PROBLEMS

1-31 The distance between two nitrogen atoms in a nitrogen molecule
is commonly given as 1.10 Å. What is this distance in nanometers,
micrometers, millimeters, and centimeters?

1-32   The ionic radius of the oxide ion is listed as 1.32 Å by
some crystallographers.   What is the radius in centimeters,
millimeters, micrometers, and nanometers?

1-33   What is the weight in milligrams of 200.59 g of mercury?
What is this weight in kilograms?

1-34   How many milliliters of water may be placed in a hollow
cylinder 8 cm long and 12.6 mm in diameter?

1-35   What is the volume, in milliliters, of a hollow cylinder 25 cm
long and 35.6 mm in diameter?

1-36   Calculate the volume occupied by 26.71 g of a substance whose
density is 2.321 mg/cc.

1-37   What is the weight of 2.718 liters of a gas whose density is
1.896 mg/cc?

1-38   A bottle of sulfuric acid contains 2 lb of the acid.   How
many pounds of HNO_3 can be stored in a bottle of the same volume?
The density of sulfuric acid is 1.84 g/ml, and the density of
nitric acid is 1.38 g/ml.

1-39   The five-pint bottle has been a standard container for concentrated acids for many years.   What is the approximate weight of
glacial acetic acid contained in a 5-pint bottle?   The density of
glacial acetic acid is 1.05 g/ml.

1-40   Which is the better source of a liquid to be used in an experiment, a supplier who advertises the product at $12.00 per lb or
another who offers it at $14.00 per liter?   The density of the
liquid is 0.782 g/ml.

1-41 Glass tubing and rod are normally sold in 4-ft lengths, and the diameter is designated in millimeters. Calculate the volume of a 4-ft length of 8-mm rod.

1-42 A reaction was carried out in a 1-ft length of tubing whose outside diameter was 30 mm. If the wall thickness was 1 mm, what was the volume, in milliliters, available for the reaction?

1-43 What is the weight of a package of 6-mm glass rod which contains 145 pieces 4 ft long?

1-44 Calculate the number of pieces of 10-mm glass tubing 4 ft long in a package weighing 34.6 lb. Assume the wall thickness of the tubing to be 1 mm.

1-45 A certain molecular species was found to exist for 3 msec. What was the life of the species in seconds and microseconds?

1-46 The density of difluoramine at $-80.5^{\circ}C$ was measured in a pycnometer with a volume of 0.0343 ml. A value of 1.587 g/ml was observed. Calculate the weight of NHF_2 contained in the pycnometer.
        [E. A. Lawton and J. Q. Weber, J. Am. Chem. Soc., 85, 3596 (1963)]

1-47 What is the quantity of fluorine, in grams, in a cylinder containing ¼ lb?

1-48 What is the weight in grams of a 51.5% suspension of sodium hydride in mineral oil if 1.00 lb of sodium hydride is present?

1-49 How many liters of acetone are present in a 1-gal jug? How many milliliters?

1-50 How many milliliters of pyridine are there in a 1-pint bottle? How many liters?

1-51 How many grams of sodium hypochlorite are present in 1 lb of a 5.0% solution? How many grams of water?

1-52 What is the weight of a 5.0% solution of sodium hypochlorite if it contains 63 g of water?

1-53 A bottle of concentrated hydrochloric acid has a density of 1.18 g/cc. A 25.0-ml sample of this acid is found to contain 10.50 g of HCl. What is the percent by weight of HCl in the bottle?

1-54 The freezing point of cyclohexane is $6.59^{\circ}C$. What is its freezing point on the Fahrenheit scale?

1-55 While looking at a thermometer outside your house, a cold air front moves in. The temperature drops from 81° to 54° F. What is the corresponding temperature drop in degrees Celsius?

1-56 How many calories of heat must be removed from a sample of lead weighing 0.435 g to lower its temperature from 61.2 to $48.3^{\circ}C$? The specific heat of lead is 0.0305 cal/g-deg.

1-57 A kilogram of alcohol at $27^{\circ}C$ was mixed with 450 g of alcohol at another temperature. The final temperature was $25.2^{\circ}C$. What was the initial temperature of the 450-g sample of alcohol? The specific heat of alcohol is 0.581 cal/g-deg.

Chapter 2

ATOMS, GRAM ATOMS, AND SYMBOLS

2-1 LAW OF DEFINITE PROPORTIONS

A compound always contains the same elements in the same pro-
portion by mass. This statement is called the law of definite
proportions, or the law of constant composition. The relative
proportions by mass of the elements in a compound may be expressed
in a variety of ways, two of which are percent and ratio. Thus,
all samples of pure silver iodide contain 45.95% silver and 54.05%
iodine. This composition may also be represented as the mass ratio
iodine/silver = 1.176. Such a ratio is the mass of one element
equivalent to 1.000 g of another element. A ratio has no units, since
the units cancel.

Example 2-1
In three experiments, silver was converted to silver iodide. Samples
of silver weighing 9.89577, 10.44012, and 9.52800 g were converted
to 21.53744, 22.72226, and 20.73713 g of silver iodide, respectively.
Calculate the percent composition of each sample of iodide to
four significant figures.
        [G. P. Baxter and O. W. Lundstedt, J. Am. Chem. Soc., 62,
        1832 (1940)]
    The first step is to determine the mass of iodine in each
sample of silver iodide.

Mass of silver iodide	21.54 g	22.72 g	20.74 g
Mass of silver	9.90 g	10.44 g	9.53 g
	11.64 g	12.28 g	11.21 g

Divide the mass of each element by the total mass of the compound, and multiply by the ratio 100. Use logarithms to obtain the proper number of significant figures.

$$\text{Percent silver} = \frac{\text{g silver}}{\text{g compd.}} \times 100$$

$$\frac{9.896 \text{ g}}{21.54 \text{ g}} \times 100 = 45.95\%$$

$$\frac{10.44 \text{ g}}{22.72 \text{ g}} \times 100 = 45.95\%$$

$$\frac{9.528 \text{ g}}{20.74 \text{ g}} \times 100 = 45.94\%$$

$$\text{Percent iodine} = \frac{\text{g iodine}}{\text{g compd.}} \times 100$$

$$\frac{11.64 \text{ g}}{21.54 \text{ g}} \times 100 = 54.05\%$$

$$\frac{12.28 \text{ g}}{22.72 \text{ g}} \times 100 = 54.05\%$$

$$\frac{11.21 \text{ g}}{20.74 \text{ g}} \times 100 = 54.05\%$$

Check by adding the percent of silver and iodine; the sum should equal 100.

Example 2-2
Use the data of Example 2-1 to calculate the iodine-to-silver ratios to four significant figures.

The ratio is obtained by dividing the mass of iodine by the mass of silver in each sample.

$$\frac{\text{Iodine}}{\text{Silver}} = \frac{11.64 \text{ g}}{9.896 \text{ g}} = \frac{12.28 \text{ g}}{10.44 \text{ g}} = \frac{11.21 \text{ g}}{9.528 \text{ g}} = 1.176$$

When symbols are used, the iodine/silver ratio in silver iodide is represented by I/Ag, since the formula of silver iodide is AgI.

The mass ratio of chlorine to calcium in calcium chloride is 2Cl/Ca, since the formula of calcium chloride is $CaCl_2$. The mass ratio of sulfur to silver in silver sulfide is represented by S/2Ag, because the formula is Ag_2S.

2-2 COMBINING MASSES; GRAM EQUIVALENTS

From the examination of the composition of a large number of compounds it has been found that a number that is related to the combining ratio of the element with other elements can be assigned to each element. If the number 126.9044 g is assigned to iodine, silver has a combining mass of 107.870 g. It is not necessary to know the formulas of the compounds involved in the determination of combining masses.

Example 2-3
Baxter and Lundstedt determined the iodine/silver ratio in silver iodide to be 1.176433. Assuming the combining mass of iodine to be 126.9044 g, calculate the combining mass of silver to five significant figures.

[J. Am. Chem. Soc., 62, 1829 (1940)]

$$\frac{\text{Comb. mass iodine}}{\text{Comb. mass silver}} = 1.1764$$

$$\frac{126.90}{\text{Comb. mass silver}} = 1.1764$$

$$\text{Comb. mass silver} = \frac{126.90}{1.1764} = 107.87 \text{ g}$$

Five-place logarithms should be used.

Example 2-4
In a similar way, the combining mass of chlorine can be calculated

from the fact that the silver chloride/silver ratio was found to be
1.328667 by T. W. Richard and R. C. Wells.

[J. Am. Chem. Soc., 27, 459 (1905)]

Mass silver chloride = mass silver + mass chlorine

$$\frac{\text{Mass silver chloride}}{\text{Mass silver}} = 1.3287$$

$$\frac{\text{Comb. mass silver} + \text{comb. mass chlorine}}{\text{Comb. mass silver}} = 1.3287$$

Comb. mass chlorine = 1.3287 x 107.87 - 107.87

= 35.45 g

The combining masses of silver, iodine, and chlorine are equal
to their atomic weights because one atom of silver combines with
óne atom of chlorine or one atom of iodine. The combining mass of
calcium calculated from the mass of chlorine with which it combines
is 20.04 g. The atomic weight is 40.08. The atomic weights and
the combining mass are related by a factor which is a small integer,
called the valence of calcium in the compound.

Example 2-5
Calculate the combining mass of selenium from the silver/silver
selenide ratio, 0.732081, in silver selenide.

[J. Am. Chem. Soc., 84, 4185 (1962)]

Mass silver selenide = mass silver + mass selenium

$$\frac{\text{Mass silver}}{\text{Mass silver selenide}} = 0.73208$$

$$\frac{107.87}{107.87 + \text{comb. mass selenium}} = 0.73208$$

comb. mass selenium = 39.44 g

The combining mass of selenium in silver selenide is one-half
the atomic weight because the formula is Ag_2Se.

2-3 GRAM EQUIVALENT

The term gram equivalent weight or gram equivalent is
encountered in several different connections in chemistry. It is
numerically equal to the combining mass, and one definition is:
     The mass of a substance which contains, will react with, or
     will displace, 7.9997 g of oxygen or its equivalent.
The direct determination of the combining mass or equivalent weight
of an element with oxygen is often difficult, and indirect methods
must be used. Problems 2-5 and 2-6 illustrate this point.

2-4 ATOMS

All atoms of an element have the same atomic number, that is,
they have the same positive electric charge on the nucleus. They
also have the same number and distribution of electrons. Sodium
atoms, for example, have a nuclear charge of 11, and they have
11 electrons outside the nucleus. All atoms of sodium also have the
same mass, 3.82×10^{-23} g. Several other elements, such as fluorine,
gold, iodine, manganese, and tantalum, also have only one type of
atom. Most of the other elements, however, have atoms that differ
in mass. All of the atoms of neon, for example, have the atomic
number 10, but they do not all have the same mass. Atoms of an
element with different masses are called isotopes. The isotopes of
an element have the same atomic number but different masses. Neon,
for example, has three isotopes.

2-5 ATOMIC WEIGHT

The mass of the atoms of an element compared to the mass of the
atoms of the carbon isotope designated carbon-12 is the atomic
weight. That is, the mass of ^{12}C has been assigned the value of

12.000000 on this relative scale, and atomic weights are relative
masses. A gold atom, for example, has a mass of 196.967 compared to
the mass of ^{12}C taken as 12.000000. For an element that has
isotopes, the atomic weight is the average relative mass of the
atoms. Note that the word mass is used when dealing with the mass
of a single isotope of an element, while the word weight is used to
describe the average mass of a mixture of isotopes.

2-6 GRAM ATOM

A quantity of an element equal in mass in grams to its atomic
weight is called a gram atomic weight. The term gram atom is used
more frequently because of its brevity. The gram atomic weight of
an element is the mass of 6.022×10^{23} atoms (Avogadro's number):
that is, an Avogadro's number of atoms of an element has a mass
equal to the atomic weight of the element expressed in grams. A
gram atom of calcium weighs 40.08 g, a gram atom of lead weighs
207.19 g, and a gram atom of boron weighs 10.811 g, as examples.
Each of these quantities contains 6.022×10^{23} atoms of the
respective element.

Example 2-6
What is the mass of a single atom of gold?
The atomic weight of Au, from Table D-3, is 196.967. (Tables
D-1 to D-4 will be found in Appendix D.) From the definition
given above, one gram atom of Au has a mass of 196.967 g. One
gram atom of an element contains 6.022×10^{23} atoms. Therefore,
to calculate the mass of one atom, divide the gram atomic weight by
6.022×10^{23}.

$$\frac{197.0 \text{ g}}{6.022 \times 10^{23}} = 3.272 \times 10^{-22} \frac{\text{g}}{\text{atom}}$$

(Note that 196.967 is rounded off to 197.0.)

Atoms, Gram Atoms, and Symbols

31

$$\frac{g/g\text{-}atom}{atom/g\text{-}atom} = \frac{g}{g\text{-}atom} \times \frac{g\text{-}atom}{atom} = \frac{g}{atom}$$

Example 2-7
What is the mass, in grams, of 0.212 g atom of Si?
The atomic weight of Si is 28.1, so one gram atom of Si has a mass of 28.1 g.

$$0.212 \times 28.1 = 5.96 \text{ g}$$

$$g\text{-}atom \times \frac{g}{g\text{-}atom} = g$$

Example 2-8
How many gram atoms of magnesium are present in 1.2151 g?
Divide the quantity present by the mass in grams of 1 g-atom to give the number of gram atoms present.

$$\frac{1.2151}{24.312} = 0.049980 \text{ g-atom}$$

$$\frac{g}{g/g\text{-}atom} = g \times \frac{g\text{-}atom}{g} = g\text{-}atom$$

(Five-place logarithms were used.)

Example 2-9
How many sodium atoms are present in 0.152 g of the metal sodium?
First calculate the number of gram atoms of Na in 0.161 g, and then multiply by the number of atoms in 1 g atom.

$$\frac{0.161}{23.0} = 0.00700 = 7.00 \times 10^{-3} \text{ g-atom}$$

$$\frac{g}{g/g\text{-}atom} = g \times \frac{g\text{-}atom}{g} = g\text{-}atom$$

$$7.00 \times 10^{-3} \times 6.02 \times 10^{23} = 42.14 \times 10^{20}$$

$$= 4.21 \times 10^{21} \text{ atoms}$$

$$g\text{-}atom \times \frac{atom}{g\text{-}atom} = atom$$

The two steps may be combined

$$\frac{0.161}{23.0} \times 6.02 \times 10^{23} = 4.21 \times 10^{23}$$

$$\frac{g}{g/g\text{-atom}} \times \frac{atom}{g\text{-atom}} = g \times \frac{g\text{-atom}}{g} \times \frac{atom}{g\text{-atom}} = atom$$

2-7 SYMBOLS

Symbols are convenient abbreviations used to represent elements. A uniform system of symbols has been agreed upon by the International Union of Pure and Applied Chemistry. An alphabetical list of the elements and their symbols is given in Table D-3.

When chemists discuss or write about their work, they often use symbols to represent the individual atoms of the elements. Used in this way, symbols represent the combination, rearrangement, or separation of atoms.

In the laboratory, however, chemists use symbols to represent gram atoms of elements: the quantities of elements used in experiments. Since a gram atom of an element contains 6.022×10^{23} atoms, the symbol of an element also represents Avogadro's number of atoms. The various uses of symbols are summarized below, where gold (Au) is used as an example.

Symbol	Au
Name of the element	Gold
1 atom	1 atom of gold
1 gram atom	197.0 g gold (1 g-atom)
Avogadro's number of atoms	6.022×10^{23} atoms of gold

2-8 DETERMINATION OF ATOMIC WEIGHTS

The atomic weights in Table D-3 are from the Report of the International Commission on Atomic Weights (1969). The values, except for a few, have been obtained from a direct comparison of the exact mass of nuclides or from the precise determination of combining masses or ratios. The first method is physical, the second chemical.

The term nuclide is used to designate a species of atom with a particular nuclear composition, i.e., a definite number of protons and neutrons. Elements that have no isotopes are said to be mononuclidic. Their mass values obtained from the mass spectrograph are identical with their relative atomic weights. Iodine is such an element, and its weight is reported as 126.9044 amu. The atomic mass unit (amu) is equal to the mass of the ^{12}C isotope of carbon divided by 12.00000.

Atomic weights of polynuclidic elements (elements possessing isotopes) can be calculated from isotopic abundance measurements and mass values of the isotopes. The uncertainties in the relative atomic weights arise entirely from limitations in the accuracy of the abundance measurements.

Example 2-10

The absolute abundances of the isotopes of silicon have been determined: ^{28}Si, 92.21%; ^{29}Si, 4.70%; ^{30}Si, 3.09%. The respective nuclidic masses are 27.97693, 28.97649, 29.97376 amu. Calculate the atomic weight of silicon to four significant figures.

$$
\begin{array}{rcl}
27.98 \times 0.9221 &=& 25.80 \text{ amu} \\
28.98 \times 0.0470 &=& 1.36 \text{ amu} \\
29.97 \times 0.0309 &=& \underline{0.93 \text{ amu}} \\
&& 28.09 \text{ amu}
\end{array}
$$

The accepted value of the atomic weight of silicon is 28.086.

Atomic weights can be calculated from combining masses or from combining ratios determined by careful measurements of masses involved in chemical reactions. When atoms combine in a 1:1 ratio, the atomic weights are identical with the combining masses. The combining mass of chlorine calculated in Example 2-4 gives the atomic weight, since one atom of chlorine combines with one atom of silver.

Example 2-11

Calculate the atomic weight of bromine from the fact that 6.23696 g of silver combined with bromine to form 10.85722 g silver bromide.

[G. P. Baxter, J. Am. Chem. Soc., 28, 1322 (1906)]

$$\begin{array}{r} 10.85722 \text{ g silver bromide} \\ -6.23696 \text{ g silver} \\ \hline 4.62026 \text{ g bromine} \end{array}$$

This gives the quantity of bromine equivalent to 6.23696 g of silver.

$$107.87 \times \frac{4.6203}{6.2370} = 79.908 \text{ g}$$

comb. mass Ag   combined mass Br

$$\text{g silver} \times \frac{\text{g bromine}}{\text{g silver}} = \text{g bromine}$$

Atomic weight of Br = 79.908

The grams of silver in one combining mass are multiplied by the grams of bromine equivalent to 1.0000 g of silver. The result is the mass of bromine equivalent to one combining mass (equivalent weight) of silver.

2-9 LAW OF DULONG AND PETIT

When atoms combine in ratios other than 1:1, the equivalent weights may not equal the atomic weights. The equivalent weight and the atomic weight are related by an integer, the valence. The value

of the integer may be obtained from an application of the law of
Dulong and Petit, which states that the produce of the atomic weight
and the specific heat of an element is approximately equal to 6.3.

   Specific heat x atomic weight = 6.3

This law enabled chemists of a century ago to decide upon the
correct atomic weight when the combining capacity (valence) of an
element was not known.

Example 2-12

Calculate the atomic weight of cadmium from the fact that 3.39038 g
of the metal was obtained from 8.21123 g of cadmium bromide.

   [G. P. Baxter, M. R. Grose, and M. H. Hartmann, J. Am.

   Chem. Soc., 38, 857 (1916)]

   First calculate the combining mass of Cd from the mass of
bromine with which it was combined.

   8.21123 g cadmium bromide
   -3.39038 g cadmium
   _____
   4.82085 g bromine

$$\text{Equiv. weight Cd} = 79.909 \times \frac{3.3904}{4.8208} = 56.199 \text{ g}$$

equiv. weight
bromine

$$\text{g Br} \times \frac{\text{g Cd}}{\text{g Br}} = \text{g Cd}$$

Applying Dulong and Petit's law:

$$\text{Approximate atomic weight} = \frac{6.3}{\text{sp. heat}} = \frac{6.3}{0.055} = 114$$

Therefore

   Atomic weight = 2 x equivalent weight

   56.199 x 2 = 112.40

(The accepted atomic weight of cadmium is 112.40.)

Example 2-13

Calculate the atomic weight of selenium from the silver/silver selenide ratio, 0.732081. The formula of silver selenide is Ag_2Se.
   [J. Am. Chem. Soc., 84, 4185 (1962)]

$$\frac{2Ag}{Ag_2Se} = 0.732081$$

$$\frac{2 \times 107.870}{2 \times 107.870 + Se} = 0.732081$$

$$\frac{215.74}{215.74 + Se} = 0.732081$$

$$215.74 \times 0.732081 \times Se = 215.74$$

$$Se = \frac{215.74 - 215.74 \times 0.73208}{0.73208} = 78.953 \ g/g\text{-atom}$$

Additional problems concerned with the determination of atomic weights are Probs. 4-31 to 4-34, 4-65 to 4-68, 5-28, 5-29, 5-59 and 5-60.

PROBLEMS

2-1 A sample of 7.505 g cadmium chloride gave 4.602 g cadmium by electrolysis. Calculate the percent composition of cadmium chloride.
   [G. P. Baxter and M. L. Hartmann, J. Am. Chem. Soc., 37, 113 (1915)]

2-2 Calculate the cadmium/chloride ratio in cadmium chloride from the data in Prob. 2-1.

2-3 What is the zinc/chlorine ratio if 5.94448 g of zinc chloride gave 2.85136 g of zinc by electrolysis?
   [G. P. Baxter and J. H. Hodgens, J. Am. Chem. Soc., 43, 1242 (1921)]

2-4 What is the percent zinc and chlorine in zinc chloride if the zinc/chlorine ratio is 0.92195?
   [G. P. Baxter and J. H. Hodgens, J. Am. Chem. Soc., 43, 1242 (1921)]

2-5 W. A. Noyes and H. C. P. Weber found that 0.47989 g of hydrogen combined with 16.88423 g of chlorine to form hydrogen chloride. Use the combining mass of chlorine from Example 2-4 to determine the combining mass of hydrogen.

[J. Am. Chem. Soc., 30, 13 (1908)]

2-6 Calculate the combining mass of sulfur from the silver sulfide/silver ratio, 1.1486.

[J. Am. Chem. Soc., 84, 4183 (1962)]

2-7 What is the mass, in grams, of a single atom of sodium? What is the mass in milligrams?

2-8 What is the average mass of an atom of tin? What is meant by the term "average mass" in this problem?

2-9 How many atoms of zinc are present in a sample of the metal which weighs 51.8 g?

2-10 Calculate the number of gram atoms of carbon in a diamond weighing 6.31 g.

2-11 A solution was made by dissolving 0.00612 g of sulfur in 1 ml of CS_2. How many atoms of S were present in 1 liter of solution? (Assume that the volume of CS_2 did not change with the addition of S.)

2-12 How many gram atoms of gold are present in a cube of the element measuring 2.00 in. on an edge? How many atoms of gold are in the cube?

2-13 What is the mass in grams of a sample of sulfur which contains one trillion atoms?

2-14 Calculate the number of sodium atoms in a rectangular block of
the element 8.36 x 10^{-3} mm long, 2.04 x 10^{-4} mm wide, and
80.3 Å high.

2-15 How many atoms of uranium are there in a uranium sphere whose
radius is 1.00 in.?

2-16 How many mercury atoms are present in 36.0 ml of the element
at 20°C?

2-17[*] The element iodine is mononuclidic, and its atomic weight was
been assigned the value 126.9044 based on direct determination of
its mass in a mass spectrograph. Calculate a value for the atomic
weight of silver from the I/Ag ratio, 1.176433.

2-18[*] Use the value of the atomic weight of silver from Prob. 2-37
to calculate the atomic weight of chlorine from the ratio
AgCl/Ag = 1.328667.

2-19[*] The absolute abundances of the gallium isotopes were determined
as 60.16% ^{69}Ga and 39.84% ^{71} Ga. Calculate the atomic weight of
gallium. The nuclidic masses are 68.9257 and 70.9249 amu,
respectively.

2-20[*] Calculate the atomic weight of krypton from the following
data:

---

[*]The data for the starred problems in this section are from
Report of the International Commission on Atomic Weights (1961)
unless otherwise noted. Isotope abundances are in atom percent.
Nuclide masses are from Handbook of Chemistry and Physics,
45th ed., Chemical Rubber Co., Cleveland, 1964

Isotopes	Abundance, atom %	Mass, amu
^{78}Kr	0.35	77.9204
^{80}Kr	2.27	79.9164
^{82}Kr	11.56	81.9135
^{83}Kr	11.55	82.9141
^{84}Kr	56.90	83.9115
^{86}Kr	17.37	85.9106

2-21 Calculate the atomic weight of cadmium from the data in Prob. 2-1.

2-22* The absolute abundance ratio ^{14}N/^{15}N in atmospheric nitrogen has been determined to be 272 ± 0.3. The nuclidic masses are ^{14}N = 14.003074 and ^{15}N = 15.000108. Calculate the atomic weight of nitrogen.

2-23 Use the data in Prob. 2-26 and Dulong and Petit's rule to calculate the atomic weight of copper.

SUPPLEMENTARY PROBLEMS

2-24 E. W. Morley carefully determined the mass ratio of hydrogen to oxygen in water as 0.12596. Use the combining mass of hydrogen calculated in Prob. 2-5 to determine the combining mass of oxygen. Compare your result with the value given in the definition of equivalent weight, Sec. 2-3.

[Z. Physik. Chem. (Leipzig), 20, 417 (1896)]

---

*See footnote on page 38.

2-25 In a very early experiment Dumas and Stas determined the composition of water by mass as O, 88.864%; H, 11.136%. Use the value 7.9997 as the equivalent weight of oxygen to calculate the equivalent weight of hydrogen and compare your answer with the latest accepted value of the atomic weight.

[Ann. Chem., 8, 189 (1843)]

2-26 What is the combining mass of copper if the copper oxide/copper ratio is 1.125181?

[J. Am. Chem. Soc., 84, 4185 (1962)]

2-27 The silver telluride/silver ratio was found to be 1.59145. What is the combining mass of tellurium?

[J. Am. Chem. Soc., 84, 4189 (1962)]

2-28 What is the mass of an atom of Ta in milligrams?

2-29 What is the mass, in grams, of 1.63 g-atom copper?

2-30 How many gram atoms of beryllium are contained in 32.1 g of the metal?

2-31 How many atoms of silicon are present in 1.00 g of the element? In 1.00 mg? In 1.00 μg? In 1.00 γ?

2-32 How many atoms of aluminum are there in a cube of the metal 1.0 x 10^{-3} mm on an edge?

2-33 How many atoms of Pt are present in a piece of platinum wire 10.0 mm long and 1.00 mm in diameter?

2-34 How many gram atoms of potassium are present in a sample containing 1.62 x 10^{18} atoms?

2-35 How many gram atoms of neon are present in a sample containing 7.86×10^4 atoms?

2-36 Calculate the number of boron atoms in a cylinder of the element 2.54×10^{-4} cm in diameter and 32.2 mm long.

2-37 How many gram atoms of lead are there in a lead sphere 22.8 mm in diameter?

2-38 Determine the number of gram atoms of mercury in 36.0 ml of the element at $20^{\circ}C$.

2-39 How many gallium atoms are present in 2.16 cc of the element at $40^{\circ}C$?

2-40 The absolute isotope abundance ratio of chlorine $^{35}Cl/^{37}Cl$ = 3.1272. Calculate the atomic weight of chlorine and compare the result with that obtained in Prob. 2-18. The masses of ^{35}Cl and ^{37}Cl are 34.96885 and 36.96590 amu, respectively.

[W. R. Shields et al., J. Am. Chem. Soc., 84, 1519 (1962)]

2-41 Calculate the atomic weight of sulfur from the data in Prob. 2-6.

2-42[*] The mass ratio of gallium to gallium oxide was found to be 0.74396. Calculate the atomic weight of gallium and compare the result with the atomic weight calculated in Prob. 2-19.

2-43[*] The isotope abundances of bromine were found to be 50.537% ^{79}Br and 49.46% ^{81}Br. The isotopic masses are respectively 78.9183 and 80.9163. Calculate the atomic weight of bromine. Compare the result with Example 2-11.

---

*See footnote on page 38.

2-44 Calculate the atomic weight of zinc from the data in Prob.2-3.

2-45* Calculate the atomic weight of tellurium from the data in
Prob. 2-27.

2-46* The isotopic abundances of strontium are ^{84}Sr, 0.560;
^{86}Sr, 9.870; ^{87}Sr, 7.035; ^{88}Sr, 82.535%. The respective masses are
83.9134, 85.9094, 86.9089, 87.9056. What is the atomic weight of
strontium?

2-47* The abundances of ^{17}O and ^{18}O in atmosphere oxygen are
0.0374% and 0.2039%, respectively. The nuclidic masses are
^{16}O = 15.994915, ^{17}O = 16.999134, and ^{18}O = 17.999160. Calculate
the atomic weight of oxygen.

---

*See footnote on page 38.

Chapter 3

MOLECULES, MOLES, AND FORMULAS

3-1 MOLECULAR WEIGHTS

When two or more atoms combine, larger particles called molecules are formed. The molecular weight is the average mass of the molecules of a substance compared to the mass of a ^{12}C atom. The molecular weight is the sum of the weights of the atoms in the molecule. The molecular weight of hydrogen chloride is 36.461, for example, since each molecule contains one hydrogen atom and one chlorine atom. The molecular weight of water is 18.0153, since each molecule contains two hydrogen atoms and one oxygen atom.

3-2 THE MOLE

The molecular weight of a substance expressed in grams is called the gram molecular weight of the substance. A gram molecular weight of a substance contains 6.022×10^{23} molecules. It is common practice to use the term mole in place of gram molecular weight. The mole has recently been adopted as a base unit of the International Systems of Units. The mole is defined as the amount of substance of a system containing as many elementary entities as there are atoms in 0.012 kg of carbon-12. The entities may be atoms, molecules, ions, electrons or any specified group of particles. The symbol for mole is mol. A mole of hydrogen chloride weighs 36.461 g and

contains 6.022×10^{23} hydrogen chloride molecules. A mole of water
contains 6.022×10^{23} water molecules and weighs 18.0153 g. The
molecular weight of ammonia NH_3 is 17.036; one mole weighs 17.036 g
and contains 6.022×10^{23} molecules.

Example 3-1

What is the mass in grams of 2.00 moles of carbon tetrachloride CCl_4?
Determine first the mass of 1 mole.

CCl_4 1 x 12.0   =   12.0 g
             4 x 35.4   =   141.6 g
                            ─────────
                            153.6 g/mole

             2.00 x 154 =   308 g

             $\overbrace{mole} \times \dfrac{g}{\underbrace{mole}}$ = g

Example 3-2

How many moles of carbon dioxide, CO_2, are present in 6.28 g?
Determine the mass of 1 mole.

CO_2 1 x 12.0   =   12.0 g
            2 x 16.0   =   32.0 g
                          ─────────
                          44.0 g/mole

Divide the mass in grams by the number of grams in 1 mole.

$\dfrac{6.28}{44.0}$ = 0.143 mole

$\dfrac{g}{g/mole}$ = $\cancel{g} \times \dfrac{mole}{\cancel{g}}$ = mole

Example 3-3

What is the average mass of a molecule of methane, CH_4?

CH_4 1 x 12.0   =   12.0 g
            4 x 1.0   =   4.0 g
                          ─────────
                          16.0 g/mole

$$\frac{16.0}{6.02 \times 10^{23}} = 2.66 \times 10^{-23} \text{ g/molecule}$$

$$\frac{\text{g/mole}}{\text{molecules/mole}} = \frac{\text{g}}{\text{mole}} \times \frac{\text{mole}}{\text{molecule}}$$

Example 3-4

How many molecules of boron trichloride, BCl_3, are present in a
sample weighing 0.133 g?

First calculate the number of moles of BCl_3 present.

BCl_3 $1 \times 10.8 = 10.8$ g

          $3 \times 35.4 = \underline{106.2}$ g

                  117.0 g/mole

$$\frac{0.133}{117} = 0.00114 = 1.14 \times 10^{-3} \text{ moles}$$

$$\frac{\text{g}}{\text{g/mole}} = \text{g} \times \frac{\text{moles}}{\text{g}}$$

$$1.14 \times 10^{-3} \times 6.02 \times 10^{23} = 6.85 \times 10^{20} \text{ molecules}$$

$$\text{mole} \times \frac{\text{molecules}}{\text{mole}} = \text{molecules}$$

3-3 FORMULAS

Formulas are combinations of symbols which represent compounds.
In speaking and writing of their work, chemists use formulas to
represent molecules. The nature and number of atoms in a molecule
are also apparent from the formula of a compound.

Just as a symbol may be used to represent one gram atom of an
element, so a formula is often used to represent one mole of a
compound. The senses in which formulas are used in the laboratory
are illustrated in the following:

Formula	H_2SO_4
Name of compound	Sulfuric acid
One molecule	One molecule of sulfuric acid
Nature and number of atoms	Two hydrogen atoms, one sulfur atom, and four oxygen atoms combined in some manner
One mole	98.0 g sulfuric acid
Avogadro's number of molecules	6.022×10^{23} molecules of sulfuric acid

A formula indicates the number of atoms of each element in a molecule, and it is understood that the atoms are joined by one means or another. The exact arrangement of the atoms is not apparent from the formula. For example, the formula of water is usually written H_2O, but each of the hydrogen atoms is in fact joined to the oxygen atom.

The number of atoms of an element in a molecule is indicated by a subscript, for example, H_2, F_2, NH_3, SO_2, and N_2F_4. The subscript implies that the atoms in the molecule are joined together.

A coefficient is used to represent a number of atoms or molecules which are not joined, for example, 2Ne, $3H_2$, and $2NH_3$. Two neon atoms, three hydrogen molecules (each containing two atoms), and two ammonia molecules (each containing one nitrogen and three hydrogen atoms) are thus represented.

If a compound is ionic and does not exist in the molecular state, its formula represents the relative number of ions of each type in the crystal. Thus $CaCl_2$ means that in a crystal of calcium chloride there are two chloride ions for each calcium ion present. Moreover, in one mole of calcium chloride there are one mole of calcium ions and two moles of chloride ions.

The nature of the bonds in a compound is not apparent from the formula. The properties of the compound give evidence of the type of bonds present.

3-4 DETERMINATION OF FORMULA

A formula represents an actual substance of a definite chemical composition. This chemical composition, and hence the formula, is determined by experiments. If a chemist says that the formula of sulfuric acid is H_2SO_4, he is saying that someone has determined the composition of sulfuric acid by experiment and has reported the results. Of course, with even a limited knowledge of chemistry one can predict from theory what the formula of a compound is likely to be. A certain knowledge of the composition of a substance is based on experiment however.

3-5 COMPOSITION FROM FORMULAS

When a known compound is prepared and analyzed, the analysis is compared with that calculated from the formula of the expected compound. If a sample of lithium nitride, Li_3N, for example, was made, the results of the analysis obtained in the laboratory would be compared with the theoretical composition calculated as follows;

$$Li_3N \quad 3 \times 6.94 \ = \ 20.82 \text{ g Li/mole } Li_3N$$
$$1 \times 14.01 \ = \ \underline{14.01 \text{ g N/mole } Li_3N}$$
$$34.83 \text{ g } Li_3N/\text{mole}$$

$$\frac{20.82}{34.83} \times 100 \ = \ 59.77\% \text{ Li}$$

$$\frac{14.01}{34.83} \times 100 \ = \ 40.22\% \text{ N}$$

The formula Li_3N represents the weight of 1 mole. The symbols with their subscripts denote the quantity of each element in 1 mole of the compound. The ratio 20.82/34.83 is the grams of lithium per gram of compound. Multiplying this ratio by 100 changes the basis from 1 g of Li_3N to 100 g of Li_3N.

Example 3-5

Recently, W. L. Jolly, K. D. Maguire, and D. Rabinovich prepared a compound believed to be S_4N_3Cl. The analysis of the compound was reported as N, 21.02%; S, 62.09%; Cl, 17.58%. Do these results agree with the theoretical composition of S_4N_3Cl?

[Inorg. Chem., 2, 1304 (1963)]

$$4 \times 32.06 = 128.24 \text{ g S/mole } S_4N_3Cl$$

$$3 \times 14.01 = 42.03 \text{ g N/mole } S_4N_3Cl$$

$$\underline{1 \times 35.45 = 35.45 \text{ g Cl/mole } S_4N_3Cl}$$

$$205.72 \text{ g/mole } S_4N_3Cl$$

$$\frac{128.2}{205.7} \times 100 = 62.31\% \text{ S in } S_4N_3Cl$$

$$\frac{\text{g S}}{\text{g } S_4N_3Cl} \times 100 = \frac{\text{g S}}{100 \text{ g } S_4N_3Cl} = \% \text{ S in } S_4N_3Cl$$

$$\frac{42.03}{205.7} \times 100 = 20.43\% \text{ N in } S_4N_3Cl$$

$$\frac{35.45}{205.7} \times 100 = 17.23\% \text{ Cl in } S_4N_3Cl$$

The compound prepared was S_4N_3Cl, because the composition found by analysis agrees well with the calculated composition.

Note that, in the problem illustrated, four significant figures may be carried throughout the calculations. The use of logarithms is recommended for problems with four significant figures, because a slide rule is limited to three.

3-6 FORMULAS FROM COMPOSITION

The elemental composition of a compound may be used to calculate a formula which denotes the relative number of atoms of each type in the molecule. Such formulas are called empirical, or simplest, formulas.

There are three steps necessary to determine the empirical formula of a compound from the percent composition.

1. Divide the percent composition of each element by its atomic weight. This gives the number of gram atoms of each element in 100 g of compound.

2. Take the _smallest_ number from step one and divide this into
all the other results from step one. This gives the ratio of gram
atoms of the elements in 100 g of compound.

3. If in step two a fraction is obtained for any element,
multiply all the numbers obtained in step two by the lowest common
denominator to clear of fractions.

Example 3-6

What is the simplest formula of the compound, prepared by J. H.
Junkins and his group, which contained 49.7% niobium and 50.2%
fluorine?

[J. Am. Chem. Soc., 74, 3464 (1952)]

The first step is to find the number of gram atoms of each
element in 100 g of the compound.

$$\frac{49.7 \text{ g Nb}/100 \text{ g compd.}}{92.9 \text{ g Nb}/\text{g-atom Nb}} = 0.536 \frac{\text{g-atom Nb}}{100 \text{ g compd.}}$$

$$\frac{50.2 \text{ g F}/100 \text{ g compd.}}{19.0 \text{ g F}/\text{g-atom F}} = 2.64 \frac{\text{g-atom F}}{100 \text{ g compd.}}$$

Next find the simplest ratio of the gram atoms of the elements
in 100 g of compound.

$$\frac{0.536}{0.536} = 1.00 \qquad \frac{2.64}{0.536} = 4.93 \frac{\text{g-atom F}}{\text{g-atom Nb}}$$

Since a gram atom of any element contains 6.02×10^{23} atoms,
the ratio can be expressed as $1 \times 6.02 \times 10^{23}$ atoms of Nb to
$5 \times 6.02 \times 10^{23}$ atoms of F. Hence for each atom of niobium there are
five atoms of fluorine. The simplest formula is NbF_5.

Example 3-7

In a recent paper, Guenther described an interesting method of
preparation of a compound of titanium carried out at the I. I. T.
Research Institute, Chicago. Potassium chloride and titanium(IV)
chloride were allowed to react in fused antimony trichloride. The

analysis of the product was K, 23.1%; Ti, 14.2%; Cl, 62.7%. What
is the simplest formula of the compound?

[K. F. Guenther, Inorg. Chem., 3, 923-924 (1964)]

K $\dfrac{23.1}{39.1}$ = 0.592 $\dfrac{\text{g-atom K}}{100 \text{ g compd.}}$ $\dfrac{0.592}{0.297}$ = 1.99   2

Ti $\dfrac{14.2}{47.9}$ = 0.297 $\dfrac{\text{g-atom Ti}}{100 \text{ g compd.}}$ $\dfrac{0.297}{0.297}$ = 1.00   1

Cl $\dfrac{62.7}{35.4}$ = 1.77 $\dfrac{\text{g-atom Cl}}{100 \text{ g compd.}}$ $\dfrac{1.77}{0.297}$ = 5.96   6

The simplest formula is K_2TiCl_6. Note that the slide rule is
quite appropriate for the calculations in this problem, since the
data are limited to three significant figures.

Example 3-8
Recently a new fluoride of oxygen was prepared by mixing the proper
ratio of oxygen and fluorine at 60 K. This compound contains 32.1%
F and 67.9% O. What is the empirical formula of the compound?

[A. G. Streng and A. V. Grosse, J. Am. Chem. Soc., 88,
169 (1966)]

F $\dfrac{32.1}{19.0}$ = 1.69 $\dfrac{\text{g-atom F}}{100 \text{ g}}$ $\dfrac{1.69}{1.69}$ = 1 x 2 = 2

O $\dfrac{67.9}{16.0}$ = 4.24 $\dfrac{\text{g-atom O}}{100 \text{ g}}$ $\dfrac{4.24}{1.69}$ = 2.50 x 2 = 5

The simplest formula is O_5F_2

Molecular formulas. Simplest formulas represent the relative number
of atoms of each element in the compound in question. The molecular
weight of the compound must be determined in order to learn if the
formula also indicates the total number of atoms in the molecules
of the compound.

Example 3-9
A compound contains 92.3% carbon and 7.7% hydrogen. Calculation of
the simplest formula gives CH.

$$\frac{92.3}{12.0} = 7.7 \; \frac{\text{g-atom C}}{100 \text{ g}} \qquad \frac{7.7}{7.7} = 1 \text{ C}$$

$$\frac{7.7}{1.00} = 7.7 \; \frac{\text{g-atom H}}{100 \text{ g}} \qquad \frac{7.7}{7.7} = 1 \text{ H}$$

At least two compounds have this simple atomic ratio. One has a molecular weight of 26.0; the other has a molecular weight of 78.0. The mass represented by CH is 13.0. The two molecular formulas are C_2H_2 and C_6H_6.

$$\frac{26.0}{13.0} = 2 \qquad 2 \text{ x CH} = C_2H_2 \qquad \frac{78.0}{13.0} = 6 \qquad 6 \text{ x CH} = C_6H_6$$

Example 3-10

At the University of Washington, the reaction of N_2F_4 with $S_2O_6F_2$ was carried out at room temperature. A gas, boiling at $-2.5^{\circ}C$, was collected. Its analysis gave N, 9.48%; S, 20.9%; F, 38.0%. Find the simplest formula.

[M. Lustig and G. H. Cady, Inorg. Chem., 2, 388 (1963)]

Since the oxygen content was not reported, we must obtain it by difference:

    9.5%   100.0%
   20.9%   68.4%
   38.0%   _____
   _____   31.6% oxygen
   68.4%

Note the application of the rule of significant figures in the addition; the number 9.48 is changed to 9.5.

N $\frac{9.48}{14.0} = 0.676 \; \frac{\text{g-atom N}}{100 \text{ g compd.}} \qquad \frac{0.676}{0.651} = 1.04$

S $\frac{20.9}{32.1} = 0.651 \; \frac{\text{g-atom S}}{100 \text{ g compd.}} \qquad \frac{0.651}{0.651} = 1.00$

F $\frac{38.0}{19.0} = 2.00 \; \frac{\text{g-atom F}}{100 \text{ g compd.}} \qquad \frac{2.00}{0.651} = 3.07$

O $\frac{31.6}{16.0} = 1.98 \; \frac{\text{g-atom O}}{100 \text{ g compd.}} \qquad \frac{1.98}{0.651} = 3.04$

The simplest formula is NSO_3F_3. This is called the simplest formula because the same analytical data would be obtained from $N_2S_2O_6F_6$, $N_3S_3O_9F_9$, etc. The molecular weight was found by experiment to be 150.8.

What is the molecular formula of the compound? The molecular weight calculated from the simplest formula is 151.1.

$$
\begin{array}{rl}
NSO_3F_3 & 14.0 \\
& 32.1 \\
3 \times 16.0 = & 48.0 \\
3 \times 19.0 = & 57.0 \\
\hline
& \overline{151.1}
\end{array}
$$

Therefore the molecular formula is NSO_3F_3.

PROBLEMS

<u>3-1</u> What is the mass in grams of 2.21 moles of phosphorus trichloride, PCl_3?

<u>3-2</u> Calculate the number of moles of water in 23.2 ml.

<u>3-3</u> Calculate the number of phosphorus and oxygen atoms in 16.8 g of orthophosphoric acid, H_3PO_4.

<u>3-4</u> Calculate the number of molecules of sulfur dioxide, SO_2, in 3.61 g. How many atoms of oxygen are present?

<u>3-5</u> Calculate the mass in grams of a molecule of hydrogen bromide, HBr.

<u>3-6</u> How many grams of each of the following compounds is represented by:

a. $2SO_3$ b. $4.56\ HNO_3$
c. $9.33 \times 10^{-8}\ KClO_3$ d. $1.68\ Mg_3N_2$
e. $4CuSO_4 \cdot 5H_2O$

3-7 How many calcium ions and how many chloride ions are present in 2.12 moles of $CaCl_2$?

3-8 What is the mass in grams of 3.01×10^{20} molecules of nitrogen trifluoride, NF_3?

3-9 What is the mass ratio potassium/chlorine in potassium chloride?

3-10 What is the mass ratio zirconium/chlorine in zirconium tetrachloride?

3-11 In a study of the use of SF_4 as a fluorinating agent at the DuPont laboratories, two compounds believed to be $KAsF_6$ and Na_2SiF_6 were prepared. The samples contained 32.9% As and 14.8% Si, respectively. Compare these results with the calculated composition.
[E. L. Meuterties, J. Inorg. Nucl. Chem., 18, 2497 (1961)]

3-12 A method for preparation of $GaCl_3$ is described by Johnson and Haskew. Analysis of their product gave Ga, 39.42 and 39.50%; Cl, 60.48 and 60.45%. How well do these results compare with the calculated composition?
[Inorg. Syn., 1, 26 (1939)]

3-13 Two methods of reduction of molybdenum pentachloride to molybdenum tetrachloride were investigated at the laboratories of the Climax Molybdenum Company of Michigan.
    Method 1. Analysis Cl, 58.98; Mo, 40.26
    Method 2. Analysis Cl, 58.69; Mo, 41.04
Compare the results of the two methods.
[M. L. Larson and F. W. Moore, Inorg. Chem., 3, 286 (1964)]

3-14 Compare the analytical results for the following compounds with the calculated values.

Compound	K,%	M,%*	F,%
$KMnF_3$	25.5	37.4	36.6
$KFeF_3$	25.6	36.6	37.4
$KCoF_3$	25.7	37.9	36.6
$KNiF_3$	25.0	37.9	35.8
$KCuF_3$	23.9	39.6	35.1
$KZnF_3$	24.1	40.5	35.0

*M = Mn, Fe, Co, Ni, Cu, and Zn, respectively.
Source: D. J. Mackin, R. L. Martin, and R. S. Nyholm, J. Chem.
Soc., 1963, 1493.

3-15 What is the simplest formula of the compound containing
7.63% magnesium and 91.78% arsenic?
       [K. Pigon, Helv. Chim. Acta., 44, 30 (1961)]

3-16 A mixture of Ir metal powder and strontium oxide was heated
in air at the chemistry laboratories of the University of Connecticut.
The compound formed contained 39.77% Sr and 44.52% Ir according to
analysis performed. What is the formula of the compound?
       [J. J. Randall, L. Katz, and R. Ward, J. Am. Chem. Soc.,
       79, 266 (1957)]

3-17 What is the simplest formula of a compound which contained
37.7% B, 47.5% N, and 13.9% H?
       [G. H. Dahl and R. Schaeffer, J. Am. Chem. Soc., 83,
       3033 (1961)]

3-18 The reaction of trisilylamine $(SiH_3)_3N$ with ammonia gave a
compound containing 68.8% Si, 17.8% N, and 6.39% H. The rest is
impurities. What is the simplest formula for the compound?
       [R. Wells and R. Schaeffer, J. Am. Chem. Soc., 88,
       37 (1966)]

3-19   At the I.I.T. Research Institute, the reaction of thallium(I)
chloride with titanium(IV) chloride was carried out in molten
antimony trichloride. The product formed was analyzed with the
following results. Tl, 61.2; Ti, 7.2; Cl, 31.6. What is the
simplest formula of the compound formed?

[K. F. Guenther, Inorg. Chem., 3, 923 (1964)]

3-20   Titanium metal was treated with hydrogen fluoride at 225° for
48 hr. The solid product contained 44.52% Ti and 53.18% F. What
is the simplest formula?

[E. L. Mueterties, J. Inorg. Nucl. Chem., 18, 148 (1961)]

3-21   A hydrated chloride of cerium was analyzed and found to contain
37.70% cerium, 28.41% chlorine, and 33.89% water (by difference).
What is the formula of the hydrate?

[L. M. Dennis and W. H. Magee, J. Am. Chem. Soc., 16,
600 (1894)]

3-22   A compound containing 24.0% cobalt, 14.4% ammonia (present as
ammonium ion), and 59.9% chlorine was prepared at the University of
Oklahoma. It was allowed to stand in contact with air for several
days until the color was violet and the weight constant. The loss
of mass from a sample of the violet hydrate on dehydration was
13.0 ± 0.2%. What was the formula of the violet hydrate?

[N. Fogel, C. C. Lin, C. Ford, and W. Grindstoff, Inorg.
Chem., 3, 720-721 (1964)]

3-23   An oxide of manganese prepared at a Bureau of Mines laboratory
contained 69.64% manganese. What is the simplest formula of the
oxide?

[R. L. Orr, J. Am. Chem. Soc., 76, 859 (1954)]

3-24   Recently, Maercker and Roberts synthesized an organic compound
containing 74.84% C, 6.86% H, and the rest O. What is the simplest

formula for this compound?

[J. Am. Chem. Soc., 88, 1742 (1966)]

3-25 The compound described in Prob. 3-46 was treated with hydrogen
fluoride and the product analyzed. The results were reported as
follows: S, 58.0; F, 22.0; N, 19.0; H, 0.7. What is the simplest
formula of the compound?

[O. Glemsler and E. Wyszomirski, Chem. Ber., 94,
1443 (1961)]

3-26 A compound containing 84.2% boron and 15.7% hydrogen was
prepared in the laboratories at Indiana University. Its molecular
weight was calculated from its vapor density as 76.7. What is its
molecular formula?

[D. F. Gaines and R. Schaeffer, Inorg. Chem., 3,
438 (1964)]

3-27 A white crystalline material melting at 122° was prepared at
Duke University. An analysis gave B, 23.9; N, 31.1; F, 42.9;
H, 2.2. The molecular weight was found to be 132. What is the
formula of the compound?

[K. Niedenzu, Inorg. Chem., 1, 943 (1962)]

3-28 A complex metal organic compound was found to contain 60.3%
carbon, 4.3% hydrogen, 33.4% tungsten, and the remainder oxygen.
Its molecular weight was found to be 564. What is its formula?

[D. P. Tate et al., J. Am. Chem. Soc., 86, 3261 (1964)]

3-29 Boron trifluoride and ammonia gases were mixed and allowed to
stand for several hours at 0°. The solid that formed analyzed
B, 12.8; NH_3, 19.9. The molecular weight was determined by the
freezing-point-depression method as 83. What is the formula of the
compound?

[W. A. Jenkins, J. Am. Chem. Soc., 78, 5500 (1956)]

3-30 What is the formula of the compound with the composition
28.76% phosphorus, 17.25% fluorine, 32.40% chlorine, and the
remainder oxygen? Its molecular weight was determined to be 225.
[M. M. Crutchfield, C. F. Callis, and J. R. Van Wazer,
Inorg. Chem., 3, 282 (1964), footnote 8]

SUPPLEMENTARY PROBLEMS

3-31 What is the mass in grams of 0.818 mole of lithium bromide,
LiBr?

3-32 How many moles of calcium oxide, CaO, are present in 4.82 g?

3-33 How many molecules of dinitrogen monoxide, N_2O, are present
in a sample weighing 0.752 g?

3-34 What is the mass in grams of sulfur represented by each of
the following:
   a. S_8 b. $2 S_2$
   c. $0.25 S_8$ d. $1.2 S_2$
   e. $8.08 \times 10^{-5} S_8$

3-35 What is the mass in grams of each of the following?
   a. 3 Au   b. $2.28 F_2$
   c. $6.51 \times 10^{-3} P_4$ d. 7.43 Pb
   e. $2.19 N_2$

3-36 How many magnesium ions and how many bromide ions are present
in 0.233 g of $MgBr_2$?

3-37 Calculate the mass ratio aluminum/oxygen in aluminum oxide.

3-38 Thallium(III) fluoride was prepared at Manchester University
by the reaction of thallium(I) iodate with bromine trifluoride, and

the ratio $TlF_3/TlIO_3$ was determined as 0.690. Compare this value
with the theoretical ratio.

[A. A. Woolf, J. Chem. Soc., 1954, 4695]

3-39 The compound silver tetrafluoroaurate was prepared and the
$AgAuF_4/Au$ ratio was determined at 1.969. Compare this value with
the theoretical ratio.

[A. A. Woolf, J. Chem. Soc., 1954, 4695]

3-40 To obtain an infrared spectrum of the white solid $Na_2N_2O_3$, a
sample was prepared at the Mellon Institute in Pittsburg.
Analysis found: N, 22.36; Na, 35.40; O, 40.00. Judging from
these data alone, was a pure sample of $Na_2N_2O_3$ prepared?

[P. D. Feltham, Inorg. Chem., 3, 900 (1964)]

3-41 The reaction of fluorine with V_2O_5 at 475° was investigated at
the University of New Hampshire. Do the analytical results agree
with the composition of the expected product of the reaction,
VOF_3? Analysis found: V, 42.2, 41.1, 40.7; F, 45.0.

[H. M. Haendler et al., J. Am. Chem. Soc., 76,
2177 (1954)]

3-42 A finely powdered mixture of ammonium fluoride and copper(II)
fluoride was subjected to a pressure of 100,000 psi at 25°. The
compound isolated after this treatment analyzed NH_3, 20.6%;
Cu, 36.7%; F, 42.7%. Do these data support the proposed formula
$(NH_4)_2CuF_4$?

[D. S. Crocket and R. A. Grossman, Inorg. Chem., 3,
645 (1964)]

3-43 What is the simplest formula of the compound, prepared at the
University of Tennessee, which contained 2.97% boron and 97.66%
iodine?

[W. J. McDowell and C. W. Keenan, J. Am. Chem. Soc., 78,
2069 (1956)]

3-44   Bromine vapor was passed over a mixture of zirconium dioxide
and carbon at 560°.   A white product containing 22.30% zirconium and
77.68% bromine was obtained.   What is its simplest formula?

[R. C. Young and H. G. Fletcher, Inorg. Syn., 1, 49 (1939)]

3-45   Direct combination of tungsten and chlorine led to the
production of a compound, the composition of which was determined
to be W, 46.28%; Cl, 53.59%.   What is the simplest formula of the
compound?

[R. E. McCarley and T. M. Brown, Inorg. Chem., 3, 1233 (1964)]

3-46   What is the simplest formula of a compound with the composition
N, 21.02; S, 62.09; Cl, 17.58?

[W. L. Jolly, K. D. Maguire, and D. Rabinovich, Inorg. Chem., 3, 1304 (1963)]

3-47   A sample of a fluoride of thorium prepared by J. C. Warf
contained 75.37% Th and 24.60% F.   What is its simplest formula?

[J. Am. Chem. Soc., 74, 1864 (1952)]

3-48   A yellow solid melting at 147° was prepared and analyzed.   Its
composition was found to be Re, 74.2; F, 7.61; and the remainder
oxygen.   What is its formula?

[A. Engelbrecht and A. V. Grosse, J. Am. Chem. Soc., 76, 2042 (1954)]

3-49   A series of complex chlorides of rhenium of the general formula
$M_x ReCl_6$ were made at the University of North Carolina.   Calculate
the formula of the potassium salt from the analytical results.
Found: Re, 39.25; Cl, 43.88.

[C. W. Horner, F. M. Collier, Jr., and S. Y. Tyree, Inorg. Chem., 3, 1388 (1964)]

3-50 An unstable thioiodide of phosphorus was prepared by stirring
a solution of the elements at room temperature for 65 hr. Analysis
Found; I, 80.31; S, 10.01; and the remainder phosphorus. What is
the simplest formula you would assign to the compound? Consult the
original paper after you have reached a decision.

      [A. H. Cowley and S. T. Colren, Inorg. Chem., 3,
      780 (1964)]

3-51 Iodine was added to liquid chlorine cooled with a dry
ice-acetone mixture. Orange crystals separated, and they were found
to have the composition I, 54.5%; Cl, 45.7%. What was the compound?

      [H. S. Booth and W. C. Morris., Inorg. Syn., 1, 168 (1939)]

3-52 What is the formula of the fluoride which was prepared at the
University of New Hampshire from TiO_2 and found to contain 38.6%
titanium?

      [H. M. Haendler et al., J. Am. Chem. Soc., 76,
      2177 (1954)]

3-53 What is the simplest formula of the compound which resulted
from the reaction of zirconium with zirconium tetrabromide if its
analysis gave Zr, 27.63; Br, 72.41?

      [E. M. Larsen and J. J. Leddy, J. Am. Chem. Soc., 78,
      5984 (1956)]

3-54 A compound prepared at Nagoya University in Japan had the
composition I, 78.9; Te, 13.1; and the remainder potassium. What
was its formula?

      [D. Nakamura, K. Ito, and M. Kubo, J. Am. Chem. Soc.,
      84, 163 (1962)]

3-55 What is the simplest formula of the compound which contained
4.03% boron and 94.5% iodine?

      [W. C. Schumb, E. L. Gamble, and M. Banus, J. Am. Chem.
      Soc., 71, 3225 (1949)]

3-56 What is the formula of the bromide of tungsten prepared by
treating a sample of WBr_2 with excess bromine at 50° for 2 weeks?
Analysis found: W, 43.22;   Br, 56.30

        [R. E. McCarley and T. M. Brown, J. Am. Chem. Soc., 84,
        3216 (1962)]

3-57 What is the formula of the fluoride of rhodium prepared at
the Argonne National Laboratory in Chicago? Analysis found:
Rh, 47.8;   F, 51.8.

        [C. L. Chernick, H. H. Classen, and B. Weinstock,
        J. Am. Chem. Soc., 83, 3165 (1961)]

3-58 The reaction of boric acid, H_3BO_3, with urea, $(NH_2)_2CO$, at
600° produced a solid containing 31.6% boron, 20.5% nitrogen, and
the remainder oxygen, This solid was then treated with ammonia at
1650° and another compound, containing 44.2% boron and 56.0%
nitrogen was formed. What are the simplest formulas of the two
compounds?

        [T. E. O'Connor, J. Am. Chem. Soc., 84, 1753 (1962)]

3-59 Calculate the simplest formula of the compound prepared by
Krebs, Müller, and Zürn. Analysis found: P, 51.8;   Cd, 47.5.

        [Z. Anorg. Allgem. Chem., 285, 25-26 (1956)]

3-60 Aqueous solutions of $N_2H_4 \cdot$ 2HCl and $HgCl_2$ were mixed and a
yellow precipitate was formed. Analysis found: Hg, 79.96;
Cl, 14.2;   N, 5.7. Calculate the simplest formula of the compound.

        [K. Brodersen, Z. Anorg. Allgem. Chem., 285, 7 (1956)]

3-61 Potassium meta arsenate $(KAsO_3)$ and potassium fluoride were
mixed and heated to $800^\circ C$. The product formed contained 35.5% K,
33.7% As, 8.63% F, and the remainder oxygen. Calculate the simplest
formula of the compound.

        [N. K. Duff and A. K. Gupta, Z. Anorg. Allgem. Chem.,
        285, 93 (1956)]

3-62 What is the simplest formula of the compound formed along with $(CH_3)_2AlH$ by the reaction of $(CH_3)_3B$ with $LiAlH_4$? Analysis found: Li, 19.75; B, 28.9; 28.5; C, 36.6, 32.6; H, 16.68, 17.14.

> [T. Wartik and H. I. Schlesinger, J. Am. Chem. Soc., 75, 838 (1953)]

3-63 What is the formula of the compound from which the following data were obtained?

Sample	Mass, mg	H, mg	CH_3, mg	Al, mg
1	64.5	1.196	33.82	29.94
2	28.7	0.491	13.95	12.72
3	118.9	2.089	61.20	54.84

Source: T. Wartik and H. I. Schlesinger, J. Am. Chem. Soc., 75, 838 (1953).

3-64 Calculate the formula of the very explosive compound obtained from the reaction of silver nitrate with potassium amide in liquid ammonia. Analysis found: Ag, 87.4; N, 11.2.

> [E. C. Franklin, J. Am. Chem. Soc., 27, 834 (1905)]

3-65 A hydrate of Ag_2TeO_4 (0.7638 g) was heated to remove the water of hydration. The loss in mass was 0.0605 g. What is the formula of the hydrate?

> [E. B. Hutchins, Jr., J. Am. Chem. Soc., 27, 1164 (1905)]

3-66 A mineral has been found crystallized from a small pond in Antartica; the pond contains an extremely large amount of salt. Analysis gave Ca, 17.5; Cl, 32.7; H_2O, 49.2. The mineral has been found in no other location and has been named antarcticite. What is the formula of the mineral?

> [T. Torii and J. Ossaka, Science, 149, 975 (1965)]

3-67 A sulfide of scandium prepared at the RCA Laboratories in
Princeton, New Jersey contained 48.6% scandium and 51.0% sulfur.
What is its simplest formula?
        [J. P. Dismukes and J. G. White, Inorg. Chem., 3,
        1221 (1964)]

3-68 Calculate the simplest formula of a binary compound of sulfur
which contains 42.46% chlorine.
        [R. C. Brasted and J. S. Pond, Inorg. Chem., 4,
        1164 (1965)]

3-69 A series of silicon oxychlorides was prepared at M.I.T. The
analytical and molecular weight data of three of the series are
listed below. What is the formula of each compound?

Compound	%Cl	%Si	Mol. Weight
1	73.2	23.3	483
2	71.0	24.1	697
3	70.1	24.8	911

Source: W. C. Schumb and R. A. Lefever, J. Am. Chem. Soc.,
76, 2092 (1954).

3-70 A compound prepared at the University of Florida gave the
following analytical results: F, 45.29; N, 16.86; and the
remainder phosphorus. The molecular weight was found to be 250.
Another compound of phosphorus had a molecular weight of 328 and
contained 45.74% F and 16.83% N. What are the formulas of the
two compounds?
        [T. J. Mao, R. D. Dresdner, and J. A. Young, J. Am. Chem.
        Soc., 81, 1021 (1959)]

3-71   The reaction of fluorine with $(FCO)_2O_2$ gave a product containing 15.0% C, 40.1% O, and 45.7% F. The molecular weight
obtained from the average vapor density measurements was 82.   What
is the formula of the product?

   [R. L. Cauble and G. H. Cady, J. Am. Chem. Soc., 89,
   5161 (1967)]

3-72   A red solid prepared by the decomposition of diboron
tetrachloride gave the following analytical data: B, 24.3 and 24.7;
Cl, 73.9 and 74.4.   Molecular weight determinations were 536, 511,
516, 517.   What is the formula of the compound?

   [E. D. Schrom and G. Urry, Inorg. Chem., 2, 405 (1963)]

Chapter 4

CHEMICAL EQUATIONS

4-1 EQUATIONS

Just as symbols are used to represent elements and formulas to represent compounds, so equations are used to represent chemical reactions. What has been said previously about symbols and formulas applies to equations. Thus the equation

$$4HCl + O_2 \rightarrow 2H_2O + 2Cl_2$$

is a shorthand designation for the following statement:

Four molecules of hydrogen chloride reacted with one molecule of oxygen to form two molecules of water and two molecules of chlorine.

Chemists use equations in this manner to discuss chemical reactions. On the other hand we can say:

Four moles of hydrogen chloride reacted with one mole of oxygen to give two moles of water and two moles of chlorine.

In this sense, equations are used in the laboratory to calculate the relative masses of reactants and products in chemical reactions. These are summarized in the following table (page 66).

The total mass of the products is equal to the total mass of the reacting substances(177.6 g), an illustration of the <u>law of conservation of mass</u>. Also, note carefully that the number of atoms of each element is the same before and after the reaction.

65

4 HCl	+ O_2	→ $2H_2O$	+ $2Cl_2$

1. Hydrogen chloride + oxygen   → water   + chlorine
2. 4 molecules   + 1 molecule   → 2 molecules + 2 molecules
3. 4 moles   + 1 mole   → 2 moles   + 2 moles
4. $4 \times 6.022 \times 10^{23}$ + 6.022×10^{23} → $2 \times 6.022 \times$ + $2 \times 6.022 \times$
   molecules   molecules 10^{23} 10^{23}
                                       molecules   molecules
5. 4×36.4 g   + 32.0 g   → 2×18.0 g + 2×70.8 g

4-2 BALANCING EQUATIONS

Chemists and others who require a knowledge of chemistry must remember a vast number of chemical reactions. It is not wise to attempt to remember complete equations such as we have just discussed. The reactions are committed to memory and the equations are written and balanced as needed.

For example, oxygen was first prepared by Priestley, who heated mercury(II) oxide; metallic mercury was the other product. This statement can be depicted by the equation

$$Hg + O_2 = HgO$$

Since there are two O's on the left and only one on the right, the equation is not "balanced." However, the equation

$$2Hg + O_2 = 2HgO$$

is balanced.

There are several types of equations which you must learn to balance; some are easier to balance than others. It would be convenient if a simple set of rules would enable one to balance all equations to be met, but this is not the case. The method to be used depends upon the type of reaction which is represented by the equation.

Chemical reactions may be classified in several ways. Two such classifications are useful for balancing equations. The first is a classification according to whether or not changes in oxidation state (transfer of electrons) occur. The second is according to the conditions under which the reactions occur. Some reactions occur in the absence of a solvent or in an indifferent solvent, i.e., one that serves only to disperse the reactants. Other reactions occur only in solution.

4-3 TYPES OF CHEMICAL EQUATIONS

A. Simple: No change in oxidation number occurs.

    1. Anhydrous

$$TiO_2(s) + CCl_4(g) \rightarrow TiCl_4(\ell) + CO_2(g)$$

    2. In aqueous solution

       a. Precipitation

$$AgNO_3(aq) + KCl(aq) \rightarrow AgCl(s) + KNO_3(aq)$$

       b. Neutralization

$$HCl(aq) + NaOH(aq) \rightarrow H_2O + NaCl(aq)$$

       c. Gas formation

$$SiO_2(s) + 4HF(aq) \rightarrow SiF_4(g) + 2H_2O$$

       d. Complex formation

$$AgCl(s) + 2NH_3(aq) \rightarrow Ag(NH_3)_2Cl(aq)$$

B. <u>Redox</u>:   Changes in oxidation numbers occur

    1. Anhydrous

$$2KClO_3(s) \rightarrow 2KCl(s) + 3O_2(g)$$

    2. In aqueous solution

$$16HCl + 2KMnO_4 \rightarrow 5Cl_2 + 2MnCl_2 + 2KCl + 8H_2O$$

A set of rules which enables one to balance all of the simple (A) equations and most of the anhydrous redox equations (B.1) follows. The application of these rules is often called <u>balancing by inspection</u>.

4-4 RULES FOR BALANCING SIMPLE EQUATIONS

    1. Balance the metals present first.

    2. Balance the nonmetals except hydrogen and oxygen.

    3. Balance hydrogen and oxygen.

    4. Check all elements.

    5. Repeat in the same order until all are balanced.

<u>Example 4-1</u>

$$Na_2O_2 + H_2O = NaOH + O_2$$

    <u>Rule 1</u> $Na_2O_2 + H_2O = 2NaOH + O_2$

    <u>Rule 2</u> Does not apply.

    <u>Rule 3</u> The hydrogen atoms balance:

$$Na_2O_2 + 2H_2O = 2NaOH + O_2$$

The oxygen atoms are now balanced.

    <u>Rule 4</u> The hydrogen atoms are no longer balanced.

    <u>Rule 5</u> $2Na_2O_2 + 2H_2O = 4NaOH + O_2$

The equation is balanced.

Example 4-2

$KClO_3 = KCl + O_2$

> Rule 1 $KClO_3 = KCl + O_2$
>
> Rule 2 $KClO_3 = KCl + O_2$
>
> Rule 3 $2KClO_3 = KCl + 3O_2$
>
> Rule 4   The K and Cl are not balanced.
>
> Rule 5 $2KClO_3 = 2KCl + 3O_2$

Example 4-3

$ZrO_2 + C + Br_2 = ZrBr_4 + CO$

> Rule 1 $ZrO_2 + C + Br_2 = ZrBr_4 + CO$
>
> Rule 2 $ZrO_2 + C + 2Br_2 = ZrBr_4 + CO$
>
> Rule 3 $ZrO_2 + C + 2Br_2 = ZrBr_4 + 2CO$
>
> Rule 4   The C atoms are not balanced.
>
> Rule 5 $ZrO_2 + 2C + 2Br_2 = ZrBr_4 + 2CO$

Example 4-4

$C_2H_2(g) + O_2(g) \rightarrow CO_2(g) + H_2O(g)$

> Rule 2 $C_2H_2 + O_2 \rightarrow 2CO_2 + H_2O$

Rule 3   The H's are balanced. There is an odd number of O's on the right side and an even number on the left. It is useful in such cases to double the coefficients of the elements already balanced and proceed as before.

$$2C_2H_2 + O_2 \rightarrow 4CO_2 + H_2O$$

The H's are not balanced.

$$2C_2H_2 + O_2 \rightarrow 4CO_2 + 2H_2O$$

Only the O's remain unbalanced.

$$2C_2H_2 + 5O_2 \rightarrow 4CO_2 + 2H_2O$$

4-5 BALANCING REDOX EQUATIONS

Sometimes redox equations which occur in the absence of a
solvent (B.1) may be quite difficult to balance by inspection. When
several substances are involved, the key ratio required to balance
the equation may not be readily apparent. In these circumstances it
is wise to use the change in oxidation number method.

4-6 OXIDATION NUMBER METHOD

1. Determine the oxidation number of each element present
(oxidation numbers are discussed in Sec. 14-1).

(0)   (0)   (+3)(-2)

Al + S → Al_2S_3

2. Indicate the changes in oxidation number.

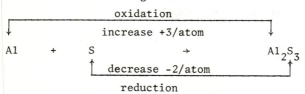

3. Multiply the substances oxidized and reduced by the proper
coefficients to make the increase in oxidation number equal to the
decrease.

increase 2 x 3 = 6

decrease 3 x 2 = 6

$2Al$ + $3S$ → Al_2S_3

4. Continue as for simple equations.

5. Check.

Example 4-5

Balance the equation Mg + B_2O_3 → MgO + B

$$\underline{\text{Rule 1}} \quad \overset{(0)}{\text{Mg}} \quad + \quad \overset{(+3)(-2)}{\text{B}_2\text{O}_3} \quad \rightarrow \quad \overset{(+2)(-2)}{\text{MgO}} \quad + \quad \overset{(0)}{\text{B}}$$

$$\underline{\text{Rule 2}} \quad \text{Mg} \quad \overset{\text{oxidation}}{\underset{\text{increase +2/atom}}{\overbrace{}}} + \quad \text{B}_2\text{O}_3 \quad \rightarrow \quad \text{MgO} \quad + \quad \text{B}$$
decrease -3/atom
reduction

$$\underline{\text{Rule 3}} \quad 3\text{Mg} + \text{B}_2\text{O}_3 \rightarrow 3\text{MgO} + 2\text{B}$$

It is not necessary to multiply B_2O_3 by 2 (the increase in oxidation number of Mg) because there are two B's in B_2O_3, and the total decrease in oxidation number is -6.

Example 4-6

Balance the equation $\text{Ba} + \text{NH}_3 \rightarrow \text{Ba}_3\text{N}_2 + \text{H}_2$

$$\underline{\text{Rule 1}} \quad \overset{(0)}{\text{Ba}} \quad + \quad \overset{(-3)(+1)}{\text{NH}_3} \quad \rightarrow \quad \overset{(+2)(-3)}{\text{Ba}_3\text{N}_2} \quad + \quad \overset{(0)}{\text{H}_2}$$

$$\underline{\text{Rule 2}} \quad \text{Ba} \quad \overset{\text{oxidation}}{\underset{\text{increase +2/atom}}{}} + \quad \text{NH}_3 \rightarrow \text{Ba}_3\text{N}_2 \quad + \quad \text{H}_2$$
decrease -1/atom
reduction

$$\underline{\text{Rule 3}} \quad 3\text{Ba} + 2\text{NH}_3 \rightarrow \text{Ba}_3\text{N}_2 + \text{H}_2$$

Example 4-7

Balance the equation $\text{NpCl}_4 + \text{NH}_3 \rightarrow \text{NpCl}_3 + \text{HCl} + \text{N}_2$

$$\underline{\text{Rule 1}} \quad \overset{(+4)(-1)}{\text{NpCl}_4} \quad + \quad \overset{(-3)(+1)}{\text{NH}_3} \quad \rightarrow \quad \overset{(+3)(-1)}{\text{NpCl}_3} \quad + \quad \overset{(+1)(-1)}{\text{HCl}} \quad + \quad \overset{(0)}{\text{N}_2}$$

$$\underline{\text{Rule 2}} \quad \text{NpCl}_4 \quad + \quad \text{NH}_3 \quad \overset{\text{oxidation}}{\underset{\text{increase +3/atom}}{\overbrace{}}} \rightarrow \text{NpCl}_3 + \text{HCl} + \text{N}_2$$
decrease -1 atom
reduction

$\underline{\text{Rule 3}}$ Since there are two N's in N_2, there must be at least 2NH_3 in the balanced equation. This makes the total increase $2(+3)$.

$$6NpCl_4 \quad + \quad 2NH_3 \quad \rightarrow \quad 6NpCl_3 \quad + \quad HCl \quad + \quad N_2$$

<u>Rule 4</u> $6NpCl_4 \quad + \quad 2NH_3 \quad \rightarrow \quad 6NpCl_3 \quad + \quad 6HCl \quad + \quad N_2$

<u>Example 4-8</u>

Balance the equation $CuF_2 \quad + \quad NH_3 \quad \rightarrow \quad Cu_3N \quad + \quad NH_4F \quad + \quad N_2$

<u>Rule 1</u>
$$\overset{(+2)(-1)}{CuF_2} \quad + \quad \overset{(-3)(+1)}{NH_3} \quad \rightarrow \quad \overset{(+1)(-3)}{Cu_3N} \quad + \quad \overset{(-3)(+1)(-1)}{NH_4F} \quad + \quad \overset{(0)}{N_2}$$

<u>Rule 2</u> $CuF_2 \quad + \quad NH_3 \quad \rightarrow \quad Cu_3N \quad + \quad NH_4F \quad + \quad N_2$

oxidation increase +3/atom

decrease -1/atom reduction

<u>Rule 3</u> The total increase is 2(+3) because there are two N's in N_2.

$$6CuF_2 \quad + \quad NH_3 \quad \rightarrow \quad 2Cu_3N \quad + \quad NH_4F \quad + \quad N_2$$

<u>Rule 4</u> To balance the F's:

$$6CuF_2 \quad + \quad NH_3 \quad \rightarrow \quad 2Cu_3N \quad + \quad 12NH_4F \quad + \quad N_2$$

To balance the N's and H's:

$$6CuF_2 \quad + \quad 16NH_3 \quad \rightarrow \quad 2Cu_3N \quad + \quad 12NH_4F \quad + \quad N_2$$

Balancing redox equations which occur in solution is discussed in Sec. 14-4.

4-7 THEORETICAL YIELD OR CALCULATED YIELD

It is possible to calculate the quantity of a product that can be obtained from a reaction by assuming that the reaction is the only one involved and that all of the product is collected. This calculated quantity is called the <u>theoretical yield</u>. Because side reactions may occur and some product may be lost in the laboratory manipulations, the <u>actual yield</u> is often less than the theoretical one. The ratio of the actual yield to the theoretical yield is represented by the <u>percent yield</u>.

Example 4-9

How many grams of magnesium oxide can be prepared from 121.5 g of
magnesium metal by the reaction

$$2Mg \ + \ O_2 \ = \ 2MgO$$

Check to be sure the equation is balanced.

A. The Mole Method

1. $2Mg \ + \ O_2 \ = \ 2MgO$

    2 g atoms + 1 mole → 2 moles

    In general, 2 g atoms of magnesium will produce 2 moles of
magnesium oxide, but in this problem we have 122 g of magnesium.
How many gram atoms have we?

$$\frac{122 \ g}{24.3 \ g/g \ atom} \ = \ 5.00 \ g \ atom \ Mg$$

According to the equation, 5.00 g atoms of Mg will produce 5.00 moles
of MgO

        MgO   24.3

                16.0

                40.3 g/mole

        5.00 moles x 40.3 g/mole = 202 g

This is the theoretical yield of the reaction.

2. The mole ratio of MgO to Mg from the equation is

$$\frac{2MgO}{2Mg} \ = \ 1$$

then

$$\frac{moles \ of \ MgO}{5.00 \ Mg} \ = \ 1$$

    moles of MgO = 5.00

Since 1 mole of MgO weighs 40.3 g, the total weight produced is

    5.00 x 40.3 = 202 g.

B. The Factor Method

    $2Mg \ + \ O_2 \ \rightarrow \ 2MgO$

    2 x 24.3 g   2 x 40.3 g

        48.6 g   80.6 g

    According to the equation, 48.6 g of Mg will yield 80.6 g

MgO. Since we started with 122 g of Mg in this problem, we must
multiply 122 by the factor.

$$\frac{80.6 \text{ g MgO}}{48.6 \text{ g Mg}}$$

from the equation

$$122 \times \frac{80.6}{48.6} = 202 \text{ g (theoretical yield)}$$

In a series of reactions it is not necessary to calculate the
quantities of intermediate materials prepared, provided the reactions
are quantitative, i.e., no side reactions occur.

Example 4-10

Sulfuric acid is manufactured by a process approximately represented
by the reactions:

$$S + O_2 \rightarrow SO_2$$

$$2SO_2 + O_2 \rightarrow 2SO_3$$

$$SO_3 + H_2O \rightarrow H_2SO_4$$

What weight of sulfuric acid could be obtained from 3.20 g of
sulfur if all the reactons were quantitative?

According to the balanced equations, 1 g atom S produces 1 mole
SO_2; 1 mole SO_2 gives 1 mole SO_3. (The second equation may be
written $SO_2 + \frac{1}{2} O_2 \rightarrow SO_3$.) One mole SO_3 gives 1 mole H_2SO_4.
One g atom S gives 1 mole H_2SO_4, through the series of reactions.

$$\frac{3.20}{32.0} = 0.100 \text{ g atom S was present}$$

therefore 0.100 mole H_2SO_4 was produced

$$0.100 \times 98.0 = 9.80 \text{ g } H_2SO_4$$

$$\text{mole} \times \frac{g}{\text{mole}} = g$$

Example 4-11

A solution containing $CaCl_2$ was added to a solution of $AgNO_3$. The
$AgNO_3$ solution was prepared by dissolving 1.00 g of Ag in nitric

acid. What was the weight of calcium chloride in the solution if all of it reacted with the silver nitrate solution to form AgCl?

The equations are

$$Ag(s) + 2HNO_3(aq) \rightarrow AgNO_3(aq) + NO_2(g) + H_2O(\ell) \qquad (1)$$

$$2AgNO_3(aq) + CaCl_2(aq) \rightarrow 2AgCl(s) + Ca(NO_3)_2(aq) \qquad (2)$$

2 moles $AgNO_3$ required in Reaction (2)

Therefore, 2 g atoms Ag required in Reaction(1)

$$2Ag(s) + 4HNO_3(aq) \rightarrow 2AgNO_3(aq) + 2NO_2(g) + 2H_2O(\ell)$$

In the series of reactions, 2 g atoms Ag are equivalent to 1 mole $CaCl_2$. The g atom Ag/mole $CaCl_2$ ratio = 2.

$$\frac{1.00}{108} = 0.00928 \text{ g atom Ag present}$$

$$\frac{g}{g/g \text{ atom}} = \cancel{g} \times \frac{g \text{ atom}}{\cancel{g}} = g \text{ atom}$$

0.00928 g atom Ag $\rightarrow$ 0.00464 mole $CaCl_2$

$$CaCl_2 \quad 1 \times 40.0 = 40.0$$
$$2 \times 35.4 = \underline{70.8}$$
$$110.8 \text{ g/mole}$$

$$0.00464 \times 111 = 0.514 \text{ g } CaCl_2$$

If it is known that all of an element present is transformed quantitatively from one substance to another, it is not necessary to know what the other products were in order to calculate a theoretical yield. In Example 4-10 all of the sulfur was eventually transformed to sulfuric acid, S $\rightarrow$ H_2SO_4. It was not necessary to know the reaction sequence to calculate the quantity of H_2SO_4 it is theoretically possible to obtain from a given quantity of sulfur.

Example 4-12

How many grams of AgCl can be made from 0.579 g of $SeOCl_2$?

1 mole $SeOCl_2$ $\rightarrow$ 2 moles AgCl, assuming all of the chlorine in $SeOCl_2$ is retained in the AgCl.

$$SeOCl_2 \quad 1 \times 79.0 = 79.0 \qquad\qquad AgCl \quad 107.9$$
$$1 \times 16.0 = 16.0 \qquad\qquad\qquad\quad \underline{35.4}$$
$$2 \times 35.4 = \underline{70.8} \qquad\qquad\qquad\quad 143.3 \text{ g/mole}$$
$$165.8 \text{ g/mole}$$

$$\frac{0.578}{166} = 0.00347 \text{ mole SeOCl}_2$$

$$\frac{g}{g/\text{mole}} = \cancel{g} \times \frac{\text{mole}}{\cancel{g}} = \text{mole}$$

3.47×10^{-3} mole $SeOCl_2$ would give $2 \times 3.47 \times 10^{-3}$ mole AgCl, and $7.94 \times 10^{-3} \times 143 = 1.00$ g AgCl

$$\cancel{\text{mole}} \times \frac{g}{\cancel{\text{mole}}} = g$$

When one element in a reactant is incorporated into two or more products, all of the products must be known.

It is possible to calculate the quantity of silver nitrate which can be prepared from a quantity of silver without consideration of the reactions involved. It is not possible to make a similar calculation in the case of nitric acid. Not all of the nitrogen in the nitric acid is transformed to silver nitrate.

$$Ag + 2HNO_3 \rightarrow AgNO_3 + NO_2 + H_2O$$

Two moles of HNO_3 are required to make 1 mole of $AgNO_3$.

Example 4-13

What weight of copper sulfate can be obtained by the action of 2.941 g of hot concentrated sulfuric acid on an excess of copper?

$$Cu + 2H_2SO_4 \rightarrow CuSO_4 + SO_2 + 2H_2O$$

2 moles $H_2SO_4 \rightarrow$ 1 mole $CuSO_4$

H_2SO_4
$\quad$ 2 x 1.0 = 2.0
$\quad$ 1 x 32.1 = 32.1
$\quad$ 4 x 16.0 = $\underline{64.0}$
$\qquad\qquad\qquad$ 98.1 g/mole

$CuSO_4$
$\quad$ 1 x 63.5 = 63.5
$\quad$ 1 x 32.1 = 32.1
$\quad$ 4 x 16.0 = $\underline{64.0}$
$\qquad\qquad\qquad$ 159.6 g/mole

$$\frac{2.943}{98.1} = 0.0300 \text{ mole } H_2SO_4$$

$$\frac{g}{g/\text{mole}} = \cancel{g} \times \frac{\text{mole}}{\cancel{g}} = \text{mole}$$

According to the equation

$$3.00 \times 10^{-2} \text{ mole } H_2SO_4 \rightarrow 1.50 \times 10^{-2} \text{ mole } CuSO_4$$

$$1.50 \times 10^{-2} \times 160 = 2.40 \text{ g } CuSO_4$$

4-8 ACTUAL YIELD, PERCENT YIELD

When side reactions occur or when material is lost during transfer and separation of the products, the actual quantity of product is less than that calculated. The actual yield is often expressed as a percent of the calculated yield.

Example 4-14

Calculate the percent yield of oxygen if 0.0240 g resulted from heating 0.303 g of KNO_3.

$$2KNO_3 \rightarrow 2KNO_2 + O_2$$

1. Check to be certain that the equation is balanced before starting any calculations.

2. Calculate the theoretical yield.

$$2 \text{ moles } KNO_3 \rightarrow 1 \text{ mole } O_2$$

KNO_3	1 x 39.1 = 39.1	O_2	2 x 16.0 = 32.0
	1 x 14.0 = 14.0		g/mole
	3 x 16.0 = 48.0		
	101.1 g/mole		

$\frac{0.303}{101} = 0.00300$ mole KNO_3, which according to the equation, decomposed as follows:

$$3.00 \times 10^{-3} \text{ mole } KNO_3 \rightarrow 1.50 \times 10^{-3} \text{ mole } O_2$$

The calculated yield is

$$1.50 \times 10^{-3} \times 32.0 = 48 \times 10^{-3} = 4.80 \times 10^{-2} \text{ g}$$

3. Percent yield $= \dfrac{\text{actual yield}}{\text{calculated yield}} \times 100$

$$= \frac{2.40 \times 10^{-2}}{4.80 \times 10^{-2}} = 50.0\%$$

Example 4-15

Calculate the percent yield if the reaction of 64.0 g of $NaBH_4$ with I_2 produced 15.0 g of BI_3.

> [H. C. Anderson and L. H. Belz, J. Am. Chem. Soc., 75, 4828 (1953)]

1. Balanced equation

$NaBH_4 + 4I_2 \rightarrow BI_3 + NaI + 4HI$

2. 1 mole $NaBH_4 \rightarrow$ 1 mole BI_3

$NaBH_4$ 1 x 23.0 = 23.0

              1 x 10.8 = 10.8

              4 x 1.0 = $\underline{4.0}$

                         $\overline{37.8}$ g/mole

BI_3 1 x 10.8 = 10.8

            3 x 126.9 = $\underline{380.7}$

                        $\overline{391.5}$ g/mole

$\dfrac{64.0}{37.8}$ = 1.69 mole $NaBH_4$

Calculated yield: 1.69 x 392 = 662 g BI_3

3.   Percent yield: $\dfrac{15.0}{662}$ x 100 = 2.27%

PROBLEMS

4-1 Balance the following equations:

a. $Fe(s) + H_2O(g) \rightarrow Fe_3O_4(s) + H_2(g)$

b. $Na(s) + H_2O(\ell) \rightarrow H_2(g) + NaOH(s)$

c. $CO(g) + H_2(g) \rightarrow CH_3OH(\ell)$

d. $CaH_2(s) + H_2O(\ell) \rightarrow Ca(OH)_2(s) + H_2(g)$

e. $BaO_2(s) \rightarrow BaO(s) + O_2(g)$

f. $KNO_3(s) \rightarrow KNO_2(s) + O_2(g)$

g. $KNO_3(s) + K(s) \rightarrow K_2O(s) + N_2(g)$

h. $Ti(OH)_3(s) \rightarrow TiO_2 + H_2O + H_2(g)$

i. $Na_2CO_3(s) + FeCr_2O_4(s) + O_2(g) \rightarrow$
$$Fe_2O_3(s) + Na_2CrO_4(s) + CO_2(g)$$

j. $Fe_2O_3(s) + CO(g) \rightarrow Fe_3O_4(s) + CO_2(g)$

k. $Fe_3O_4(s) + CO(g) \rightarrow FeO(s) + CO_2(g)$

l. $CS_2(\ell) + O_2(g) \rightarrow CO_2(g) + SO_2(g)$

m. $Al_4C_3(s) + F_2(g) \rightarrow AlF_3(s) + CF_4(g)$

n. $Na_2Cr_2O_7(s) + HCl(g) \rightarrow CrO_2Cl_2(g) + NaCl(s) + H_2O(\ell)$

o. $Fe_2O_3(s) + CO(g) \rightarrow Fe(\ell) + CO_2(g)$

p. $CuFeS_2(s) + O_2(g) \rightarrow Cu(s) + FeO(s) + SO_2(g)$

q. $Al(s) + Fe_2O_3(s) \rightarrow Fe(s) + Al_2O_3(s)$

r. $CS_2(g) + Cl_2(g) \rightarrow CCl_4(g) + S_2Cl_2(g)$

s. $CaC_2(s) + H_2O(\ell) \rightarrow Ca(OH)_2(s) + C_2H_2(g)$

t. $SiH_4(g) + O_2(g) \rightarrow SiO_2(s) + H_2O(\ell)$

u. $PbS(s) + O_2(g) \rightarrow PbO(s) + SO_2(g)$

v. $PbO(s) + C(s) \rightarrow Pb(s) + CO_2(g)$

w. $PbS(s) + PbO(s) \rightarrow Pb(\ell) + SO_2(g)$

x. $PbS(s) + PbSO_4(s) \rightarrow Pb(\ell) + SO_2(g)$

y. $NH_4NO_2(s) \rightarrow N_2(g) + H_2O(g)$

z. $NH_4NO_3(s) \rightarrow N_2O(g) + H_2O(g)$

4-2 The next eight equations appeared in journal articles in recent years. Balance them.

a. $BI_3 + NH_3 \rightarrow B_2(NH)_3 + NH_4I$
  [J. Am. Chem. Soc., 78, 2069 (1956)]

b. $NpF_3 + Si \rightarrow NpSi_2 + SiF_4$

c. $NpCl_4 + NH_3 \rightarrow NpCl_3 + HCl + N_2$

d. $Np_3P_4 + HCl \rightarrow NpCl_4 + PH_3$

[J. Am. Chem. Soc., 75, 1232 (1953)]

e. $CH_4 + SO_2 \rightarrow CS_2 + H_2O + CO_2$

[Chem. Abstr., 53, 18713 g (1959)]

f. $B_2Cl_4 + O_2 \rightarrow B_2O_3 + BCl_3$

g. $B_2Cl_4 + H_2O \rightarrow H_3BO_3 + HCl + H_2$

[J. Am. Chem. Soc., 80, 6153-6155 (1958)]

h. $UF_6 + HBr \rightarrow UF_5 + HF + Br_2$

4-3 How many moles of sulfur dioxide and how many moles of carbon dioxide can be formed by the reaction of six moles of carbon disulfide with oxygen?

4-4 How many millimoles of fluorine are required to react with two moles of Al_4C_3 if the products are aluminum fluoride and carbon tetrafluoride?

4-5 How many grams of diborane (B_2H_6) can be produced from 0.100 mM $LiAlH_4$, according to the reaction

$BF_3 + LiAlH_4 \rightarrow B_2H_6 + LiF + AlF_3$

(Balance the equation first.)

> In the next five problems reactions from recent
> research papers are used. In each case, balance
> the equation and then calculate the theoretical
> yield of the designated product from the quantity
> of starting material given.

4-6 $P_2I_4S_2 \rightarrow P_4S_7 + PI_3 + I_2$
       0.262 g   ?

[A. H. Cowley and S. T. Cohen, Inorg. Chem., 3, 780 (1964)]

4-7 $Si_3Cl_8 + NH_3 \rightarrow Si_3N_4H_4 + NH_4Cl$
      1.65 g   ?
      [M. Billy, Compt. Rend., 251, 1639 (1960)]

4-8 $ScO_3 + B \xrightarrow{1800-1850^{\circ}} ScB_2 + BO$
      2.72 g   ?
      [G. V. Samsonov, Dok. Akad. Nauk, SSSR, 133 1344 (1960)]

4-9 $UO_3 + CBr_4 \xrightarrow{110^{\circ}} UOBr_3 + COBr_2 + Br_2$
      0.0265 g   ?
      [J. Prigent, Ann. Chim. (Paris), 5, 65 (1960)]

4-10 $BiS_2 + O_2 \rightarrow Bi_2O_3 + SO_2$
      16.5 g   ?
      [M. S. Silverman, Inorg. Chem., 3, 1041 (1964)]

4-11 In the following reaction, 0.273 g of NF_2SO_3F gave 1.85 mM of
nitrite ion. Was this quantitative?

$NF_2SO_3F + 4OH^- \rightarrow NO_2^- + SO_3F^- + 2F^- + 2H_2O$

      [M. Lustig and G. H. Cady, Inorg. Chem., 2, 388 (1963)]

4-12 As part of a research project at Iowa State University, the
composition of tungsten(IV) chloride was determined by hydrogen
reduction at 650°. The tungsten metal formed was weighed as such,
and the chlorine was converted to hydrogen chloride by reaction
with hydrogen. The hydrogen chloride was collected in a known
amount of standard sodium hydroxide solution. Write equations for
the reactions which occurred.

      [R. E. McCarley and T. M. Brown, Inorg. Chem., 3,
      1233 (1964)]

4-13 Tl_2CO_3 was fused with V_2O_5, and one product obtained was found
to contain 67.12% Tl, 16.33% V, and the remainder oxygen. The other

product of the reaction was CO_2. Write an equation for the reaction.
[T. Carnelley, J. Chem. Soc., 26, 334 (1873)]

4-14 A 4:11 atomic weight mixture of Bi and S was subjected to a
pressure of 50 kbars and a temperature of 1250° for 5 to 10 min.
The product consisted of gray, needlelike crystals. Analysis
found: Bi, 74.0; S, 23.0. When the compound was oxidized with
pure oxygen, the residue contained 90.0% Bi and no sulfur. The
mass loss because of oxidation was 14.7%. When heated in vacuo to
300°, the initial product decomposed to Bi_2S_3, as evidenced by the
X-ray diffraction pattern of the residue. Write three equations
for the reactions which occurred.
[M. S. Silverman, Inorg. Chem., 3, 1041 (1964)]

4-15 Boron trifluoride and dioxygen difluoride were allowed to
react at 126°. Excess boron trifluoride was removed, and it was
found that 7.17 mM of BF_3 had been consumed and 3.14 mM of fluorine
had been formed. Are these results consistent with the claim that
the compound O_2BF_4 (dioxygenyl tetrafluoroborate) had been formed?
[I. J. Solomon et al., Inorg. Chem., 3, 457 (1964)]

4-16 The compound described in Prob. 4-15 gave 6.80, 7.21, and
3.40 mM of BF_3, O_2, and F_2, respectively, when allowed to warm to
room temperature and decompose. Write the equation for the
decomposition.

4-17 A chocolate-colored substance (0.4512 g) prepared by the
reaction of KNH_2 on mercury(II) iodide in liquid ammonia gave
0.0218 g of nitrogen when decomposed and 0.4982 g of mercury(II)
sulfide by reaction with H_2S. What is the formula of the highly
explosive, chocolate-colored compound?
[E. C. Franklin, J. Am. Chem. Soc., 27, 870 (1905)]

4-18 Xenon (4.25 mM) was mixed with a large excess of O_2F_2 at
-118°, and from a yellow solid that formed, 0.7052 g of XeF_2 was

sublimed. Assume that oxygen was the other product of the decomposition. What is the formula of the yellow solid consistent with
these observations?

[S. I. Morrow and A. R. Young, Inorg. Chem., 4,
759 (1965)]

4-19 Ammonium chloride and carbon tetrachloride were heated to
475° for an hour. Hydrogen chloride gas and a yellow solid containing
36.6% C, 1.2% H, and 62.2% N were formed. Write an equation for
the reaction which occurred.

[D. D. Cubbicciotti and W. M. Ratimer, J. Am. Chem. Soc.,
70, 3509 (1949)]

4-20 CCl_2F_2 gas was passed over WO_2 at 525°, and a solid mixture of
WOF_4 and $WOCl_4$ and CO gas were the products. The solid mixture
contained W, 59.1%; F, 12.0%; Cl, 22.7%; O, 6.2%. Calculate the
composition of the solid mixture. Write an equation for the reaction
which occurred.

[A. D. Webb and H. A. Young, J. Am. Chem. Soc., 72,
3358 (1950)]

4-21 A mixture of mercury(II) chloride, indium, and aluminum was
heated to 325°. A white solid (In, 40.51; Al, 9.69; Cl, 49.74)
and mercury were the products. In a typical preparation, 3.6061 g
of mercury(II) chloride, 0.761 g of indium, and 0.1792 g of aluminum
were employed. Were any of the reactants in excess? What was the
theoretical yield?

[R. J. Clark, E. Griswold, and J. Kleinberg, J. Am. Chem.
Soc., 80, 4766 (1958)]

4-22 Protactinium metal was prepared by the reduction of 0.100 mg
of protactinium(IV) fluoride with barium metal at 1400°. How many
grams and how many moles of protactinium could be prepared
assuming a 100% yield?

[P. A. Sellers et al., J. Am. Chem. Soc., 76, 5937 (1954)]

4-23 How many grams of neptunium nitride can be made from 1.00 mg
of Np by the following reactions, assuming 100% yield?

$$2Np + 3H_2 \rightarrow 2NpH_3$$

$$NpH_3 + NH_3 \rightarrow NpN + 3H_2$$

  [I. Sheft and S. Fried, J. Am. Chem. Soc., 75,
  1236 (1953)]

4-24 Copper(II) chloride reacted with lithium borohydride in
diethyl ether:

$$CuCl_2 + 2LiBH_4 \rightarrow Cu + 2LiCl + B_2H_6 + H_2$$

In one experiment 3.58 mM of $CuCl_2$ and 7.15 mM of $LiBH_4$ gave 3.17 mM
B_2H_6. What was the mole percent yield of the product?

  [T. J. Klinger, Inorg. Chem., 3, 1058 (1964)]

4-25 A sample of copper (0.1350 g) was dissolved in sulfuric acid,
and the solution was evaporated to dryness. $CuSO_4$ (0.1340 g) was
collected. What was the percent yield? The other products of the
reaction were SO_2 and H_2O.

  [C. Baskerville, J. Am. Chem. Soc., 17, 904 (1895)]

4-26 What was the percent yield if 1.7395 g of lead reacted with
oxygen to give 1.8650 g of lead(II) oxide?

  [F. J. Brisbie, J. Chem. Soc., 93, 161 (1908)]

4-27 A 43.92 g sample of $CF_3CF_2CF_2COOAg$ was treated with 65.7 g
of bromine. The products were a solid, a liquid, and a gas. The
liquid had a boiling point of 12°C and contained 14.56% carbon by
analysis. Its molecular weight (gas-density method) was found to be
249. The solid product weighed 25.42 g. The gas was assumed to be
carbon dioxide and was changed to barium carbonate. The weight of
barium carbonate represented a 99.5% yield, assuming the gas to be
CO_2. Write an equation for the reaction. Was either of the reactants
in excess?

[M. Hauptschein, E. A. Nodiff, and A. V. Grosse, J. Am.
Chem. Soc., 74, 1348 (1952)]

4-28 Twenty-five grams of the compound prepared in Prob. 3-45 was
placed in a sealed tube with aluminum foil and heated to 475° for
18 hr. The aluminum chloride formed was removed by sublimation;
the other product contained 56.45% W and 43.10% Cl. Based on the
weight of aluminum used, a 100% yield of the product was obtained.
If the starting tungsten compound was 2.0 g in excess, how many
grams of final product was formed?

4-29 In the preparation of the compound described in Prob. 3-61,
32.5 g of $KAsO_3$ and 11.6 g KF were used. The yield of product was
25.0 g. Was either of the reactants in excess? Calculate the
percent yield of product.

4-30 Chlorine was passed over 0.1878 g of $NaNO_3$ heated in a boat in
a combustion tube. What was the percent yield of NaCl if 0.1289 g
was prepared along with NO_2 and O_2?
          [E. F. Smith and J. G. Hibbs, J. Am. Chem. Soc., 17,
          682 (1895)]

4-31 Calculate the atomic weight of nitrogen from the fact that
12.46620 g of $AgNO_3$ gave 7.91609 g of Ag by reduction.
          [O. Honigschmid, E. Zintl, and P. Thilo, Z. Anorg.
          Allgem. Chem., 163, 65 (1924)]

4-32 Calculate the atomic weight of calcium from the conversion
of $CaCl_2$ to AgCl in which the calcium chloride/silver chloride ratio
was 0.387200.
          [J. Am. Chem. Soc., 84, 4184 (1962)]

4-33 The atomic weight of vanadium was determined from the mass of
NaCl obtained from the reaction of $NaVO_3$ with HCl gas. What is the

atomic weight of V if 4.8564 g of $NaVO_3$ produced 2.3277 g of NaCl?

[D. J. McAdam, Jr., J. Am. Chem. Soc., 32, 1603 (1910)]

4-34 T. W. Richards and C. R. Hoover determined the atomic weight
of sodium from the mass of silver (9.68023 g) necessary to prepare
sufficient silver nitrate to react with the quantity of sodium
bromide made by the reaction of hydrobromic acid with 4.75555 g
sodium carbonate. Using the accepted atomic weights of silver,
carbon, and oxygen, calculate the atomic weight of sodium from
these data.

[J. Am. Chem. Soc., 37, 106 (1915)]

SUPPLEMENTARY PROBLEMS

4-35- Balance:

a. $NH_3(g) + O_2(g) \rightarrow NO(g) + H_2O(g)$

b. $NO_2(g) + H_2O \rightarrow HNO_3 + NO(g)$

c. $HNO_3(\ell) \rightarrow NO_2(g) + O_2(g) + H_2O(\ell)$

d. $N_2H_4(\ell) + H_2O_2(\ell) \rightarrow N_2(g) + H_2O(g)$

e. $Ca_3(PO_4)_2(s) + SiO_2(s) + C(s) \rightarrow CaSiO_3 + CO(g) + P_2(g)$

f. $PCl_3(s) + H_2O(\ell) \rightarrow H_3PO_3(aq) + HCl(g)$

g. $Ca_3P_2(s) + H_2O(\ell) \rightarrow PH_3(g) + Ca(OH)_2(s)$

h. $NaH_2PO_4(s) \rightarrow Na_3P_3O_9(s) + H_2O(g)$

i. $Cu_2S(s) + O_2(g) \rightarrow Cu(s) + SO_2(s)$

j. $HClO_3(\ell) \rightarrow ClO_2(g) + O_2(g) + H_2O(g)$

k. $LiH(s) + H_2O(\ell) \rightarrow LiOH(s) + H_2(g)$

l. $Ca(OH)_2(aq) + H_3PO_4(aq) \rightarrow Ca_3(PO_4)_2(s) + H_2O$

m. $Na_2O(s) + H_2O(\ell) \rightarrow NaOH(s)$

n. $CaO(s) + P_4O_{10}(s) \rightarrow Ca_3(PO_4)_2(s)$

o. $P_4O_{10}(s) + H_2O(\ell) \rightarrow H_3PO_4(\ell)$

p. $MnO_2(s) + HCl(aq) \rightarrow MnCl_2(aq) + H_2O + Cl_2(g)$

q. $KBF_4(s) + Na(\ell) \rightarrow KF(s) + NaF(s) + B(s)$

r. $Al_2O_3(s) + Mg(\ell) \rightarrow MgO(s) + Al(s)$

s. $H_3BO_3(s) \rightarrow H_2O + B_2O_3(s)$

t. $NH_3(g) + CuO(s) \rightarrow Cu(s) + H_2O(g) + N_2(g)$

u. $H_2SO_4(aq) + CaF_2(s) \rightarrow CaSO_4(s) + HF(aq)$

v. $PbS(s) + H_2O_2(aq) \rightarrow PbSO_4(s) + H_2O$

w. $Cl_2(g) + HgO(s) \rightarrow HgCl_2(s) + Cl_2O(g)$

x. $Pb(NO_3)_2(s) \rightarrow PbO(s) + NO_2(g) + O_2(g)$

y. $Zn_3As_2(s) + HCl(aq) \rightarrow ZnCl_2(aq) + AsH_3(g)$

z. $I_2 + HNO_3(aq) \rightarrow HIO_3(aq) + NO_2(g) + H_2O$

4-36 How many millimoles of CrO_2Cl_2 will result from the reaction of HCl with 2.68 g of sodium dichromate if the other products are sodium chloride and water?

4-37 How many grams of $VOCl_3$ can be made from 100 g of V_2O_5 by the reaction

$$V_2O_5 + Cl_2 \rightarrow VOCl_3 + O_2$$

4-38 At $-45°$, $CuCl_2$ reacted with $LiBH_4$ according to the equation

$$2CuCl_2 + 4LiBH_4 \rightarrow 2CuBH_4 + 4LiCl + B_2H_6 + H_2$$

In one experiment, 2.30 mM $CuCl_2$ and 4.54 mM $LiBH_4$ gave 0.98 mM B_2H_6. Was either of the reactants in excess? What was the percent yield of B_2H_6?

[T. J. Klinger, Inorg. Chem., 3, 1058 (1964)]

4-39 Zirconium metal was heated in an atmosphere of chlorine. The product of the reaction was analyzed and found to contain 39.0% Zr and 60.6% Cl. One sample of the product weighing 12.0 g was mixed with 0.35 g of zirconium metal and heated to 460°. The product of this reaction contained 45.9% Zr and 53.7% Cl. Write equations for the reactions which occurred.

[H. L. Schlafer and H. W. Wille, Z. Anorg. Allgem. Chem., 327, 253 (1964)]

4-40 Ten grams of high-purity silicon, a 90% yield, was made by the reaction of sodium with $SiCl_4$. What other product was formed? How many grams of $SiCl_4$ were used?

[B. Kamenas and D. Ordenic, Z. Anorg. Allgem. Chem., 321, 113 (1963)]

4-41 A sample of a compound containing 56.45% tungsten and 43.10% chlorine was heated to $450\text{-}500^\circ$ in a sealed, evacuated tube. Tungsten(V) chloride and a compound containing 72.11% tungsten and 27.67% chlorine were formed. What is the equation for the reaction which occurred?

[R. E. McCarley and T. M. Brown, Inorg. Chem., 3, 1233 (1964)]

4-42 Fluorine gas was passed over tin(II) sulfide at 500°, and the solid fluoride that was formed contained 61.00% tin. Assuming the other product was SF_6, write the equation for the reaction.

[H. M. Haendler et al., J. Am. Chem. Soc., 76, 2178 (1954)]

4-43 Carbon tetrachloride vapor was passed over a sample of Gd_2O_3 in a tube furnace at 650°, and phosgene and a solid product containing 40.2% chlorine were formed. Write an equation for the reaction.

[J. F. Miller, S. E. Miller, and R. C. Himes, J. Am. Chem. Soc., 81, 4450 (1959)]

4-44 The oxide of tungsten, WO_3, was reduced with hydrogen gas. A sample of tungsten weighing 2.49330 g was obtained from 3.14520 g of WO_3.   What was the percent yield?

   [G. E. Thomas, J. Am. Chem. Soc., 21, 373 (1899)]

4-45 A sample of the compound (0.47 mM prepared in Prob. 3-17
was heated at 205° for 150 min. Hydrogen (1.44 mM) and 0.36 mM
$B_3N_3H_6$ were collected. What was the molecular formula of the
original compound if the yield of $B_3N_3H_6$ was 77%?

4-46 What was the percent yield in the reaction
   $BaPt(CN)_4 \cdot 4H_2O + ZnSO_4 \rightarrow ZnPt(CN)_4 + BaSO_4 + H_2O$
   0.6029 g   0.4261 g
      [L. A. Levy, J. Chem. Soc., 101, 1084 (1912)]

4-47 Refer to Prob. 3-17. The molecular weight of the compound
was 86.5. A sample weighing 0.47 mM was heated to 205° for 150 min,
and 1.44 mM of hydrogen and 0.36 mM of $B_3N_3H_6$ were obtained. Write
an equation for the reaction. What was the mole percent yield
of $B_3N_3H_6$?
      [G. H. Dahl and R. Schaeffer, J. Am. Chem. Soc., 83,
      3033 (1961)]

4-48 What was the percent yield in the following reaction if
1.07 g of $Co_2(CO)_8$ gave 0.30 g of $[Co_2S(CO)_5]_n$? Assume n = 1.
   $Co_2(CO)_8 + S \rightarrow CO + [Co_2S(CO)_5]_n$
      [L. Marko, G. Bor, and G. Almasy, Chem. Ber., 94,
      847 (1961)]

4-49 Trifluoroacetic anhydride $(CF_3CO)_2O$ was prepared by treating
100 g of trifluoroacetic acid CF_3COOH with 100 g of P_4O_{10}. The
anhydride collected weighed 65.6 g. What was the percent yield?
(The P_4O_{10} removes water to form H_3PO_4.)

4-50 When stibene, SbH_3, was heated at pressures greater than
2/3 atm, it decomposed explosively into the elements. If the
yield was 99.90% of the theoretical, how much hydrogen, in grams,

was obtained from 0.8231 g of stibene?

[W. Jolly et al., J. Inorg. Nucl. Chem., 14, 190 (1960)]

4-51 A sample of UCl_3 (10.2 g) was obtained, along with $ZnCl_2$, from the reaction of UCl_4 with zinc at 450° for 18 hr. If the yield was 86.0% of the theoretical, how much UCl_4, in grams, was used in the reaction?

[H. S. Young, Chem. Abstr., 53, 18710 i (1959)]

4-52 A compound, believed to be CdO_2, was heated, and cadmium(II) oxide and oxygen gas were formed. The loss in mass was 11.5%. Does this loss in mass substantiate the belief that the compound was CdO_2?

[C. W. W. Hoffmann, R. C. Ropp, and R. W. Mooney, J. Am. Chem. Soc., 81, 3830 (1959)]

4-53 How many grams of titanium dioxide would be required as the starting material in the preparation of 100 g of the fluoride described in Prob. 3-52?

4-54 How many grams of a 3.00% hydrazoic acid solution, HN_3, was required to make 14.7 g of potassium azide, KN_3, from potassium hydroxide if the yield was 91.5%?

[A. W. Browne, Inorg. Syn., 1, 80 (1939)]

4-55 In the following reaction, is either of the reactants in excess? What is the percent yield?

$$B_{10}H_{14} + 2(CH_3CH_2)_3NBH_3 \rightarrow [(CH_3CH_2)_3NH]_2B_{12}H_{12} + 3H_2$$
    1.727 g   5.688 g   4.6 g

[N. N. Greenwood and J. H. Morris, Proc. Chem. Soc., 1963, 338]

4-56 A 40-mg sample of XeF_2 reacted with water and gave 0.25 mM Xe, 0.12 mM O_2, and 0.55 mM HF. Which of the quantities reported was in excess of the quantity predicted by the equation for the reaction which presumably occurred?

> [L. V. Streng and A. G. Streng, Inorg. Chem., 4, 1370 (1965)]

4-57 A sample of the hydrate described in Prob. 3-65 (0.4847 g) was dissolved in water, and a solution containing chloride ion was added. A white precipitate of silver chloride formed. Calculate the weight of AgCl to be expected.

4-58 A sample of the amide prepared in Prob. 3-64 weighing 0.549 g reacted with HCl to yield silver chloride. How many grams of silver chloride would be expected?

4-59 Calculate the theoretical yield of $B_{10}H_{13}Na$ to be expected from the reaction of 1.56 g of $B_{10}H_{14}$ with 2.784 g of a 60.62% sodium hydride solution in n-propyl ether. The reaction was

$$B_{10}H_{14} + NaH \rightarrow B_{10}H_{13}Na + H_2$$

> [R. J. F. Polchak, J. H. Norman, and R. E. Williams,
> J. Am. Chem. Soc., 83, 3381 (1961)]

4-60 Diboron tetrachloride (1.400 mM) and an excess of boron tribromide were allowed to stand at room temperature. The diboron tetrabromide (0.3663 g) which formed was distilled away. What was the percent yield?

> [G. Urry et al., J. Am. Chem. Soc., 76, 5297 (1954)]

4-61 LaF_3 (22.83 mg) was reduced to the metal with Li. What was the percent yield if 15.29 mg of La was formed?

> [J. G. Stites, M. R. Salutsky, and B. D. Stone, J. Am.
> Chem. Soc., 77, 240 (1955)]

4-62 $SiCl_4$ (65 g) reacted with H_2S to form HCl and 26 g of the solid, $HSSiCl_3$. What was the percent yield of the silicon compound?

> [W. C. Schumb and W. J. Bernard, J. Am. Chem. Soc., 77,
> 862 (1955)]

4-63 A sample of gold(III) fluoride was prepared at Manchester
University by the reaction of bromine trifluoride with gold. The
AuF_3/Au ratio was found to be 1.287. How does this compare with the
theoretical AuF_3/Au ratio?

[A. S. Woolf, J. Chem. Soc., 1954 4695]

4-64 A sample of beryllium bromide weighing 6.04443 g was dissolved
in water, and to the resulting solution an excess of silver nitrate
was added. The precipitate of AgBr which formed was collected and
carefully dried; it weighed 13.44565 g. By using the accepted
values of the atomic weights of Ag and Br, from Table D-3, calculate
a value for the atomic weight of beryllium.

[G. E. F. Lundell, J. Am. Chem. Soc., 70, 3531 (1948)]

4-65 T. W. Richards and G. S. Forbes dissolved silver in nitric
acid and collected the $AgNO_3$ by evaporation. In one experiment, of
a series of six, 4.60825 g Ag gave 7.25706 g $AgNO_3$. Calculate the
atomic weight of silver from the equation

$Ag + HNO_3 \rightarrow AgNO_3 + NO_2 + H_2O.$
[J. Am. Chem. Soc., 29, 808 (1907)]

4-66 Calculate the atomic weight of iron from the fact that
0.574244 g of $FeBr_3$ gave 1.00000 g of AgBr.

[J. Am. Chem. Soc., 84, 4185 (1962)]

4-67 Calculate the atomic weight of calcium if 1.03950 g of $CaCl_2$
gave 2.68467 g of AgCl.

[O. Honigschmid and K. Kempter, Z. Anorg. Allgem. Chem., 195,
1 (1931)]

4-68 What is the atomic weight of vanadium calculated from the
observation that 4.42279 g of $VOCl_3$ gave 10.97187 g of AgCl?

[H. V. A. Broscoe and H. F. V. Little, J. Chem. Soc., 105
1310 (1941)]

Chapter 5

GASES

The introductory study of gases, i.e., substances in the gaseous
state, involves the relationships between four measurable quantities:
mass, volume, temperature, and pressure. These are the four
properties sufficient to define a state.

One of the outstanding characteristics of substances in the
gaseous state is the sensitivity of the volume to changes in pressure
and temperature. The quantitative relationships between the volume
of a given mass of a gas and the pressure and temperature are
described by two physical laws.

5-1 BOYLE'S LAW

This law may be stated:
    The volume of a given mass of gas, at constant
    temperature, varies inversely with the pressure.
The volume is an inverse function of the pressure.*

---

* The SI unit of pressure is the pascal, symbol Pa. This is defined
as a force per unit area of 1 newton (N) per square meter. The most
commonly used unit is the atmosphere: 1 atmosphere (atm) =
101,325 N m^{-2}. Pressure is also measured in millimeters of Hg:
1 atmosphere (atm) = 760 mm Hg, or 760 torr. The units mm Hg and
torr do not belong to the International System of Units.

$$V \propto 1/P \qquad \text{or} \qquad V = k/P$$

The law may be stated another way:

The product of the pressure and volume of a given

mass of a gas is constant at a fixed temperature.

$PV = k$

It follows from this that $P_1V_1 = P_2V_2 = P_3V_3$, etc = k.

Example 5-1

A sample of gas was collected under a pressure of 740 mm in a 200-ml vessel. What volume would the same sample of gas occupy at 760 mm?

Method 1: Since the pressure was increased, the volume would decrease. The original volume must be multiplied by the factor less than 1, that is, 740/760.

$$200 \times \frac{740}{760} = 195 \text{ ml}$$

$$\text{ml} \times \frac{\text{mm}}{\text{mm}} = \text{ml}$$

Method 2: Let P_1 = 740 mm, V_1 = 200 ml, P_2 = 760 mm, and V_2 = the volume at P_2.

$$P_1V_1 = P_2V_2$$

$$740 \times 200 = 760 \, V_2$$

$$V_2 = \frac{740 \times 200}{760} = 195 \text{ ml}$$

5-2 GAY-LUSSAC'S LAW (CHARLES'LAW)

At a constant pressure, the volume of a given mass of a gas increases 1/273 of the volume at 0°C for each 1° rise in temperature.

$$V_t = V_o + \frac{1}{273} V_o t$$

where V_t is the volume at some temperature t°C and V_o is the volume at 0°C.

$$V_t = V_o \left(1 + \frac{t}{273} \right)$$

$$V_t = V_o \left(\frac{273 + t}{273} \right)$$

$$V_t = \frac{V_o}{273} (273 + t)$$

if

$$T = 273 + t$$

then

$$V_t = \frac{V_o}{273} T$$

where T is the absolute temperature (more correctly the thermodynamic temperature) in degrees Kelvin (K).

$$V = k'T$$

The volume is directly proportional to the absolute temperature. For a given quantity of a gas at constant pressure: $V_1 = k'T_1$, $V_2 = k'T_2$, $V_3 = k'T_3$, etc.

$$\frac{V_1}{T_1} = \frac{V_2}{T_2} = \frac{V_3}{T_3}$$

Example 5-2

What was the volume of a quantity of a gas at 27°C if its volume was 400 ml at 0°C? The pressure remained constant.

Method 1: Since the temperature increased, the volume increased; and the original volume must be multiplied by the factor greater than 1, that is, 300/273.

The temperature must be in degrees Kelvin,

$$27°C + 273 = 300 \text{ K}$$

$$0°C + 273 = 273 \text{ K}$$

$$400 \times \frac{300}{273} = 438 \text{ ml}$$

$$ml \times \frac{K'}{K'} = ml$$

Method 2: Let $V_1 = 400$ ml, $T_1 = 0 + 273 = 273$ K, $T_2 = 27 + 273 = 300$ K and

V_2 = the volume at T_2

$$\frac{V_1}{T_1} = \frac{V_2}{T_2}$$

$$\frac{400}{273} = \frac{V_2}{300}$$

$$V_2 = \frac{400 \times 300}{273} = 438 \text{ ml}$$

5-3 EQUATION OF STATE

Boyle's law and Charles' law may be combined to give what is called an __equation of state__. Suppose a given mass of a gas has a volume V_1 and a pressure P_1. Suppose that the pressure is changed to P_2 while the temperature remains constant at T_1. The volume will assume a new value V'_1. Then

$$P_1 V_1 = P_2 V'_1$$

$$V'_1 = \frac{P_1 V_1}{P_2}$$

Suppose further that the pressure is now kept constant at P_2 and the temperature is changed from T_1 to T_2. The volume will change to a new value, V_2. Then

$$\frac{V'_1}{T_1} = \frac{V_2}{T_2}$$

but

$$V'_1 = \frac{P_1 V_1}{P_2}$$

$$\frac{P_1 V_1}{P_2 T_1} = \frac{V_2}{T_2}$$

$$\frac{P_1 V_1}{T_1} = \frac{P_2 V_2}{T_2}$$

By similar arguments

$$\frac{P_1V_1}{T_1} = \frac{P_2V_2}{T_2} = \frac{P_3V_3}{T_3} = \ldots\ldots$$

or, in general

$$\frac{PV}{T} = \text{constant}$$

Example 5-3

What volume will be occupied by a sample of gas at 40°C and 720 mm if its volume at 20°C and 750 mm was 250 ml?

Method 1: Since the pressure decreased, the volume would increase and the original volume should be multiplied by the factor greater than 1, that is 750/720. The temperature increased; therefore the volume would increase. The original volume should be multiplied by the factor greater than 1, or $(273+40)/(273+20)$

$$250 \times \frac{750}{720} \times \frac{313}{293} = 278 \text{ ml}$$

Method 2: Let $P_1 = 750$ mm, $V_1 = 250$ ml, $T_1 = 20 + 273 = 293$ K, $P_2 = 720$ mm, $T_2 = 40 + 273 = 313$ K, and V_2 = volume at P_2 and T_2.

$$\frac{P_1V_1}{T_1} = \frac{P_2V_2}{T_2}$$

$$\frac{750 \times 250}{293} = \frac{720 \; V_2}{313}$$

$$V_2 = \frac{750 \times 250 \times 313}{293 \times 720} = 278 \text{ ml}$$

5-4 QUANTITY OF A GAS

Previously we have considered the relationship of volume, pressure, and temperature for a given quantity of a gas. The volume of a gas is obviously related to the quantity present. We would now like to introduce quantity as a variable. In the general equation

$$\frac{PV}{T} = \text{constant}$$

the value of the constant depends upon the mass of the gas present.
According to Avogadro's law, equal volumes of gases under the same
conditions of temperature and pressure contain the same number of
molecules. We have previously stated that <u>one mole</u> of any substance
contains the same number of molecules. We can then introduce a
universal constant R, the <u>gas constant</u>, which has the same value
for all gases.

$$\frac{PV}{T} = R$$

or PV = RT, where V is the volume occupied by 1 mole, the <u>molar</u>
<u>volume</u>, at the pressure P and the temperature T. For <u>n</u> moles of
gas,

$$PV = nRT$$

$$n = \frac{m}{M}$$

where <u>m</u> is the mass in grams and M is the gram molecular weight
(grams per mole).

In this derivation we have assumed that all gases exhibit the
same behavior and that Boyle's and Charles' laws accurately describe
this behavior for all gases. For real gases these laws can only be
regarded as approximations, which can be applied at relatively low
pressures and moderately high temperatures. It is useful to
postulate an imaginary gas or <u>ideal gas</u>. An <u>ideal gas</u> is a gas
to which Boyle's and Charles' laws are strictly applicable. The
equation PV = <u>n</u>RT is called the ideal-gas equation.

5-5 EVALUATION OF THE GAS CONSTANT

$$R = \frac{PV}{nT}$$

$$\frac{\text{pressure x volume}}{\text{No. of moles x temp. in K}}$$

At standard temperature and pressure (STP), defined as 0°C and 1 atm (760 mm or torr), one mole of a gas occupies 22.41 liters.

For one mole of an ideal gas:

$$R = \frac{1 \times 22.41}{1 \times 273.2} = 0.08205 \text{ liter-atm/K-mole}$$

Example 5-4

What volume will be occupied by 0.115 g of xenon at 37°C and 0.250 atm? (Assume the ideal-gas law applies.)

$$PV = \frac{mRT}{M}$$

$P = 0.250$ atm

$m = 0.155$ g

$M = 131.3$ g/mole

$R = 82.05$ ml-atm/K-mole

$T = 37 + 273 = 310$ K

$$V = \frac{mRT}{PM} = \frac{0.155 \times 82.0 \times 310}{0.250 \times 131} = 120 \text{ ml}$$

Example 5-5

What pressure will be exerted by 0.623 g of hydrogen at 42°C in a volume of 58.4 ml?

$$PV = \frac{mRT}{M}$$

$V = 58.4$ ml

$m = 0.623$ g

$M = 2 \times 1.008 = 2.016$ g/mole

$R = 82.05$ ml-atm/K-mole

$T = 42 + 273 = 315$ K

$$P = \frac{mRT}{VM} = \frac{0.623 \times 82.0 \times 315}{58.4 \times 2.02}$$

$$= 136 \text{ atm}$$

5-6 MOLECULAR WEIGHTS OF GASES

One of the most useful applications of the ideal-gas law is the calculation of molecular weights of gases.

Example 5-6

E. W. Morley found that 43.35 liters of hydrogen weighed 3.716 g at 0.0°C and 724.4 mm. Calculate the molecular weight.

[Am. Chem. J., 18, 162 (1896)]

Method 1: First correct the volume to standard pressure. The corrected volume will be smaller.

$$43.35 \times \frac{724.4}{760.0} = 41.31 \text{ liters}$$

At standard conditions 1 mole of a gas occupies 22.41 liters.

$$M = 3.716 \times \frac{22.41}{41.31} = 2.015 \text{ g}$$

$$g \times \frac{\text{liter/mole}}{\text{liter}} = \frac{g \times \text{liter}}{\text{liter} \times \text{mole}} = g/\text{mole}$$

Method 2:

$$PV = \frac{mRT}{M}$$

$$M = \frac{mRT}{PV}$$

The value of R to be used is 0.08205 because the volume is in liters. Pressure must be expressed in atmospheres and the temperature in degrees Kelvin.

$$M = \frac{3.716 \times 0.08205 \times 273.2}{(724.4/760.0) \times 43.35} = 2.012 \text{ g}$$

$$\frac{g \times \frac{\text{liter-atm}}{\text{K-mole}}}{\text{atm} \times \text{liter}} = \frac{g \times \text{liter} \times \text{atm} \times K}{\text{atm} \times \text{liter} \times K \times \text{mole}} = g/\text{mole}$$

Gases   101

<u>Example 5-7</u>

A sample of a new compound in a bulb whose capacity was 268 cc
weighed 0.0516 g. The pressure was 17.9 mm, and the temperature
was 69°C. Calculate the molecular weight of the compound.

[G. Urry, A. G. Garrett, and H. I. Schlesinger, <u>Inorg.</u>
<u>Chem.</u>, <u>2</u>, 396 (1963)]

Method 1: Correct the volume to standard conditions. The
volume will <u>decrease</u> because of an increase in pressure and a
decrease in temperature.

$$268 \times \frac{17.9}{760} \times \frac{273}{342} = 5.04 \text{ cc}$$

Here 0.0516 g occupies 5.04 cc (ml); one mole will occupy 22,414 ml.

$$M = 0.0516 \times \frac{22,414}{5.04} = 230 \text{ g/mole}$$

$$g \times \frac{\text{ml/mole}}{\text{ml}} = \frac{g \times \text{ml}}{\text{ml} \times \text{mole}} = \text{g/mole}$$

Method 2:

$$M = \frac{mRT}{PV}$$

R has the value 82.05 when V is in milliliters.

$$M = \frac{0.0516 \times 82.05 \times 342}{(17.9/760) \times 268} = 230 \text{ g/mole}$$

$$\frac{g \times \frac{\text{ml-atm}}{\text{K-mole}} \times K}{\text{atm} \times \text{ml}} = \frac{g \times \text{ml} \times \text{atm} \times K}{\text{atm} \times \text{ml} \times K \times \text{mole}} = \text{g/mole}$$

Note:

$$\frac{17.9}{760}$$

$$\frac{\text{mm}}{\text{mm/atm}} = \frac{\text{mm} \times \text{atm}}{\text{mm}} = \text{atm}$$

5-7 GAS DENSITY AND IDEAL-GAS EQUATION

The ideal-gas equation may be rearranged:

$$PV = \frac{m}{M} RT$$

$$M = \frac{m}{V} \frac{RT}{P}$$

Density d has been defined as m/V (Sec. 1-3); the equation may then be written

$$M = d \frac{RT}{P} \quad \text{or} \quad d = \frac{PM}{RT}$$

Example 5-8

Calculate the density of silane, SiH_4, at 27°C and 190 mm.

SiH_4

$$\begin{array}{r} 1 \times 28.1 = 28.1 \\ 4 \times 1.0 = \underline{4.0} \\ \overline{32.1} \text{ g/mole} \end{array}$$

$$d = \frac{(190/760) \times 32.1}{82.0 \times 300} = 3.26 \times 10^{-4} \text{ g/ml}$$

$$\frac{\text{atm} \times \frac{g}{\text{mole}}}{\frac{\text{ml-atm}}{\text{K-mole}} \times K} = \text{g/ml}$$

Example 5-9

The density of a gas measured at 20.3°C and 749 mm was 2.66 g/liter. Calculate the molecular weight.

> [Gmelin's Handbuch der Anorganischen Chemie, System No. 9, 211 (1953)]

$$M = d \frac{RT}{P} = 2.66 \times \frac{0.0820 \times 293}{749/760} = 65.0 \text{ g/mole}$$

$$\frac{g}{\text{liter}} \times \frac{\frac{\text{liter-atm}}{\text{K-mole}} K}{\text{atm}} = \text{g/mole}$$

5-8 DALTON'S LAW OF PARTIAL PRESSURES

In a mixture of gases, each gas exerts the pressure it would exert if it alone occupied the volume, and the total pressure is equal to the sum of the partial pressures.

$$P_{total} = P_1 + P_2 + P_3 + \ldots$$

Example 5-10

A sample of argon was collected over water in a 200-ml container. The total pressure in the container at 27°C was 763 mm. What was the weight of argon? The measured pressure was the total pressure of a mixture of argon and water vapor. The pressure due to water vapor in equilibrium with water at 27°C = 26.7 mm (Appendix D).

Total pressure   763 mm

Partial pressure H_2O $\underline{27\ mm}$

Partial pressure Ar   736 mm

$$PV = \frac{m}{M} RT$$

$$m = \frac{PVM}{RT}$$

$P = \frac{736}{760}$ atm, $V = 200$ ml, $M = 39.9$ g/mole

$R = 82.0$ ml-atom/K-mole and $T = 27 + 273 = 300$ K

$$m = \frac{736/760 \times 200 \times 39.9}{82.0 \times 300} = \frac{736 \times 200 \times 39.9}{760 \times 82.0 \times 300}$$

$$= 0.314\ g$$

5-9 GAY-LUSSAC'S LAW OF COMBINING VOLUMES

In chemical reactions involving gases, the volumes are in the ratio of whole numbers. In the balanced equation of the reaction, the whole numbers are represented by the coefficients. For example:

$$4HCl(g) + O_2(g) \rightarrow 2H_2O(g) + 2Cl_2(g)$$

Four volumes of hydrogen chloride will react with one volume of oxygen to form two volumes of water vapor and two volumes of chlorine. If 100 ml of HCl reacts, 25 ml of oxygen will be consumed and 50 ml of $H_2O(g)$ and 50 ml of Cl_2 will be formed.

Example 5-11

How many milliliters of ammonia gas will be formed when 50 ml of nitrogen reacts with an excess of hydrogen?

$$N_2 + 3H_2 \rightarrow 2NH_3$$
$$1 \text{ vol} \rightarrow 2 \text{ vols}$$
$$50 \times \frac{2}{1} = 100 \text{ ml}$$

5-10 MOLAR VOLUME

In Chap. 4 we saw that a formula may be used to represent 1 mole of a substance. If the substance is a gas, the formula may be used to represent one gram molecular volume. A gram molecular volume is the volume occupied by one mole of a gas. At 0°C and 1 atm the gram molecular volume of an ideal gas is 22.4 liters.

Example 5-12

How many milliliters of oxygen (measured at STP) can be prepared by the decomposition of 2.65 g of KNO_3?

$$\begin{array}{r} KNO_3 \quad\quad 39.1 \\ 14.0 \\ \underline{48.0} \\ 101.1 \end{array}$$

$$2KNO_3 \rightarrow 2KNO_3 + O_2$$
$$2 \text{ moles} \rightarrow 1 \text{ mole or } 22.4 \text{ liters}$$
$$1 \text{ mole} \rightarrow 0.5 \text{ mole or } 11.2 \text{ liters}$$

$$\frac{2.65}{101} = 0.0262 \text{ mole } KNO_3 \rightarrow \frac{0.0262}{2} = 0.0131 \text{ mole } O_2$$

$$\frac{g}{g/mole} = g \times \frac{mole}{g} = mole$$

$$1.31 \times 10^{-2} \times 2.24 \times 10 = 2.94 \times 10^{-1} = 0.294 \text{ liter}$$

5-11 GRAHAM'S LAW OF DIFFUSION

The rates of diffusion of gases are inversely proportional to the square root of their densities. The law may be expressed as

$$\frac{r_1}{r_2} = \sqrt{\frac{d_2}{d_1}}$$

where r_1 and r_2 are the rates of diffusion of two gases having densities d_1 and d_2, respectively.

From the ideal-gas equation

$$d = \frac{m}{V} = \frac{PM}{RT}$$

At constant temperature and pressure

$$\frac{d_1}{d_2} = \frac{(P/RT)\ M_1}{(P/RT)\ M_2}$$

$$\frac{r_1}{r_2} = \sqrt{\frac{M_2}{M_1}}$$

The rates of diffusion of gases are inversely proportional to the square root of their molecular weights.

Example 5-13

A sample of nitrogen diffused through a small hole at the rate of 2.65 ml/min. At what rate would NF_3 flow through the same orifice under the same conditions?

$$\frac{r_{N_2}}{r_{NF_3}} = \sqrt{\frac{M_{NF_3}}{M_{N_2}}}$$

$$\frac{2.65}{r_{NF_3}} = \sqrt{\frac{71}{28}} = \sqrt{2.53} = 1.59$$

$$r_{NF_3} = \frac{2.65}{1.59} = 1.67 \text{ ml/min}$$

Example 5-14

A sample of xenon required 1.0 min 8.3 sec to diffuse through a small orifice. What is the molecular weight of a gas which required only 57.0 sec for diffusion under identical conditions? The time required is inversely proportional to the rate of diffusion and is therefore directly proportional to the square root of the molecular weights.

$$\frac{t_{Xe}}{t} = \sqrt{\frac{M_{Xe}}{M}}$$

$$\frac{68.3}{57.0} = \sqrt{\frac{131}{M}} = 1.20$$

$$M = \frac{131}{1.20^2}$$

$$M = \frac{131}{1.44} = 91.0$$

5-12 BEHAVIOR OF REAL GASES

The PVT behavior of most gases is represented fairly accurately by the ideal-gas equation at relatively low pressures and high temperatures. The use of the ideal-gas equation is justified in many applications because departures from ideal behavior are not great. When more accurate calculations must be made, the discrepancies must be avoided.

5-13 LIMITING DENSITIES

One method of minimizing the departures from ideal behavior in the determination of accurate molecular weights is to apply the

principle of limiting densities. We have seen that the ideal-gas
law can be written in the form

$$M = d \frac{RT}{P}$$

This can be rearranged

$$M = RT \frac{d}{P}$$

The value of M will be more accurate as the pressure assumes
smaller values. In practice a series of measurements of density is
made at different pressures. The values of d/P are plotted against
P, and the value of d/P at P = 0 is obtained by extrapolation. This
method has been used to determine the atomic weights of noble gases.

Example 5-15

Determine the molecular weight of carbon dioxide from the densities
of the gas measured at 10°C and the pressures listed.

P, atm	d, g/liter
0.68	1.29
2.72	5.25
8.14	16.32

First calculate values of d/P at the three pressures 1.90, 1.93,
2.01; then plot d/P vs P and extrapolate to P = 0.

$$\frac{d}{P} \text{ at } P = 0.68 \text{ atm} \quad \frac{1.29}{0.86} = 1.90$$

$$\frac{d}{P} \text{ at } P = 2.72 \text{ atm} \quad \frac{5.75}{2.72} = 1.93$$

$$\frac{d}{P} \text{ at } P = 8.14 \text{ atm} \quad \frac{16.32}{8.14} = 2.01$$

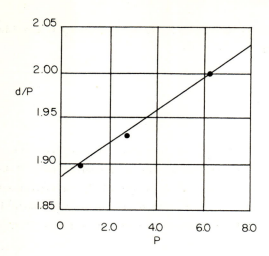

FIGURE   5-1

Substitute the value of d/P obtained by extrapolation in the equation.

$$M = RT \frac{d}{P}$$

$$= 0.0821 \times 283 \times 1.89 = 44.0 \text{ g/mole}$$

5-14   EQUATIONS OF STATE FOR REAL GASES

Several equations which correct for the deviations from ideal behavior have been proposed. One of the most useful of these equations of state is the van der Waals equation:

$$\left(P + \frac{n^2 a}{V^2}\right)(V-nb) = nRT$$

The constant a is the correction for the attractive force between molecules, and b is the correction factor for the volume occupied by the molecules of the gas. Values of the constants for several gases are given in Table 5-1.

TABLE 5-1 van der Waals Constants

Gas	a, liter^2-atm/mole^2	b, liters/mole
Helium	0.0341	0.0237
Argon	1.35	0.0322
Nitrogen	1.39	0.0391
Carbon dioxide	3.59	0.0427
Acetylene	4.39	0.0514
Carbon tetrachloride	20.39	0.1383

Example 5-16

Calculate the pressure exerted by <u>one mole</u> of ammonia at 27°C in a flask whose volume is 0.800 liter.

Since n = 1

$$\left(P + \frac{a}{V^2}\right)(V - b) = RT$$

$$\left(P + \frac{4.17}{0.8^2}\right)(0.800 - 0.0371) = 0.0820 \times 300$$

$$\left(atm + \frac{liter^2\text{-}atm \times mole^2}{liter^2 \times mole^2}\right)\left(liter - mole \times \frac{liter}{mole}\right) =$$

$$mole \times \frac{liter\text{-}atm}{K\text{-}mole} \times K$$

atm x liter = liter x atm

$$0.763P + \frac{4.17 \times 0.763}{0.640} = 24.6$$

$$P = \frac{24.6 - 4.98}{0.763} = 25.4 \text{ atm}$$

PROBLEMS

5-1 Plot the following data for the volume of Ne at various pressures and constant temperature. Plot the volume at various pressures versus the reciprocal of the pressure. Calculate the product of the pressure and volume at each pressure.

P, mm	V, liters
1	1.366
10	0.137
20	0.0685
30	0.0457
40	0.0343
50	0.0274
60	0.0228
70	0.0196
80	0.0171

Source: Landolt-Bornstein Tabellen, 5th ed., first supplement, p. 64

5-2 What is the volume of a sample of carbon dioxide at 735 mm if its volume at 723 mm is 223 ml?

5-3 A sample of gas occupied 400 ml at STP. What volume will it occupy at 740 mm and 87°C?

5-4 Calculate the molecular weight of a gas if 0.638 g occupies a volume of 223 ml at 23°C and 758 mm.

5-5 Calculate the density, in grams per liter, of nitrogen trifluoride at standard conditions.

5-6 Calculate the molecular weight for each of the following gases from the densities. Compare the results of your calculations with the molecular weight obtained from the most recent values of the atomic weights. The densities are in grams per liter at STP.

    a. Ar, 1.7824   b. He, 0.1785
    c. H_2, 0.08987   d. Kr, 3.708

e. Ne, 0.9002   f. N_2, 1.2506

g. O_2, 1.4290   h. Rn, 9.73

i. Xe, 5.851

> [N. A. Lange, Handbook of Chemistry, 10th ed., McGraw-Hill,
> New York, 1966, pp. 100-107]

5-7 Calculate the molecular weight of stibine from its density at standard conditions, which was found to be 5.685 g/liter.

> [Stock and Gutman, Chem. Ber., 37, 887 (1904)]

5-8 A mixture of B_2H_6 (3.28 mM) and PF_3 (19.21 mM) was collected in a vessel whose capacity was 62.5 ml. What was the total pressure at 25°C?

> [B. W. Parry and T. C. Bissot, J. Am. Chem. Soc., 78,
> 1526 (1956)]

5-9 A fluoride of platinum was prepared at the Argonne National Laboratory. A sample weighing 0.1418 g was placed in a thin-walled nickel bulb at a pressure of 68.81 mm and a temperature of 26°C. The volume of the bulb was 126.63 cc. What is the molecular weight of the compound?

> [B. Weinstock, J. G. Mahn, and E. C. Weaver, J. Am. Chem.
> Soc., 83, 4313 (1961)]

5-10 A compound was found to contain 64.70% carbon, 10.97% hydrogen, and 24.33% oxygen. A sample weighing 0.1396 g occupied a volume of 41.56 ml at 147°C and 658.8 mm. What is the formula of the compound?

> [E. Frankland and B. F. Duppa, J. Chem. Soc., 20, 108
> (1867)]

5-11 A pale yellow solid containing 23.0% boron and 75% chlorine was prepared from boron trichloride. A 0.0516-g sample was vaporized at 69°, and the vapor occupied a volume of 268 cc at 22.2 mm. What is the formula of the compound?

[G. Urry, W. G. Garrett, and H. I. Schlesinger, Inorg.
Chem., 2, 396-400 (1963)]

5-12 Diboron tetrachloride was treated with sodium hydroxide, and
the following reaction occurred:

$$B_2Cl_4 + 10NaOH \rightarrow 2NaBaO_3 + 4NaCl + 4H_2O + H_2$$

If 13.26 cc of hydrogen formed, how much B_2Cl_4, in grams, was
consumed?

[G. Urry et al., J. Am. Chem. Soc., 76, 5296 (1954)]

5-13 In a preparation of carbon-14-labeled methane, 2.00 mM of
barium carbonate containing 4% radioactive carbon atoms was treated
with sulfuric acid. The carbon dioxide formed was freed of water
vapor and reduced with hydrogen at 330° over a nickel-thorium
dioxide catalyst. The yield of methane, based on barium carbonate,
was 95.0%. What volume of methane, measured at 20°C and 750 mm,
was obtained?

[W. H. Beamer, J. Am. Chem. Soc., 70, 3900 (1948)]

5-14 $LiBH_4$(500 mg) reacted with 2.4247 g of iodine to produce
boron triiodide, lithium iodide, hydrogen iodide, and 133 cc of
hydrogen. What was the percent yield of hydrogen, based on $LiBH_4$,
for the equation: $2LiBH_4 + 5I_2 \rightarrow 2BI_3 + 2LiI + 2HI + 3H_2$?

[F. Klanberg and H. W. Kohischutter, Chem. Ber., 94,
786 (1961)]

5-15 A mixture of the gases N_2 and CF_3COOF was contained in a metal
trap at room temperature (23°C) and a pressure of 43.7 mm. The
CF_3COOF was decomposed by a spark to CF_4 and CO_2. The pressure was
then 58.4 mm. What was the initial partial pressure of CF_3COOF?

[G. L. Gard and G. H. Cady, Inorg. Chem., 4, 595 (1965)]

5-16 The vapor density of the compound containing 26.8% vanadium,
73.02% chlorine, was determined in 1870. The weight of the bulb
filled with air at 15° and 751 mm was 19.8945 g. Filled with the

vapor at 205° and 758.1 mm, it weighed 20.3015 g. Its capacity was
112 cc. The density of air at 0°C and 1 atm is 1.293 g/l. What is
the formula of the compound?

    [H. E. Roscoe, J. Chem. Soc., 23, 345-347 (1870)]

5-17 A small quantity of a material believed to be ammonium ozonide
was prepared at Armour Research Foundation. After standing for
90 min at 78°C, 74.1% had decomposed according to the equation:

    $4NH_4O_3 \rightarrow 2NH_4NO_3 + O_2 + 4H_2O$

How many milliliters of oxygen would be formed if 10.0 g of NH_4O_3
had been prepared and the pressure were 740 mm?

    [I. J. Solomon et al., J. Am. Chem. Soc., 84, 34 (1962)]

5-18 In a study of the thermal decomposition of potassium chlorate
it was found that, at 551°C, 97.5% of the theoretical quantity of
oxygen was released after 60 min. How many milliliters of oxygen
would result from heating 1.00 g of potassium chlorate? Assume
that the gas volume was measured at 25°C and 743.1 mm.

    [A. Glasner and L. Weidenfeld, J. Am. Chem. Soc., 74,
    2467 (1952)]

5-19 The molecular weight of a highly explosive vapor was calculated
from the density measurements at Cornell University. A sample
weighing 0.0223 g had a volume of 13.8 cc at 747.6 mm and 19°C.
What was the molecular weight of the compound?

    [L. M. Dennis and H. Isliam, J. Am. Chem. Soc., 29,
    222 (1907)]

5-20 Calculate the molecular weight of the gas whose normal density
is 1.978 g/liter.

5-21 A sample of an oxide of nitrogen weighing 0.64304 g was
passed over nickel and catalytically decomposed, giving 0.34390 g
of oxygen and 0.29920 g of nitrogen. What was the volume of the

original sample of gas measured at 27°C and 740 mm?

    [R. W. Gray, J. Chem. Soc., 87, 1616 (1905)]

5-22 What was the percent yield if 71.4 ml of hydrogen (at STP)
was obtained from 0.8588 g of SiHBr$_3$ by the reaction

    SiHBr$_3$ + 5NaOH $\rightarrow$ 3NaBr + Na$_2$SiO$_3$ + H$_2$ + 2H$_2$O
        [W. C. Schumb, Inorg. Syn., 1(1939), 41]

5-23 A compound melting at -92.6° was prepared by the action of an
electric discharge on boron trichloride. A sample weighing 0.0995 g
was found to contain 0.0878 g of chlorine and 0.0131 g of boron.
Another sample, collected in a 191.6 cc bulb, weighed 0.0810 g
at 48.9 mm and 26.5°C. What is the formula of the compound?

        [G. Urry et al., J. Am. Chem. Soc., 76, 5296 (1954)]

5-24 Gallium metal and chlorine combined to form a compound which
melted at 77° and boiled at 200°. Analysis found: Ga, 39.55;
Cl, 60.35. A sample weighing 0.5763 g occupied a volume of 223.4 cc
at a pressure of 206.7 mm and a temperature of 177.5°C. What is
the formula of the compound?

        [A. W. Laubengayer and F. B. Schirmer, J. Am. Chem. Soc.,
        62, 1578 (1940)]

5-25 The accepted value of the molar volume of an ideal gas is
22.4136 liter. Calculate the volume occupied by 1.0000 mole of
each of the following gases at STP from the densities in g/liter.

    Ar, 1.7824
    Cl$_2$.3.214
    F$_2$, 1.695
    He, 0.1785
    H$_2$, 0.08987
    Kr, 3.708
    Ne, 0.9002
    N$_2$, 1.2506

O_2, 1.4290
O_3, 2.144
Rn, 9.73
Xe, 5.851

5-26 Calculate the rate at which arsine (AsH_3) gas will diffuse through an orifice if ammonia diffuses at the rate of 0.481 liter/min under identical conditions.

5-27 Calculate the ratio of rates of diffusion through a small orifice of silane, SiH_4, and diborane B_2H_6.

5-28 Calculate an accurate atomic weight of the monatomic gas for which the following densities were obtained at 0°C by G. P. Baxter and H. W. Starkweather at Harvard.
[Proc. Natl. Acad. Sci. U. S., 14, 56 (1928)]

Pressure, mm	Density, g/liter
760.000	0.89990
506.667	0.60004
253.333	0.30009

5-29 Calculate an accurate atomic weight of argon whose densities at 0°C and various pressures are given below.
[G. P. Baxter and H. W. Starkweather, Pro. Natl. Acad. Sci. U. S., 14, 60 (1928)]

Pressure, mm	Density, g/liter
760.000	1.78364
506.67	1.18874
253.33	0.59419

Compare the result with the atomic weight calculated from the

isotopic abundances: ^{36}Ar = 0.337%, ^{34}Ar = 0.063%, and
^{40}Ar = 99.600%

<u>5-30</u> Using the van der Waals equation, calculate the pressure
exerted by 0.399 g of argon at -23°C in a volume of 300 cc.

<u>5-31</u> Use the van der Waals equation to calculate the pressure of
1.00 mole of CCl_4 in a 250-ml flask at 17°C.

SUPPLEMENTARY PROBLEMS

<u>5-32</u> Plot the following data for the volume of Ne at constant
pressure vs temperature. Determine the slope of the curve from the
graph and compare it with the slope calculated from Gay-Lussac's
law (Charles' law).

t, °C	V, Liters
400	2.47
300	2.11
200	1.74
100	1.37
0	1.01
-50	0.82
-100	0.64
-150	0.45
-183	0.33

Source: <u>Landolt-Bornstein Tabellen</u>, 5th ed., first supplement,
pp. 64-65.

<u>5-33</u> What volume will be occupied by a sample of hydrogen at
1.51 atm if the volume at 740 mm is 325 ml?

<u>5-34</u> A sample of hydrogen sulfide was collected in a 250-ml flask

at a pressure of 740 mm and 37°C. What volume would it occupy at
3.65 atm and 40°C?

5-35 Calculate the density of diborane, B_2H_6, at 40° and 1.81 atm.

5-36 Calculate the molecular weights of the following gases and
compare them with the values obtained from the latest values of the
atomic weights of the elements. Account for any discrepancies. The
densities are in grams per liter.

    a. Ammonia, 0.7708   b. Stibine, 5.685

    c. Argon, 1.782   d. Arsine, 3.50

    e. Boron trifluoride, 3.00   f. Hydrogen bromide, 3.644

    g. Hydrogen chloride, 1.639   h. Fluorine, 1.71

        [Landolt-Bornstein Tabellen, 5th ed., first supplement]

5-37 A canary-yellow liquid was found to have 29.58% vanadium,
61.27% chlorine, and the remainder oxygen. A sample weighing
1.02 g had a pressure of 780 mm at 18.6° in a bulb, the capacity
of which was 135.1 cc. What was the compound?

        [H. E. Roscoe, J. Chem. Soc., 21, 343-347 (1868)]

5-38 A compound was prepared at Cornell University by the reaction
of ammonium chloride with boron trichloride. Analysis found:
Cl, 57.62, 57.80, 57.92; B, 17.88, 17.36, 17.62; N, 22.83, 22.84.
A sample weighing 0.2182 g in a 668.5-ml bulb had a pressure of
45.0 mm at 141.1°C. What was the compound formed?

        [C. A. Brown and A. W. Laubengayer, J. Am. Chem. Soc.,
        77, 3699 (1955)]

5-39 The molecular weight of chromyl fluoride was calculated from
vapor-density measurements as 121.4. The temperature was 25° and
the pressure 500 mm. What was the density of the vapor?

        [C. A. Engelbrecht and A. V. Grosse, J. Am. Chem. Soc.,
        74, 5262 (1952)]

5-40 In a study of the hydrolysis of pentaborane, B_5H_9, a sample
(6.56 cc at STP) was shaken with 20% water in dioxane. The volume
of hydrogen collected was 78.4 cc. If the other product was ortho-
boric acid, write a balanced equation consistent with the results
of the experiment.

> [I. Shapiro and H. G. Weiss, J. Am. Chem. Soc., 76,
> 6020 (1954)]

5-41 Samples of zirconium trichloride, tribromide, and triiodide
were hydrolyzed, and the hydrogen which formed was collected.
$$2ZrX_3 + 2H_2O \rightarrow 2Zr(OH)X_3 + H_2$$

Compound	Mass, g	Volume H_2(corr.), ml
$ZrCl_3$	0.9183	50.3
$ZrBr_3$	1.2591	39.2
ZrI_3	1.3121	29.7

What was the percent yield of hydrogen in each case?

> [E. M. Larsen and J. J. Leddy, J. Am. Chem. Soc., 78,
> 5985 (1956)]

5-42 A mixture containing 1.3 mM of HNF_2 gas and equal quantity of
chlorine was led into a flask containing 5.0 g of potassium fluoride
and allowed to stand for 18 hr at room temperature. The gas $ClNF_2$
(66% yield) and the solid KF·HCl were formed. What was the volume
percent of $ClNF_2$ in the gaseous mixture present after the reaction?
What was the weight percent of $ClNF_2$ in the gas mixture after
the reaction?

> [W. C. Firth, Jr., Inorg. Chem., 4, 254 (1965)]

5-43 When pure sulfuric acid was heated to 450°, its molecular
weight, calculated from a vapor-density measurement, was found to
be 50.03. What conclusion can be drawn from this?

5-44   A sample of the compound in Prob. 5-45 weighing 0.0947 g
was subjected to a series of reactions with the eventual preparation
of 0.3140 g of silver chloride. If 1 mole of the unknown produced
1 mole of silver chloride, are the results of this experiment
consistent with the molecular weight found in Prob. 5-45?

5-45   A bulb whose capacity was 267.43 cc at 0°C was filled with
a gas at a pressure of 760 mm. The difference in weight between
the filled bulb and a tare weight of similar displacement was
0.35851 g. When the bulb was filled with oxygen under the same
conditions, the difference in weight between the bulb and tare was
0.38228 g. Calculate the molecular weight of the gas.
        [R. W. Gray, J. Chem. Soc., 87, 1608 (1905)]

5-46   A bulb filled with a gaseous element weighed 0.32280 g at
10°C and 760 mm; when filled with oxygen under the same conditions,
it weighed 0.36884 g. What is the molecular weight of the gaseous
element?
        [R. W. Gray, J. Chem. Soc., 87, 1616 (1905)]

5-47   What is the molecular weight of a gas if 503.5 ml of it weighs
1.10 g at 545.5 mm and 0°C?
        [D. R. Martin and R. E. Dial, J. Am. Chem. Soc., 72,
        853 (1950)]

5-48   The volume of 0.0808 g of a boron hydride was 128.8 ml at
17.3°C and 405.7 mm. The same sample of gas was then decomposed and
63.6 mg of boron and 197.3 ml of hydrogen (at STP) were obtained.
What is the molecular weight of the boron hydride? What is its
formula?
        [A. Stock and E. Kuss, Chem. Ber., 56, 801 (1923)]

5-49   Difluoramine (24.6 cc) was maintained in a high-pressure
ampoule for 68 hr at 0° in the presence of a catalyst. Some HNF_2

(10.0 cc) was recovered, but the remainder had been converted to N_2F_4 (5.5 cc). Calculate the percent yield of N_2F_4 assuming its formation by the reaction $5HNF_2 = 2N_2F_4 + NH_4F + HF$.

     [E. A. Lawton and J. Q. Weber, J. Am. Chem. Soc., 85,
     3597 (1963)]

5-50 The compound described in Prob. 5-24 was mixed with gallium metal and heated to 175° in a vacuum. Analysis found: Ga, 49.42; Cl, 50.41. Write an equation for the reaction which occurred.

5-51 Calculate the mass of each component present in a mixture of fluorine and xenon in a 2-liter glass flask. The partial pressure of Xe was 350 mm and the total pressure was 724 mm at 25°.

     [L. V. Streng and A. G. Streng, Inorg. Chem., 4,
     1370 (1965)]

5-52 Calculate the percent yield of product if the components of the mixture in Prob. 5-51 reacted to form 0.50 g XeF_2.

5-53 A mixture of the gases PH_3 and BF_3 weighing 1.03 g exerted a pressure of 538.0 mm in a volume of 503.51 ml at 0°. What was the mole ratio BF_3/PH_3 of the mixture?

     [D. R. Martin and R. E. Dial, J. Am. Chem. Soc., 72,
     853 (1950)]

5-54 What was the weight of $SF_5ONF_2(g)$ contained in a 200-ml glass vessel if the pressure was 56.5 mm when the vessel was immersed in a water bath maintained at 85°C?

     [W. H. Hale, Jr. and S. M. Williamson, Inorg. Chem., 4,
     1341 (1965)]

5-55 A sample of $SF_5OF(g)$ was contained in a glass vessel at 25° and a pressure of 80 mm. A quantity of N_2F_4 that was added brought the total pressure to 160 mm. The reaction that occurred produced

a variety of products, NF_3, NO, SiF_4 (by reaction with glass),
SF_6, SO_2F_2, SOF_4, SF_5ONF_2, and NO_2. The yield of SF_5ONF_2 was
40 mole% with respect to the reactant, SF_5OF. All of the SF_5OF
and N_2F_4 were consumed in the reaction. What was the mass of SF_5ONF_2
produced if the volume of the vessel was 1 liter? (The other product
of the reaction in question was NF_3.)

[W. H. Hale, Jr., and S. M. Williamson, Inorg. Chem., 4,
1342 (1965)]

5-56 What time will be required for a sample of ethane, C_2H_6, to
diffuse through an orifice if a similar sample of butane, C_4H_{10},
required 2 min 23 sec?

5-57 Calculate the rate of flow of krypton through the walls of a
porous tube if helium flows at the rate of 2.73×10^{-4} ml/sec under
identical conditions.

5-58 What will be the ratio of the rate of flow through a porous
wall of UF_6 containing ^{234}U to that of UF_6 containing ^{238}U?

5-59 Calculate an accurate molecular weight of phosphine, PH_3,
from the following data. Determine a value of the atomic weight of
phosphorus by using the accepted value for hydrogen

Pressure, atm	Density, g/liter
1	1.5307
3/4	1.1454
1/2	0.76190
1/4	0.38012

5-60 Calculate the atomic weight of oxygen from the following values
of the density of carbon dioxide at 0°C and various pressures. Use
the accepted value of the atomic weight of carbon.

Pressure, atm	Density, g/liter
1	1.97676
2/3	1.31485
1/2	0.98505
1/3	0.65596

Compare your result with that obtained in Prob. 2-47.

5-61 According to calculations made by using the van der Waals equation, what will be the pressure of 2.00 moles of nitrogen at 27°C in a 500-ml flask?

Chapter 6

CHEMICAL BONDS

The atomic theory proposes that the atoms of elements are combined in compounds. Chemists use the term chemical bond to identify the forces between atoms in compounds, and the nature of these forces is inferred from the properties of the compounds. Substances may be roughly divided into five types according to their properties. A detailed examination of their structural properties indicates that substances of a given type also have the same type of bonds.

6-1 CLASSIFICATION OF COMPOUNDS

1. Ionic, for example, NaCl, KNO_3, MgO
2. Covalent, for example, CCl_4, SiH_4, NF_3
3. Polar covalent, for example, HCl, HNO_3, $SnCl_4$
4. Metallic, for example, TiN, VC, Fe_2N
5. Adamantine, for example, SiC, BN

6-2 THE IONIC BOND

Ionic compounds are crystalline aggregates of ions, i.e., atoms, or groups of atoms, with an electric charge. The cations (+ charge) and anions (-) are arranged in a three-dimensional array approximately represented as spheres in a closely packed arrangement. Cations are formed by the loss of electrons by atoms; anions are formed by the gain of electrons by atoms. The electrostatic attraction between these ions of opposite charge leads to the formation of ionic crystals.

6-3 THE COVALENT BOND

Many compounds exist as gases, liquids, or low-melting solids. These compounds contain molecules; the attractive forces between them are relatively weak. The atoms within a molecule are joined by electron-pair or covalent bonds. In a covalent bond, atoms share pairs of electrons, attain a stable electronic configuration, and form molecules.

6-4 LEWIS STRUCTURES

Molecules of covalent compounds are represented by structural formulas called Lewis structures or electron-dot formulas. The usual symbols designate atoms, and dots are used for electrons.

6-5 RULES FOR WRITING LEWIS STRUCTURES

1. Distinguish between ionic and covalent bonds.

Example 6-1

Lithium fluoride, LiF, ionic

$$Li + F \rightarrow Li^+ + F^-$$

Electrons are transferred, not shared.

Example 6-2

Carbon tetrafluoride, CF_4, <u>covalent</u>

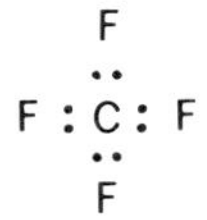

```
      F
      ..
  F : C : F
      ..
      F
```

Example 6-3

Sodium hydroxide, NaOH, <u>ionic</u>

$$Na^+ \quad O\!:\!H^-$$

Note that the bond within O-H$^-$ is covalent. Some compounds have both ionic and covalent bonds.

Example 6-4

$NaBH_4$

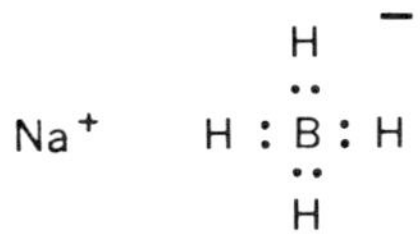

```
              H    ‾
              ..
  Na⁺    H : B : H
              ..
              H
```

2. The proper arrangement of the atoms must be known.

    a. In some cases only one arrangement is possible.

Example 6-5

HCl

```
  H : Cl
```

Example 6-6

NH_3

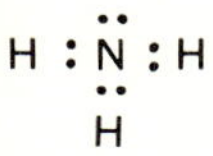

H : N̈ : H
　 H

　　b.　In other cases chemical and physical evidence must be available to support a structure.

Example 6-7

H_2O_2

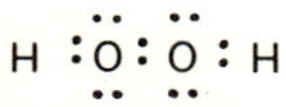

H :Ö: Ö: H   correct

Example 6-8

H_2O_2

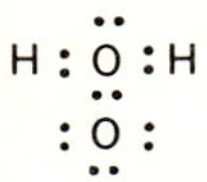

H : Ö :H   incorrect
: Ö :

　　Chemists may predict the Lewis structure of many compounds with confidence because of experience.　Students must be given more information about compounds in order to know the proper arrangement. In many cases, more than one structure is possible and even experienced chemists must have evidence to support the arrangement proposed.

　　The compound nitrosylchloride has the formula NOCl and is usually written that way.　A determination of its structure has shown that the Cl is attached to N, and not O. The N——Cl distance is 1.95 Å, and the Cl—N—O angle is 116°.　The appropriate Lewis structure is:

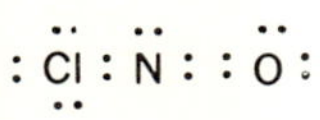

: C̈l : N̈ : : Ö :

3. The correct number of electrons must be included. The correct number of electrons for each atom is given by the periodic group number.

Example 6-9
PCl_3

$$\cdot \overset{\displaystyle ..}{\underset{\displaystyle .}{P}} \cdot \qquad 3 \; \overset{\displaystyle ..}{\underset{\displaystyle ..}{:} Cl} \cdot$$

$$5 + (3 \times 7) = 26$$

$$: \overset{..}{\underset{..}{Cl}} : \overset{..}{\underset{..}{P}} : \overset{..}{\underset{..}{Cl}} :$$
$$: \overset{}{\underset{..}{Cl}} :$$

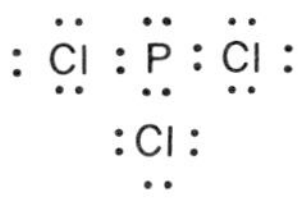

4. Arrange the electrons to provide full octets of electrons, if possible.

Example 6-10
Methane CH_4

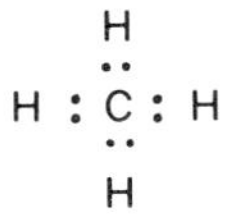

$$\begin{array}{c} H \\ H : \overset{..}{\underset{..}{C}} : H \\ H \end{array}$$

Example 6-11
NF_3

$$: \overset{..}{\underset{..}{F}} : \overset{..}{\underset{..}{N}} : \overset{..}{\underset{..}{F}} :$$
$$: \overset{}{\underset{..}{F}} :$$

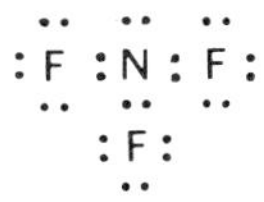

A covalent bond is usually represented by a dash (——). Students should use dots until they are thoroughly familiar with writing of structures. The use of dashes often leads to errors in the application of Rule 3.

6-6 MULTIPLE BONDS

Multiple bonds involve two or three pairs of electrons and are used when necessary to complete octets.

Example 6-12
Ethylene, C_2H_4

H : C :: C : H   A double bond (::)
    ..   ..
    H   H

Example 6-13
Nitrous acid, HNO_2

    ..
H : O : N :
    ..   ::
         O :
         ..

Example 6-14
Acetylene, C_2H_2

H : C ::: C : H   A triple bond (:::)

Example 6-15

Nitrogen, N_2

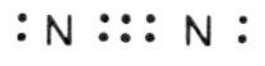

: N ::: N :

6-7 COORDINATE COVALENT BONDS

When one atom supplies both electrons of the electron pair in the formation of a covalent bond, the bond is called a coordinate covalent bond. Other names for this type of bond are coordinate bond and dative bond. Once formed, the coordinate covalent bond cannot be distinguished from a normal covalent bond.

Example 6-16

NH_3BF_3

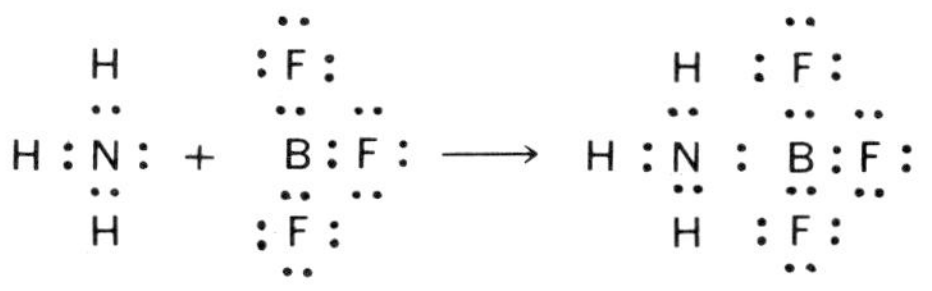

```
        ..                      ..
H      :F:              H   :F:
..     .. ..                .. ..
H:N: +  B:F:  ⟶  H:N: B:F:
..     .. ..                .. ..
H      :F:              H   :F:
        ..                      ..
```

Example 6-17

Chloric acid, $HClO_3$

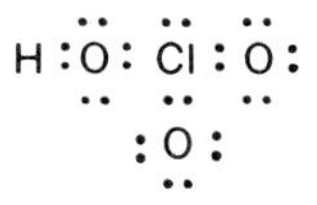

```
  ..  ..  ..
H:O: Cl:O:
  ..  ..  ..
     :O:
      ..
```

Two coordinate covalent bonds are present between Cl and the two O's not bonded to H.

6-8 DEPARTURES FROM THE OCTET RULE

    1. Less than an octet

       a. Two electrons

Example 6-18

Hydrogen

    H : H

Example 6-19

Hydrogen chloride

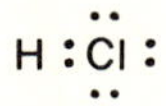

        $\overset{\cdot\cdot}{\underset{\cdot\cdot}{\text{H : Cl :}}}$

       b. Quartet

Example 6-20

Beryllium chloride vapor

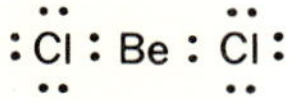

      $\overset{\cdot\cdot}{\underset{\cdot\cdot}{\text{: Cl}}}$: Be : $\overset{\cdot\cdot}{\underset{\cdot\cdot}{\text{Cl :}}}$

       c. Sextet

Example 6-21
Boron trichloride

$$
\overset{..}{:}\overset{..}{Cl}:B:\overset{..}{Cl}:\\
\quad\ :\overset{..}{Cl}:
$$

2. More than an octet

Example 6-22
Phosphorus pentachloride

```
Cl   Cl
  . .
Cl :P: Cl
  . .
   Cl
```

Example 6-23
Sulfur hexafluoride

```
    F
F.  .. .F
  : S :
F.  .. .F
    F
```

Example 6-24
Sulfur tetrafluoride

```
F.  .. .F
  : S :
F.    .F
```

6-9 RESONANCE

In the case of many molecules, more than one Lewis structure that satisfies the rules may be drawn. Since we are attempting to represent a complex system of atoms and electrons with letters of the alphabet and dots, the limitations of our symbols cause this difficulty. An accurate representation of the bonds is not possible; more than one Lewis structure is possible. These are called canonical structures.

Example 6-25
Nitric acid, HNO_3

$$H:\overset{..}{O}:N::O: \qquad H:\overset{..}{O}:N:\overset{..}{O}:$$
$$\overset{..}{:}\overset{..}{O}\overset{..}{:} \qquad\qquad :O$$

or

$$H—O—N=O \qquad H—O—N—O$$
$$\overset{|}{O} \qquad\qquad\quad \overset{\|}{O}$$

Note the positions of the double and single bonds in the two structures.

A more accurate description of the molecules describes the bonds as being neither single nor double, but bonds which lie somewhere between. They may be represented as:

$$H—O—N{-}{-}O$$
$$\overset{|}{O}$$

or more commonly

$$H-O-N-O \quad \longleftrightarrow \quad H-O-N=O$$
$$\overset{\|}{O} \qquad\qquad\qquad \overset{|}{O}$$

The two structures are called canonical forms, and the molecule is described as a resonance hybrid of the forms. The term resonance refers to this view of the bonding in molecules of this type. Resonance does not imply that the bonds are alternately double or single bonds; it is a description of the bond as lying somewhere between a single bond and a double bond.

6-10   THE SHAPE OF MOLECULES

The structure of molecules, i.e., the spatial arrangement of atoms in a molecule, is of genuine importance in chemistry. A sound understanding of the properties of compounds requires a knowledge of the structure of molecules. The spatial arrangement of atoms in the molecules of most compounds of the nontransitional elements has been determined by a variety of methods including X-ray and electron diffraction as well as infrared and nuclear magnetic resonance spectrometry. The distances between atoms and the angles between them can be determined with considerable precision.

A theoretical basis for the observed molecular structures and electron distributions, which requires an extensive knowledge of mathematics, is still under development. It is possible, however, to predict with reasonable accuracy the shapes of the molecules of compounds of the nontransitional elements with the aid of a few relatively simple principles. These principles were first proposed by Sidgwick and Powell, but they were applied and advanced by Gillespie and Nyholm. Students are urged to study the discussion of these principles [Quart. Rev. Chem. Soc., (London), 11, 339 (1957), R. J. Gillespie, J. Chem. Ed., 47, 18 (1970)].

6-11 ELECTRON-PAIR REPULSION THEORY

In a molecule, the arrangement of the pairs of electrons about
an atom, whether shared (bonding) or unshared (nonbonding), depends
only on the number of pairs.

TABLE 6-1   Arrangement of Electron Pairs in Regular Structures

No. of electron pairs	Arrangement*
2	Linear
3	Triangular plane
4	Tetrahedron
5	Trigonal bipyramid
6	Octahedron
7	Pentagonal bipyramid

*The word "arrangement" is used to designate the positions of
electron pairs. The word "shape" is used to designate the positions
of atomic nuclei.

The arrangements of electron pairs given in Table 6-1 are those
predicted if it is assumed that the electron pairs repel each other
so that they arrange themselves as far apart as possible.

The arrangements of electron pairs listed in Table 6-1 lead
to regular shapes of molecules when all pairs are bonded to other
atoms or groups which are alike. When the atoms or groups are not
alike or when some of the positions are occupied by unshared pairs
or lone pairs, deviations from regular shapes result. The regular
shapes to be expected are given in Table 6-2.

Table 6-3 shows the shapes of molecules with unshared pairs of
electrons.

6-12 DEVIATIONS FROM REGULAR SHAPES

Deviations from the regular shapes can be attributed to decreasing order of repulsion among electron pairs; unshared pair-
unshared pair>unshared pair-bond pair>bond pair-bond pair. This can
be abbreviated u-u>u-b>b-b.

TABLE 6-2

No. of electron pairs	Arrangement of electron pairs	Shape of molecule	Example
2	Linear	Cl—Be—Cl	$BeCl_2$(vapor)
3	Triangle	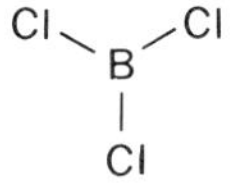	BCl_3
4	Tetrahedron	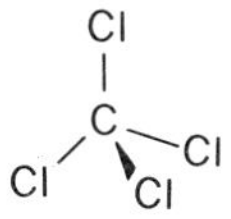	CCl_4
5	Trigonal bipyramid	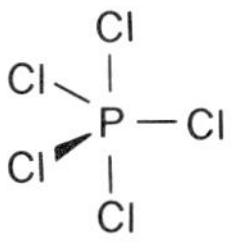	PCl_5
6	Octahedron		SF_6
7	Pentagonal bipyramid	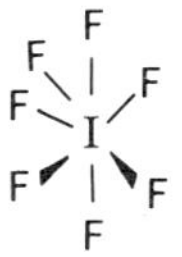	IF_7

TABLE   6-3

No. of electron pairs	No. of unshared pairs	Shape	Example
4	1	Trigonal pyramid	NH_3
	2	V shape	H_2O
5	1	Irregular tetrahedron	SF_4
	2	T shape	BrF_3
	3	Linear	ICl_2^-
6	1	Square-based pyramid	IF_5
	2	Square plane	ICl_4^-
7	1	Irregular octahedron	$(SbBr_6)^{3-}$

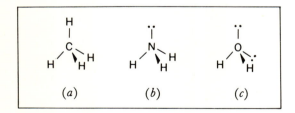

FIGURE 6-1   (a) Methane   (b) ammonia   and   (c) water molecules

Applying the principle that the shape of a molecule depends upon the number of pairs of electrons to methane leads to the tetrahedral structure. Here the H-C-H bond angle is 109°28'. In a similar fashion the ammonia molecule is expected to be a trigonal pyramid. The H-N-H bond angle observed is only 107°,however, not 109°28'. The greater u-b repulsions decrease the H-N-H angle below the regular 109°28'. The b-b repulsions are less than the u-b repulsions.

The H-O-H bond angle in water is 105° in the V-shape molecule, a lower value than that in NH_3 because there are two unshared pairs of electrons, leading to greater repulsion and a smaller angle.

When considering more complicated molecules, repulsions of electron pairs at angles greater than 90° may be ignored. Applying the principles to SF_4 leads to the prediction that the five pairs of electrons will occupy the apices of a trigonal bipyramid. The position of the unshared pair will be either axial [Fig. 6-2(a)] or equatorial [Fig. 6-2(b)]. In Fig. 6-2(a) there are three u-b repulsions at 90°, and in Fig. 6-2(b) there are only two. The second structure is predicted.

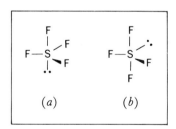

FIGURE 6-2

Three possibilities may be considered for the shape of ClF_3, as in Fig. 6-3(a), (b), and (c). The first has six u-b repulsions at 90°,the second has one u-u and six u-b repulsions at 90°, and the third has four u-b repulsions at 90°. The third is the preferred structure.

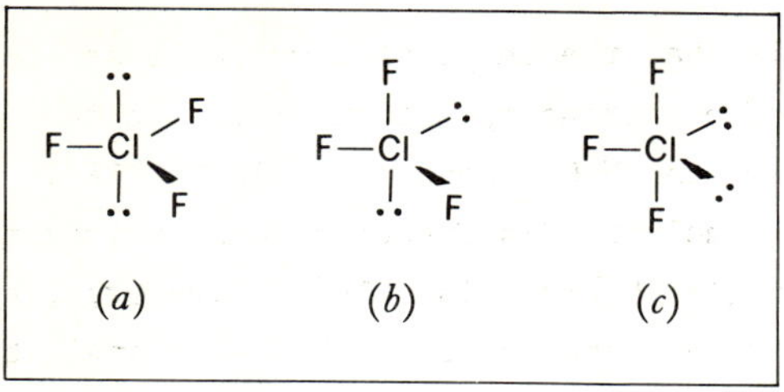

FIGURE 6-3

Example 6-26

Predict the shapes of the following molecules and ions. In each case
write the Lewis structure before attempting to decide on a shape.

    a. BeH_2

    H : Be : H

Two pairs of electrons—linear arrangement of electrons; therefore
the molecule is linear: H-Be-H.

    b. BBr_3

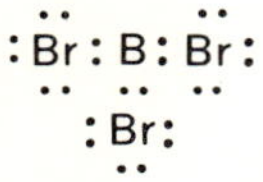

      : Br : B : Br :
          : Br :

Three pairs of electrons—trigonal plane arrangement; therefore
the molecule is a triangular plane.

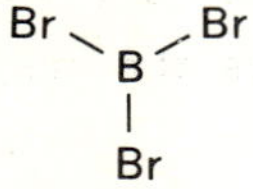

  Br   Br
      B
      |
      Br

c. H_2Te

H:Te:H

Four pairs of electrons—tetrahedral arrangement;   therefore the
molecule is V-shaped.

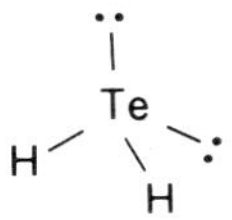

```
      ..
      |
      Te
   H-   \   -.
        \
        H
```

d. SiH_4

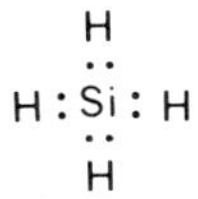

```
      H
      ..
   H:Si:H
      ..
      H
```

Four electron pairs—tetrahedral arrangement;   therefore the molecule
is tetrahedral in shape.

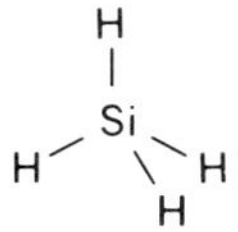

```
      H
      |
      Si
   H-   \-H
        \
        H
```

e. PH_3

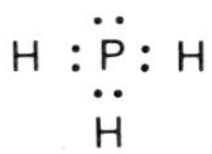

```
      ..
   H :P: H
      ..
      H
```

Four electron pairs—tetrahedral arrangement;   therefore the molecule
is trigonal pyramidal in shape.

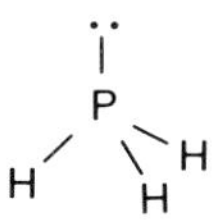

```
      ..
      |
      P
   H-   \-H
        \
        H
```

f. $TeCl_4$

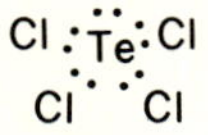

Cl :Te: Cl
Cl   Cl

Five electron pairs—trigonal bipyramidal arrangement; therefore
the molecule is a distorted tetrahedron.

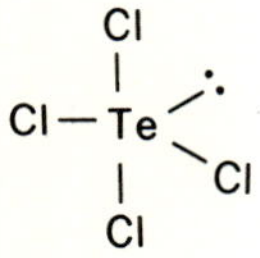

```
      Cl
      |    ..
 Cl—Te
      |    Cl
      Cl
```

g. $CsIF_4$ $Cs^+IF_4^-$

```
F .  ..  . F        —
F .  I  . F
F .  ..  . F
```

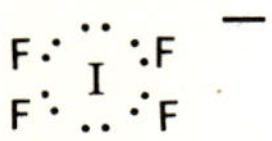

Six electron pairs—octahedral arrangement; therefore a square-
planar shape.

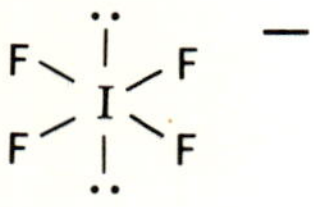

```
     ..
 F    |   F         —
  \   I  /
  /   |  \
 F    |   F
     ..
```

6-13 MOLECULES CONTAINING MULTIPLE BONDS

The shapes of molecules containing multiple bonds can also be
predicted. The electron-pair double or triple bonds are assumed to
occupy only one of the positions in the arrangements listed in
Table 6-1.

Example 6-27

a. Carbon dioxide, CO_2

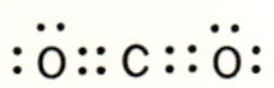

:O:: C ::O: (linear)

There are two bonds and no unshared pairs.

Chemical Bonds   141

b.   Carbonyl fluoride, COF_2

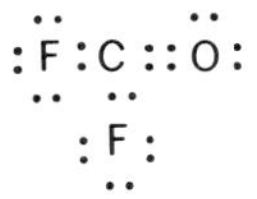

:F:C::O:

   :F: (triangular plane)

There are three bonds and no unshared pairs.

c.   Ozone, O_3

:O:O::O: (V shape)

There are two bonds and one unshared pair.

PROBLEMS

6-1   Write Lewis structures for:

  a.   Carbon tetrabromide   b.   Silane, SiH_4
  c.   Oxygen difluoride   d.   Hydrogen selenide, H_2Se
  e.   Ammonium chloride   f.   Hypochlorous acid, HOCl
  g.   Sulfuric acid   h.   Perchloric acid
  i.   Hypobromous acid   j.   Sodium sulfite
  k.   Carbon dioxide   l.   Dichlorine monoxide
  m.   Nitrogen trifluoride   n.   Phosphine, PH_3
  o.   Calcium nitrite

6-2   Write Lewis structures for:

  a.   Mercury(II) chloride   b.   Boron tribromide
  c.   Tellurium tetrachloride   d.   Antimony pentachloride
  e.   Selenium hexafluoride

6-3 Write Lewis structures for the following compounds. Include
canonical forms (resonance forms), if they are appropriate.

a. Ozone   b. Carbon monoxide
c. Formaldehyde, H₂CO   d. Lithium carbonate
e. Dinitrogen monoxide, NNO   f. Sodium nitrate
g. Sulfur dioxide, OSO   h. Hydrazoic acid, HNNN
i. Potassium nitrite   j. Sulfur trioxide

6-4 Write Lewis structures for the compounds listed below.

a. Diboron tetrachloride

B—B = 1.75 Å   B—Cl = 1.72 Å

Cl—B—Cl angle = 122°   Cl—B—B angle = 119°

    [M. J. Linevsky, E. R. Shull, D. E. Mann, and T. Wartik,

    J. Am. Chem. Soc., 75, 3287 (1953)]

b. Cyanogen (CN)₂. The molecules are symmetrical and linear.

C—C = 1.37 Å   C—N = 1.16 Å

    [L. Pauling, H. D. Springall, and K. L. Palmer, J. Am.

    Chem. Soc., 61, 933 (1939)]

c. Nitryl fluoride, NO₂F. The molecules are planar.

F—N = 1.35 Å   N—O = 1.23 Å   O—N—O angle = 125°

    [D. F. Smith, and D. W. Magnuson, Phys. Rev., 87,

    226 (1952)]

d. POCl₃

P—O = 1.45 Å   P—Cl = 1.99 Å   Cl—P—Cl angle = 103.5°

    [G. R. Badgley and R. L. Livingston, J. Am. Chem. Soc.,

    76, 261 (1954)]

e. Na₂S₂O₄

S—S = 2.389 Å

    [J. D. Dunitz, J. Am. Chem. Soc., 78, 878 (1956)]

f. Formyl fluoride, HCOF

C—F = 1.351 Å   C—O = 1.192 Å   O—C—F angle = 121.9°

[M. E. Jones, K. Hedberg, and C. Shomaker, J. Am. Chem.
Soc., 77, 5278 (1955)]

6-5 A compound prepared by the reaction of fluorine with sulfur
trioxide contained 27.0% S, 32.0% F, and the remainder oxygen.
The molecular weight calculated from gas-density measurements was
118.6. One of the fluorine atoms is bonded to oxygen. Write a
Lewis structure representing the molecules of the compound. How
is the structure related to that of sulfuric acid?

6-6 Draw diagrams representing the expected shapes of the molecules
of each of the following compounds. Name the geometrical figures
in each case.

    a. Hydrogen selenide   b. Oxygen difluoride

    c. Ammonium ion   d. Dichlorine monoxide

    e. Mercury(II) chloride   f. Boron triiodide

    g. Nitrogen trifluoride   h. Selenium tetrachloride

    i. Selenium hexafluoride   j. Arsenic pentafluoride

6-7 Predict the shapes of the molecules of the following compounds.

    a. Hypobromous acid   b. Selenium dichloride

    c. Silicon tetrafluoride   d. Tetrahydroborate ion, BH_4^-

    e. Antimony trichloride   f. Tin tetrachloride

    g. Gallium triiodide   h. Phosphorus pentafluoride

       [Inorg. Chem., 4, 1775 (1965)]

    i. Bromine trifluoride   j. Tellurium hexafluoride

6-8 Draw diagrams representing the shapes of the molecules of the
compounds listed.

    a. Hydrogen cyanide, HCN

    b. Phosgene (carbonyl chloride), $COCl_2$

    c. Sulfur dioxide

    d. Phosphorus oxychloride, $POCl_3$

e. Nitrate ion

f. Thionyl fluoride, SOF_2

g. Nitrosyl chloride, NOCl

h. Nitryl fluoride, NO_2F

6-9 A crystal structure analysis of potassium selenocyanate was
made by D. D. Swank and R. D. Willett. The C—N distance was
recorded as 1.117 Å, the Se—C distance as 1.829 Å, and the
Se—C—N angle as 178.8°. Write a Lewis structure for the compound.
Show that the structure is consistent with the shape you might
predict from electron-pair repulsions.
     [Inorg. Chem., 4, 501 (1965)]

6-10 A 14:1 mixture of F_2 and ClF_3 was heated to 350°C for 1 hr
at 250 atm. The infrared spectrum of the product was similar to
that of BrF_5. Write a Lewis structure for the expected product of
the reaction. Draw a diagram representing the shape of the molecules.
     [D. F. Smith, Science, 141, 1039 (1964)]

6-11 From an electron-diffraction investigation, Harvey and Bauer
determined the S—S distance to be 2.21 Å in the compound disulfur
decafluoride. Write a Lewis structure for the molecule and draw
a diagram representing the shape.
     [J. Am. Chem. Soc., 75, 2840 (1953)]

6-12 To a solution of 13.0 g of BCl_3 in liquid SO_2 as a solvent
at -30°, 26 g of anhydrous sodium thiocyanate, NaSCN, was added.
Sodium chloride and 7.5 g of a colorless liquid were obtained.
Analysis found: C, 19.55; N, 22.92.
     [D. B. Sowerby, J. Am. Chem. Soc., 84, 1831 (1962)]

a. What was the liquid product of the reaction?

b. Which of the starting materials was present in excess?

c. Write a Lewis structure for the liquid product.

6-13   Thionyl fluoride, SOF_2, was treated with silver(II) fluoride, and an oxygen-containing product was formed. Analysis found: S, 25.6; F, 62.0. The molecular weight from the mass spectrum was 124.

> [F. B. Dudley, G. H. Cady, and D. F. Eggers, Jr., J. Am. Chem. Soc., 78, 1554 (1956)]

   a.   Write a Lewis structure representing the molecules of the compound.

   b.   Draw a diagram representing the shape of the molecules.

6-14   The fluorination of a gaseous mixture of SNF and SNF_2 gave as a product a colorless gas. Analysis found: F, 54.4; S, 30.4. Molecular weight 1.3352 g occupied 471.3 ml at 499.3 mm and 290.5 K.

> [O. Glemser and H. Schroder, Z. Anorg. Allgem. Chem., 284, 97 (1956)]

The infrared spectrum of the compound was similar to that of OPF_3.

> [H. Richert and O. Glemser, Z. Anorg. Allgem. Chem., 307, 328 (1961)]

Predict the shape of the molecules of the compound. Compare your prediction with the results of a microwave investigation of W. H. Kirchoff and E. B. Wilson of Harvard.

> [J. Am. Chem. Soc., 84, 334 (1962)]

Chapter 7

SOLUTIONS

7-1 EXPRESSING CONCENTRATIONS OF SOLUTIONS

The method of expressing the concentration of a solution
depends upon the intended use of the solution. For many uses it
is sufficient to express the concentration as a weight percent:
the weight of solute per 100 parts by weight of solution.

Example 7-1
An experiment required 250 g of a 23.2% aqueous solution of glucose.
What weight of glucose was required? What weight of water?

$$250 \times 0.232 = 58.0 \text{ g}$$

$$\text{g solution} \times \frac{\text{g glucose}}{\text{g solution}} = \text{g glucose}$$

$$250 - 58 = 192 \text{ g } H_2O$$

Other relationships used to express very dilute concentrations
are parts per million (ppm) and parts per billion (ppb). Both are
derived from weight fractions.

If the fraction by weight of solute in solution is multiplied
by a million (10^6) the concentration is expressed as ppm, while if
the multiplier is a billion (10^9) the concentration is expressed
as ppb.

Example 7-2
A 0.500-g sample of fish was dissolved in 50 ml of concentrated nitric
acid. One ml of this solution was taken, neutralized, and diluted

147

to 25 ml. This final solution was found to contain 0.04 nanograms of mercury per milliliter of solution. Express the mercury in the fish in ppm and ppb.

$$0.04 \times 10^{-9} \; \frac{g \; Hg}{ml} \times 25 \; ml = 1 \times 10^{-9} \; g \; Hg$$

This came from 1 ml of the 50 ml solution, so

$$50 \; ml \times 1 \times 10^{-9} \; \frac{g \; Hg}{ml} = 50 \times 10^{-9} \; g \; Hg$$

There were 50×10^{-9} g of mercury in 0.500 g of fish

$$\frac{50 \times 10^{-9}}{0.500} \times 10^{9} = 100 \; ppb$$

$$= 0.10 \; ppm$$

In much of the recent work done in water pollution studies, a weight/volume unit of concentration is used. This is also expressed as parts per million or parts per billion, based on the number of milliliters of solution.

Example 7-3

The concentration of copper in salt water has been measured by atomic absorption spectroscopy. A typical result in the Atlantic Ocean, in μg-atom/liter, is 0.030. Express this in ppm and ppb.

[B. P. Faricand, R. R. Sawyer, S. G. Ungar, and S. Adler, Geochim. Cosmochim. Acta, 26, 1023 (1962)]

1 g-atom Cu = 63.55 g

$1 \; \mu g = 10^{-6} \; g$

$$\frac{1 \; \mu g\text{-atom}}{liter} = 1 \times 10^{-6} \times 63.55/10^{3} = 63.55 \times 10^{-9} = \frac{g \; solute}{ml \; solution}$$

$$= \frac{\mu g\text{-atom}}{liter} \times \frac{g}{\mu g} \times \frac{g \; solute}{g\text{-atom}} \times \frac{liter}{ml \; solution}$$

$$= \frac{g \; solute}{ml \; solution}$$

$$0.030 \; \frac{\mu g\text{-atom}}{liter} \times 63.55 \times 10^{-9} \; \frac{g \; solute/ml \; solution}{g\text{-atom}/liter}$$

$$= 1.09 \times 10^{-9} \; \frac{g \; solution}{ml \; solution}$$

$1.09 \times 10^{-9} \times 10^{9} = 1.09$ ppb

$1.09 \times 10^{-9} \times 10^{6} = 0.00109$ ppm

For chemical experiments and calculations it is usually more convenient to use solutions based on the natural unit of chemistry, the mole.

7-2 MOLAR SOLUTIONS

A molar solution (M) contains one mole of solute per liter of solution. To prepare a molar solution, the correct quantity of solute is dissolved in less than sufficient solvent to prepare the solution. When the solute has dissolved, additional solvent is added to make the proper volume of solution. The usual vessel for accurate preparation of solutions is the volumetric flask.

The symbol 3 M HNO_3 refers to a three-molar solution of nitric acid, a solution which contains 3 moles of nitric acid in 1 liter of solution. A 0.10 M H_2S solution contains 0.10 mole of H_2S per liter of solution.

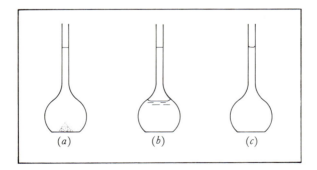

FIGURE 7-1 Preparation of solutions: (a) accurately weighed solute, (b) solvent to dissolve, (c) fill to mark.

Example 7-4

Calculate the mass of solute in grams required to make 1 liter of
1.72×10^{-2} M $HgCl_2$.

$HgCl_2$ $1 \times 200.6 = 200.6$

$2 \times 35.4 = \underline{70.8}$

271.4 g/mole

$1.72 \times 10^{-2} \times 2.71 \times 10^{2} = 4.66$ g/liter

$$\frac{mole}{liter} \times \frac{g}{mole} = \frac{g}{liter}$$

The concentration of a solution in moles per liter is referred to as
the <u>molarity</u> of the solution.

Example 7-5

What is the molarity of a solution which contains 171 g of sucrose
(sugar) in 2.00 liters of solution?

Since the molarity is the number of moles of solute in one liter
of solution, the first step is to change the solution to a basis of
one liter.

$$\frac{171 \text{ g}}{2.00 \text{ liters}} = 85.5 \text{ g/liter}$$

Next change the mass of solute in grams to the number of moles
present.

$C_{12}H_{22}O_{11}$ $12 \times 12 = 144$

$22 \times 1 = 22$

$11 \times 16 = \underline{176}$

342 g/mole

$$\frac{85.5}{342} = 0.250 \text{ mole/liter}$$

$$\frac{g/liter}{g/mole} = \frac{g}{liter} \times \frac{mole}{g} = mole/liter$$

The molarity of the solution is 0.250 M.

A more direct approach to this type of problem is to represent
<u>molarity</u> by the equation.

$$\text{Molarity} = \frac{\text{moles solute}}{\text{liters solution}}$$

$$M = \frac{171/342}{2} = 0.250 \text{ mole/liter}$$

$$\frac{g}{g/\text{mole/liter}} = g \times \frac{\text{mole}}{g} \times \frac{1}{\text{liter}}$$

It is very often much more convenient to use the smaller unit of volume, the milliliter. Since 1 ml $= \frac{1}{1000}$ liter, the molarity of a solution may be expressed in terms of milliliters and <u>millimoles</u> (mM).

$$\text{Molarity} = \frac{\text{millimoles solute}}{\text{milliliters solution}}$$

The solution in Example 7-5 contains 0.250 mM/ml. A 0.100 M H_2SO_4 solution will contain 0.100 mM of H_2SO_4 per milliliter of solution

$$\frac{\text{mole} \times 1000 \text{ mM/mole}}{\text{liter} \times 1000 \text{ ml/liter}} = \frac{\text{mM}}{\text{ml}}$$

7-3 CALCULATION OF QUANTITY OF SOLUTE

Frequently it is necessary to calculate the quantity of solute in a given volume of solution from a knowledge of its concentration.

Example 7-6

What mass of hydrogen chloride is present in 1.88 liters of 0.150 M HCl solution?

Each liter contains 0.150 mole HCl, and the volume is 1.88 liters.

$$0.150 \times 1.88 = 0.282 \text{ mole HCl}$$

$$\frac{\text{mole}}{\text{liter}} \times \text{liter} = \text{moles}$$

$$0.282 \times 36.4 = 10.3 \text{ g HCl}$$

$$\text{mole} \times \frac{g}{\text{mole}} = g$$

More directly:

Number of moles of solute = molarity x volume

Mass of solute in grams = molarity x volume x mole weight

0.150 x 1.88 x 36.4 = 10.3 g

$$\frac{mole}{liter} \times liter \times \frac{g}{mole} = g$$

Example 7-7

Calculate the quantity of potassium nitrate present in 200 ml of 0.125 M solution.

Mass solute = molarity x volume x mole weight

KNO_3 1 x 39.1 =   39.1

          1 x 14.0 =   14.0

          3 x 16.0 =   48.0

                       101.1 g/mole

200 ml = 0.200 liter

Mass solute = 0.125 x 0.200 x 101 = 2.52 g

$$\frac{mole}{liter} \times liter \times \frac{g}{mole} = g$$

If the volume is left as milliliters, then the product of molarity and volume will give the number of <u>millimoles</u> of solute. The mass of 1 mM rather than 1 mole will be required.

1 mM KNO_3 = $\frac{101}{1000}$ = 0.101 g

Mass solute = 0.125 x 200 x 0.101 = 2.52 g

$$\frac{mM}{ml} \times ml \times \frac{g}{mM} = g$$

Example 7-8

Calculate the mass of solute present in 15.2 ml of 0.212 M HBr.

HBr   1.01

        $\underline{79.91}$

        80.92 g/mole

1 mM HBr = $\frac{80.9}{1000}$ = 0.0809 g

Mass solute = 0.212 x 15.2 x 0.0809 = 0.260 g

$$\frac{mM}{ml} \times ml \times \frac{g}{mM} = g$$

Example 7-9

What is the mass of lithium chloride, in grams, in 61.8 ml of
1.03 M LiCl?

LiCl   6.9

$\dfrac{35.4}{42.3}$ g/mole

$1 \text{ mM LiCl} = \dfrac{42.3}{1000} = 0.0423 \text{ g}$

$\text{Mass LiCl} = 1.03 \times 61.8 \times 0.0423 = 2.69 \text{ g}$

$\dfrac{mM}{ml} \times ml \times \dfrac{g}{mM} = g$

Example 7-10

What is the molarity of concentrated sulfuric acid which contains
95.2% H_2SO_4? The density of the solution is 1.53 g/ml.

$\text{Mass 1 liter} = 1.53 \times 1000 = 1530 \text{ g/liter}$

$\dfrac{g}{ml} \times \dfrac{ml}{liter} = g/liter$

$\text{Mass } H_2SO_4/\text{liter} = 1530 \times 0.952 = 1450 \text{ g}$

H_2SO_4 $2 \times 1.0 = 2.0$

              $1 \times 32.1 = 32.1$

              $4 \times 16.0 = \dfrac{64.0}{98.1}$ g/mole

$\text{Molarity} = \dfrac{1450}{98.1} = 14.8 \text{ M}$

$\dfrac{g/liter}{g/mole} = \dfrac{g}{liter} \times \dfrac{mole}{g} = mole/liter$

Example 7-11

Calculate the molarity of a solution of hydrochloric acid from
the data which appeared on the label of the bottle in which it was
received from the manufacturer: Assay 36.5%, sp. gr. 1.1854. Since
the molarity is based on 1 liter of solution, base your calculations
on 1 liter.

$\text{Mass 1 liter} = 1000 \times 1.18 = 1180 \text{ g/liter}$

$\dfrac{ml}{liter} \times \dfrac{g}{ml} = g/liter$

Mass HCl/liter = 1180 x 0.365 = 430 g HCl/liter

HCl   1.0

$\dfrac{35.4}{36.4}$ g/mole

$\dfrac{430}{36.4}$ = 11.8 moles HCl/liter = 11.8 M

7-4 DILUTION

The dilution of more concentrated solutions is an everyday experience in the laboratory. The calculation of the volume of concentrated solution required to make a given volume of dilute solution is based on the fact that the quantity of solute in the two solutions is the same. The concentration (moles of solute per liter of solution) varies, but the number of moles remains the same:

No. moles solute in dilute soln. = no. moles solute in conc. soln.

$$M_d \times V_d = M_c \times V_c$$

where M_d is the molarity of the dilute solution, V_d is the volume of the dilute solution, M_c is the molarity of the concentrated solution, and V_c is the volume of the concentrated solution.

Example 7-12

Calculate the volume of 12.0 M HCl required to make 250 ml of 1.50 M HCl solution by dilution with water.

$$M_d \times V_d = M_c \times V_c$$

millimoles HCl = millimoles HCl

1.50 x 250 = 12.0 x V_c

V_c = $\dfrac{1.50 \times 250}{12.0}$ = 31.2 ml 12.0 M HCl

$$\dfrac{\dfrac{mM}{ml} \times ml}{\dfrac{mM}{ml}} = ml$$

Example 7-13

What is the concentration in moles per liter of a solution made by
adding 20.0 ml of H_2O to 15.0 ml of 2.50 M HNO_3?

$$M_d \times V_d = M_c \times V_c$$

$$M_d \times (20.0 + 15.0) = 2.50 \times 15.0$$

$$M_d = \frac{2.50 \times 15.0}{35.0} = 1.07 \text{ mM/ml}$$

$$1.07 \text{ mM/ml} = 1.07 \text{ moles/liter}$$

7-5 MOLAL SOLUTIONS

The concentration of a solution based on the volume varies with
temperature. For this and other reasons, solutions whose concentrations are based on the mass of solute and solvent are sometimes
preferred. A _molal_ solution contains 1 mole of solute per 1000 g
of _solvent_. Note carefully that the basis is 1000 g of _solvent_,
not _solution_.

$$\text{Molality} = \frac{\text{moles solute}}{1000 \text{ g solvent}}$$

Example 7-14

Calculate the molality of a 16.2% solution of alcohol in water.

A 16.2% solution contains 16.2 g of solute (alcohol) and
100.0 - 16.2 = 83.8 g of solvent (water).

Since the basis of the molal solution is 1000 g of solvent,
change the mass of solute to that basis. Multiply 16.2 by the
fraction 1000/83.8.

$$16.2 \times \frac{1000}{83.8} = 194 \text{ g}$$

This is the mass of solute which would be present in the solution
if the mass of solvent were 1000 g. Next calculate the number of
moles of solute in 1000 g of solvent.

C_2H_6O 2 x 12.0 = 24.0

         6 x 1.01 =   6.06

         1 x 16.0 = 16.0
                           46.1 g/mole

$$\frac{194}{46.1} = 4.21 \text{ moles/1000 g, or } 4.21 \text{ m(molal)}$$

Example 7-15

Calculate the molality of a concentrated HCl solution which contains
36.5% HCl.

100.0 - 36.5 = 63.5 g H_2O in 100.0 g solution

$36.5 \times \dfrac{1000}{63.5}$ = mass of HCl in 1000 g H_2O

1 mole HCl weighs 36.4 g

$$\frac{36.5 \times 1000/63.5}{36.4} = 15.7 \text{ m}$$

$$\frac{\text{g/HCl/1000 g } H_2O}{\text{g HCl/mole}} = \frac{\text{g HCl}}{1000 \text{ g } H_2O} \times \frac{\text{mole}}{\text{g HCl}} = \frac{\text{mole}}{1000 \text{ g } H_2O}$$

Compare this result with that of Example 7-11.

7-6 MOLE FRACTION

The mole fraction X of a solution may be expressed as the mole
fraction of solute or the mole fraction of solvent.

$$X \text{ solute } = \frac{n}{n + n^\circ}$$

$$X \text{ solvent } = \frac{n^\circ}{n^\circ + n}$$

where n is the number of moles of solute and n° is the number of
moles of solvent in the solution.

Example 7-16

Calculate the mole fraction of solute in a 2.50 m aqueous solution.

This solution contains 2.50 moles of solute in 1000 g of H_2O. The number of moles of H_2O is given by 1000/18.0, since 1 mole of H_2O weighs 18.0 g.

$$X = \frac{2.50}{2.50 + (1000/18.0)} = \frac{2.50}{2.50 + 55.6}$$

$$X = \frac{2.50}{58.1} = 0.0430$$

$$\frac{\_mole\widetilde{}}{\_mole\widetilde{}}$$

Example 7-17

Calculate the mole fraction of solvent in a solution containing 89.9 g of mannitol, $C_6H_{14}O_6$, in 1000 g of H_2O.
[J. C. W. Frazer, B.F. Lovelace, and T. H. Rogers, J. Am. Chem. Soc., 42, 1801 (1920)]

$$C_6H_{14}O_6 \qquad \begin{array}{rcl} 6 \times 12.0 & = & 72.0 \\ 14 \times 1.0 & = & 14.0 \\ 6 \times 16.0 & = & \underline{96.0} \\ & & 182.0 \text{ g/mole} \end{array}$$

$$H_2O \qquad \begin{array}{rcl} 2 \times 1.0 & = & 2.0 \\ 1 \times 16.0 & = & \underline{16.0} \\ & & 18.0 \text{ g/mole} \end{array}$$

$$\frac{89.9}{182.0} = 0.494 \text{ mole } C_6H_{14}O_6$$

$$\frac{1000}{18.0} = 55.6 \text{ mole } H_2O$$

$$X \text{ solvent} = \frac{55.6}{55.6 + 0.494} = \frac{55.6}{56.1} = 0.990$$

PROBLEMS

7-1 In order to determine the concentration of phosphorus present in a sample using uv spectroscopy, a solution was prepared containing 2.200 g of KH_2PO_4 per liter of solution. A stock solution was made

by diluting 20.00 ml of the first solution to 1 liter with distilled
water. Express the concentration of P in the stock solution in ppm.

[T. Hurford and D. Boltz, Anal. Chem., 40, 379 (1968)]

7-2 In analyzing for trace amounts of chromium, a standard stock
solution was prepared by dissolving 1.414 g of dry primary standard
$K_2Cr_2O_7$ in 500 ml of distilled water. How many ppm of Cr are
present in the standard solution?

[J. A. Hurlbert and C. D. Chriswell, Anal. Chem.,

43, 465 (1971)]

7-3 Calculate the mass of solute required to make 1 liter of each
of the following solutions.

a. 0.475 M sodium carbonate   f. 2.12 M sodium hydroxide
b. 0.9251 M $NaNO_3$ g. 8.5 M KCNS
c. 0.047 M copper(II) sulfate   h. 2.38 mM lead nitrate
d. 0.0531 M zinc chloride   i. 1.03 x 10^{-4}M cadmium nitrate
e. 0.896 M cesium nitrate   j. 0.0571 M aluminum sulfate

7-4 How many grams of each of the following solids must be weighed
to make the respective solutions?

a. Sodium hydroxide, 250 ml of 0.100 M NaOH
b. $CuSO_4 \cdot 5H_2O$, 300 ml of 2.00 M $CuSO_4$
c. $MnSO_4 \cdot 7H_2O$, 800 ml of 0.0020 M $MnSO_4$
d. $Na_2S_2O_3 \cdot 2H_2O$, 3.00 liters of 0.100 M $Na_2S_2O_3$
e. $NaClO_4 \cdot H_2O$, 50.0 ml of 1.25 x 10^{-4} M $NaClO_4$

7-5 Calculate the molarity of each of the following solutions:

a. 18.0% $CaCl_2$, sp. gr. 1.1578
b. 32.0% $(NH_4)_2SO_4$, sp. gr. 1.1836
c. 14.0% KBr, sp. gr. 1.1070

7-6 Calculate the volume of the concentrated reagent required to make the corresponding dilute solution in each instance.

a. 15.0 M NH_3 to make 200 ml of 1.84 M NH_3

b. 12.0 M HCl to make 300 ml of 4.0 M HCl

c. 16.0 M HNO_3 to make 50.0 ml of 0.250 M HNO_3

d. 12.3 M HCl to make 100 ml of 1.02 x 10^{-3} M HCl

e. 15.2 M HNO_3 to make 250 ml of 2.00 mM HNO_3

7-7 How many milliliters of concentrated H_2SO_4 would be required to make 20.0 liters of 6.00 M acid? The concentrated sulfuric acid contains 95.0% H_2SO_4, and its specific gravity is 1.8407.

7-8 Calculate the molarity of concentrated hydrochloric acid which contains 38.0% HCl. The specific gravity is 1.1885.

7-9 What is the molarity of concentrated nitric acid if its specific gravity is 1.4251 and it contains 71.15% HNO_3?

7-10 The assay of a bottle of aqueous ammonia reads: "Min. 23.0%, Max. 30.0%." The specific gravity is given as: "Min. 0.8957, Max. 0.9016." What are the limits of the molarity of the solution?

7-11 Calculate the molality of 32.31% nitric acid.

7-12 Calculate the molality of 11.8 M HCl. The density of the solution is 1.1854 g/ml.

7-13 The solubility of potassium iodate in water was found to be 0.0919 moles/liter. The density of the solution was 1.0160 g/ml. Calculate the molality of the saturated solution.

[W. D. Larson and J. J. Renier, J. Am. Chem. Soc., 74, 3184 (1952)]

7-14   A 27.76% solution of calcium dichromate, $CaCr_2O_7$, was found
to have a density of 1.2568 g/ml. Calculate the molality and molarity
of the solution.

> [W. H. Hartford, K. A. Lane, and W. A. Meyer, Jr., J. Am.
> Chem. Soc., 72, 3655 (1950)]

7-15   Tin(II) chloride dihydrate (112.8 g) was mixed with 22.5 g
of 49.6% HF and heated on a steam bath. When the mixture was allowed
to cool, 21.8 g of product was collected. Analysis found:
Sn, 68.7, 68.3, 68.5; F, 10.7, 10.7, 10.8; and the remainder
chlorine. The solubility of the product in water was 55.0%.

   a.   Calculate the molality of the solution.

   b.   Which of the reagents used to prepare the crystalline
product was in excess?

   c.   How many grams of tin(II) chloride dihydrate would be
required to make 100 g of a 1 m solution of the product? Assume that
the reaction and recovery of the product are quantitative.

> [W. H. Nebergall, G. Baseggio, and J. C. Muhler, J. Am.
> Chem. Soc., 76, 5353 (1954)]

7-16   Calculate the volume of oxygen, at STP, liberated by the
decomposition of hydrogen peroxide in 25.0 ml of 0.01121 M H_2O_2.
The reaction is

   $2H_2O_2(aq) \rightarrow 2H_2O + O_2(g)$

7-17   Calculate the mole fraction of solute and the mole fraction of
solvent in a 31.2% solution of sugar, $C_{12}H_{22}O_{11}$.

7-18   What is the mole fraction of mannitol, $C_6H_{14}O_6$, in 1000 g
of H_2O if the solution contains 162 g of solute?

> [J. C. W. Frazer, B. F. Lovelace, and T. H. Rogers,
> J. Am. Chem. Soc., 42, 1801 (1920)]

SUPPLEMENTARY PROBLEMS

7-19 Silver can be determined in the 0.01 to 10 ppm range using
atomic fluorescence spectroscopy. In such an experiment West and
Williams prepared a stock solution of silver by dissolving 0.9128 g
of $AgNO_3$ and diluting to 100 ml with distilled water. What was the
concentration in ppm of silver in the stock solution?

[Anal. Chem., 40, 335 (1968)]

7-20 Calculate the quantity of solute in grams per liter of each
of the following solutions, which were used in research projects
described in the literature.

    a. 0.9646 M $LiNO_3$

    b. 6.67 x 10^{-3} M $Co(NO_3)_2$

    c. 0.1511 M NaOH

    d. 2.31 x 10^{-3} M $Cu(ClO_4)_2$

    e. 1.793 x 10^{-2} M $CuCl_2$

    f. 0.45 M chromium(III) perchlorate

    g. 0.475 M sodium bicarbonate

    h. 9.93 mM $TlNO_3$

    i. 2.36 x 10^{-2} mM cadmium nitrate

    j. 2.94 x 10^{-3} M zinc chloride

7-21 Calculate the mass of solute in grams in each of the following
solutions.

    a. 200 ml of 0.1 M perchloric acid

    b. 0.125 liter of 0.16 M $NaNO_3$

    c. 1.67 liter of 7.10 x 10^{-3} M manganese(II) nitrate

    d. 50.0 ml of 8.96 x 10^{-3} M $CuCl_2$

    e. 723 ml of 9.44 x 10^{-4} M copper(II) chloride

7-22 Calculate the mass of iodine monochloride present in 5.00 ml of 6.82 x 10^{-3} M solution in acetonitrile CH_3CN.

7-23 Calculate the molarity of each of the following solutions.

   a. 200 ml of solution containing 8.31 g of $LiNO_3$
   b. 75.0 ml of solution containing 0.00641 g of $AgNO_3$
   c. 2.50 liters of solution containing 1.74 g of alcohol, C_2H_6O

7-24 Calculate the solubilities in moles per liter of the following compounds from the solubility in grams per 100 ml as reported by Wheeler, Perros and Naeser.

        [J. Am. Chem. Soc., 77, 3489 (1955)]

   a. Na_2PtF_6, 20.49 g/100 ml   b. $(NH_4)_2PtF_6$, 7.32 g/100 ml
   c. K_2PtF_6, 0.750 g/100 ml   d. $RbPtF_6$, 0.278 g/100 ml
   e. $CsPtF_6$, 0.484 g/100 ml

7-25 Glacial acetic acid is 99.7% acetic acid, and its specific gravity is 1.049. What volume of glacial acetic acid would be required to make 5.65 liters of 4.57 M CH_3COOH?

7-26 Calculate the volume of concentrated phosphoric acid required to make 10.0 liters of 6.00 M H_3PO_4. The concentrated acid contains 85.0% H_3PO_4, and its specific gravity is 1.689.

7-27 What is the molarity of a solution of hydrobromic acid whose assay is 47.0% and specific gravity 1.50?

7-28 In each of the following cases, how many ml of the concentrated acid must be taken to prepare the desired solution?

   a. 100 ml of 3 M HCl from 12 M HCl
   b. 250 ml of 0.1 M HNO_3 from 15 M HNO_3
   c. 125 ml of 0.5 M H_2SO_4 from 18 M H_2SO_4

Solutions   163

7-29 Calculate the molarity of 88.0% formic acid, HCOOH. The
specific gravity is 1.2012.

7-30 Calculate the molarity of 70.0% perchloric acid. Its specific
gravity is 1.674.

7-31 What is the molality of an 18.9% NaCl solution?

7-32 The solubility of KPF$_6$ in water at 25° C was found to be
8.35% by Sarmousakis and Low. Calculate the molality of the solution.
     [J. Am. Chem. Soc., 77, 6518 (1955)]

7-33 A saturated solution of manganese(II) sulfate was prepared
by Richards and Frapie. A sample of the solution weighing 4.4318 g
contained 1.7470 g of the salt. What was the molality of the
solution?
     [J. Am. Chem. Soc., 26, 77 (1901)]

7-34 Crystals were collected after the reaction of stannous oxide
with hydrofluoric acid. Analysis found: Sn, 75.8; F, 24.3. The
solubility of the crystals in water was 29.6% at 18°. Calculate the
molality of the solution.
     [W. H. Nebergall, J. C. Muhler, and H. G. Day,
      J. Am. Chem. Soc., 74, 1604 (1952)]

7-35 Approximately 45.0 liters of 4.360 m hydrochloric acid was
required for several experiments. How many liters of concentrated
acid (37.5% HCl) was required? The specific gravity of the concentrated acid was 1.1875.

7-36 The mole fraction of alcohol, C_2H_6O, in an aqueous solution
was 0.250. What mass of alcohol was present in a solution containing
100 g of H_2O?

Chapter 8

COLLIGATIVE PROPERTIES

8-1 COLLIGATIVE PROPERTIES

Four properties of solutions which depend only on the number
of solute particles are called colligative properties. They are
vapor-pressure lowering, boiling-point elevation, freezing-point
depression, and osmotic pressure.

8-2 VAPOR-PRESSURE LOWERING

The vapor pressure of a solution is lower than the vapor
pressure of pure solvent. The extent of the depression of the vapor
pressure is related to the composition of the solution by Raoult's
law when a nonvolatile solute is present. Raoult's law may be
represented by
$$P = P°X$$
where P is the vapor pressure of the solution, P° is the vapor
pressure of the pure solvent, and X is the mole fraction of solvent
in the solution:
$$X = \frac{n°}{n + n°}$$
where n° is the number of moles of solvent and n is the number of
moles of solute.

Example 8-1

Calculate the vapor pressure of water at 25.0° over a solution containing 171 g of sugar ($C_{12}H_{22}O_{11}$) in 900 g of H_2O.

$$P = P°X$$

$$P° = 23.8 \text{ mm (from Table D-1)}$$

$$X = \frac{n°}{n + n°}$$

$$n° = \frac{900}{18} = 50.0 \text{ moles } H_2O$$

$$n = \frac{171}{342} = 0.500 \text{ moles } C_{12}H_{22}O_{11}$$

$$P = 23.8 \times \frac{50.0}{50.5} = 23.6 \text{ mm}$$

Example 8-2

Calculate the vapor pressure at 30.0° of a 10.0% aqueous solution of glucose, $C_6H_{12}O_6$.

$$P = P°X$$

$$P° = 31.8 \text{ mm (from Table D-1)}$$

$$X = \frac{n°}{n + n°}$$

$$n° = \frac{90.0}{18.0} = 5.00 \text{ moles } H_2O$$

$$n = \frac{10.0}{180} = 0.0556 \text{ mole } C_6H_{12}O_6$$

$$P = 31.8 \times \frac{5.00}{5.06} = 31.4 \text{ mm}$$

8-3 BOILING-POINT ELEVATION

The normal boiling point of a liquid is the temperature at which its vapor pressure equals one atmosphere. It follows from

Raoult's law that the boiling point of a solution will be higher than the boiling point of pure solvent. The extent of the boiling-point elevation is given by

$$\Delta t_b = K_b m$$

where Δt_b is the amount of the temperature increase, K_b is the molal boiling-point constant that is characteristic for each solvent, and m is the molal concentration of the solution. The value of K_b for water is 0.512. See Table 8-1 for the molal boiling point elevation constants of some common solvents.

TABLE 8-1   Molal Boiling-Point Elevation Constants

Solvent	K_b	Normal boiling point, °C
Acetic acid	3.07	118.5
Benzene	2.53	80.1
Carbon disulfide	2.34	46.3
Chloroform	3.63	61.2
Dichloromethane	2.52	40.5
Water	0.512	100.0

Example 8-3

Calculate the expected boiling-point elevation of a solution containing 0.504 g of anthracene, $C_{14}H_{10}$, in 42.0 g of benzene and compare it with the value 0.175° observed by E. Beckmann.

[Z. Physik. Chem., 6, 439 (1890)]

The boiling-point elevation of a solution is given by

$$\Delta t_b = K_b m$$

$$m = \frac{moles\ solute}{1000\ g\ solvent}$$

Calculate the mass of $C_{14}H_{10}$ in 1000 g of solvent.

$$0.504 \times \frac{1000}{42.0} = 12.0\ g$$

Determine the mass of 1 mole of $C_{14}H_{10}$

$C_{14}H_{10}$ 14 x 12.0 = 168

                 10 x 1.0 = $\underline{10}$

                          $\overline{178}$ g/mole

Calculate the number of moles of $C_{14}H_{10}$ in 1000 g solvent.

$\dfrac{12.0}{178}$ = 0.0674 mole/1000 g solvent

The boiling-point elevation is then given by

Δt_b =   2.53 x 0.0674 =   0.171°

$°C$ = $\dfrac{°C}{mole/1000 \text{ g solvent}}$ x $\dfrac{mole}{1000 \text{ g solvent}}$

      = $°C$ x $\dfrac{\overline{1000 \text{ g solvent}}}{\underline{mole}}$ x $\dfrac{\underline{mole}}{\underline{1000 \text{ g solvent}}}$

8-4 FREEZING-POINT DEPRESSION

Solutions have lower freezing points than pure solvents. The extent of the freezing-point depression depends upon the <u>molal</u> concentration of the solute: it is given by the equation

Δt_f = $K_f m$

where Δt_f is the freezing-point lowering, K_f is the molal freezing-point constant of the solvent, and m is the molal concentration. K_f is also called the <u>cryoscopic constant</u>. See Table 8-2 for the molal freezing-point depression constants of some common solvents.

TABLE 8-2 Molal Freezing-Point Depression Constants

Solvent	K_f	Normal freezing point, °C
Acetic acid	3.90	16.604
Benzene	4.90	5.5
Naphthalene	6.85	80.22
Phenol	7.40	43
Water	1.86	0.00

Example 8-4

Calculate the freezing point of a solution containing 50.0 g
of ethylene glycol, $C_2H_6O_2$, in 700 g of water. The freezing-point
depression of the solution is given by

$$\Delta t_f = K_f m$$

Calculate the mass of $C_2H_6O_2$ in 1000 g of water.

$$50.0 \times \frac{1000}{700} = 71.4 \text{ g}$$

Determine the mass of 1 mole of $C_2H_6O_2$.

$$
\begin{array}{llll}
C_2H_6O_2 & 2 \times 12.0 & = & 24.0 \\
& 6 \times 1.0 & = & 6.0 \\
& 2 \times 16.0 & = & \underline{32.0} \\
& & & 62.0 \text{ g/mole}
\end{array}
$$

Calculate the moles of $C_2H_6O_2$ in 1000 g of water.

$$\frac{71.4}{62.0} = 1.15 \text{ moles/1000 g}$$

The freezing-point depression is given by

$$\Delta t_f = 1.86 \times 1.15 = 2.14°C$$

$$°C = \frac{°C}{\text{mole/1000 g solvent}} \times \frac{\text{mole}}{1000 \text{ g solvent}}$$

The predicted freezing point of the solution is -2.14°C, since the
freezing point of water is 0°C.

8-5 OSMOTIC PRESSURE

When a solution is separated from pure solvent by a semipermeable
membrane, a pressure develops in the solution. An illustration of
the existence of this pressure is given by the rise of solution in
the stem of an inverted thistle tube, containing a sugar solution
held there by a cellophane membrane, when it is immersed in water.
A semipermeable membrane is one through which solvent molecules, but
not solute molecules, can pass. The pressure thus developed is
called osmotic pressure.

The osmotic pressure π of a __dilute__ solution may be expressed by the equation

π = mRT

where m = molality

R = 0.0820

T = temperature, K

Example 8-5

Calculate the osmotic pressure of a 0.100 m aqueous solution at 20°C.

π = mRT = 0.100 x 0.0820 x 293

π = 2.40 atm

8-6 DETERMINATION OF MOLECULAR WEIGHTS

All four of the colligative properties of solutions may be used for the experimental determination of molecular weights of solutes. Cryoscopic (freezing-point depression) measurements are the most common for the usual range of molecular weights because of fewer experimental difficulties. Osmotic-pressure measurements are expecially useful for substances with very high molecular weights, such as protein and other polymeric materials.

Example 8-6

What is the molecular weight of a compound if the vapor pressure of a solution containing 0.8696 g in 29.740 g of benzene had a vapor pressure of 631.9 mm? The vapor pressure of pure benzene at the same temperature was 639.7 mm.

[M. A. Rosanoff and R. A. Dunphy, __J. Am. Chem. Soc.__, __36__, 1417 (1914)]

The mole fraction of solvent in the solution can be calculated from P = P°X

$X = \dfrac{P}{P°} = \dfrac{631.9}{639.7}$

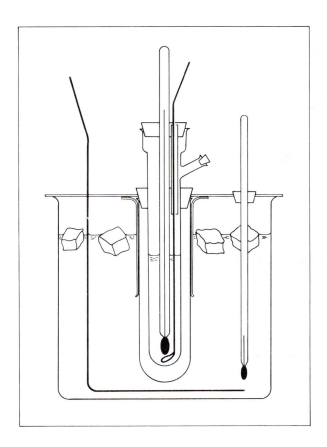

FIGURE 8-1 Apparatus for determining the freezing-point
depression from which molecular weight is calculated.

$$X = \frac{n^\circ}{n^\circ + n}$$

The molecular weight of benzene C_6H_6 is 78.0. Let M equal the molecular weight of solute

$$X = \frac{29.7/78.0}{29.7/78.0 + 0.870/M} = \frac{631.9}{639.7}$$

$$\frac{0.382}{0.382 + 0.870/M} = 0.988$$

$$\frac{0.382}{0.988} = 0.382 + \frac{0.870}{M} = 0.387$$

$$(0.387 - 0.382)M = 0.870$$

$$M = \frac{0.870}{0.005} = 174$$

Example 8-7

Calculate the molecular weight of a compound, 1.632 g of which raised the boiling point of 36.74 g of benzene 0.728°. The molarity of the solution can be calculated from $\Delta t_b = K_b m$.

[E. Beckmann, Z. Physik. Chem. (Leipzig), 6, 439 (1890)]

$$m = \frac{0.728}{2.53}$$

$$m = \frac{\text{moles solute}}{1000 \text{ g benzene}}$$

Calculate the mass of solute in 1000 g of solvent.

$$1.63 \times \frac{1000}{36.7} \text{ g solute/1000 g benzene}$$

From this, the number of moles of solute per 1000 g benzene and hence m is

$$\frac{1.63 \times 1000/36.7}{M} = \frac{1.63 \times 1000}{36.7 \times M} = m$$

$$M = \frac{1.63 \times 1000 \times 2.53}{36.7 \times 0.728} = 155$$

Example 8-8

Calculate the molecular weight of a compound if a solution containing 1.6491 g in 25.9256 g benzene had a freezing point of 5.12°C. The molality of the solution m is $\Delta t_f / K_f$ from $\Delta t_f = K_f m$.

[J. K. Ruff, Inorg. Chem., 4, 1448 (1965)] (See also Figure 8-1.)

$$\Delta t_f = 5.48 - 5.12 = 0.36°$$

$$m = \frac{0.36}{4.90}$$

Calculate the mass of solute in 1000 g solvent.

$$1.65 \times \frac{1000}{25.9} \text{ g solute/1000 g benzene}$$

The moles of solute per 1000 g benzene and hence m is

$$\frac{1.65 \times 1000/25.9}{M} = \frac{1.65 \times 1000}{25.9 \ M} = m$$

$$\frac{0.36}{4.90} = \frac{1.65 \times 1000}{25.9 \times M}$$

$$M = \frac{1.65 \times 1000 \times 4.90}{25.9 \times 0.36} = 87$$

Example 8-9

A solution containing 0.0150 g of polymeric material per gram of water had an osmotic pressure of 1.52×10^{-3} atm at 25°C. Estimate the molecular weight of the polymeric material.

$$m = \frac{\pi}{RT} = \frac{1.52 \times 10^{-3}}{0.0820 \times 298} = 6.22 \times 10^{-5}$$

$$m = 6.22 \times 10^{-5} \text{ moles/1000 g } H_2O$$

The solution contained 0.0150 g/g H_2O.

$$0.0150 \times 1000 = 15.0 \text{ g/1000 g } H_2O$$

Since 6.22×10^{-5} mole weighed 15.0 g, to find the mass of 1 mole, multiply

$$15.0 \times \frac{1}{6.22 \times 10^{-5}} = 2.41 \times 10^5 \text{ g/mole}$$

8-7 IDEAL SOLUTION

Raoult's law and the equations used to express the colligative
properties are approximate, just as the gas laws are approximate.
A solution to which Raoult's law applies accurately is called an
ideal solution. Departures from ideal behavior are pronounced when
an interaction between solute and solvent molecules occurs, when
the solute dissociates, or when a chemical reaction occurs. When
no dissociation, reaction, or appreciable interaction occurs,
solutions approach ideal behavior when they are dilute.

PROBLEMS

8-1 Calculate the vapor pressure at 20°C of an aqueous solution
containing 108.2 g mannitol $(C_6H_{14}O_6)$ in 1000 g of water. The
experimental value obtained by Frazer, Lovelace, and Rogers was
17.439 mm.

        [J. Am. Chem. Soc., 42, 1801 (1920)]

8-2 The measured pressure of a solution containing 0.5676 g of
anthracene, $C_{14}H_{10}$, in 29.740 g of benzene was 634.6 mm. How does
this value compare with the vapor pressure predicted by Raoult's
law? The vapor pressure of pure benzene at the same temperature was
639.7 mm.

        [M. A. Rosanoff and R. A. Dunphy, J. Am. Chem. Soc., 36,
        1416 (1914)]

8-3 The vapor pressure of ethyl alcohol was found to be 36.6 mm.
At the same temperature the vapor pressure of a solution of 3.9781 g

of acetamide, C_2H_5ON, in 102 g of alcohol was 35.6 mm. Compare the observed vapor pressure of the solution with that calculated from Raoult's law.

[J. Wright, J. Chem. Soc., 115, 1165 (1919)]

8-4 Make a plot of the observed vapor pressures of the solutions of oleic acid, $C_{18}H_{34}O_2$, in diethyl ether, $C_4H_{10}O$, versus the mole fraction of solvent from the data of F. H. Campbell at 30.0°C. Calculate the vapor pressure of each solution predicted from Raoult's law and plot the results on the same graph. Discuss the results.

% $(C_2H_5)_2O$	Vapor pressure, mm
4.96	96.7
7.17	119.9
14.36	217.9
23.55	303.7
37.40	414.8
61.93	535.6
100.00	642.1

Source: Trans. Faraday Soc., 11, 94 (1915).

8-5 Calculate the boiling-point elevation of a solution containing 0.5151 g of naphthalene, $C_{10}H_8$, in 60.81 g of chloroform. Compare your answer with the observation of E. Beckmann, who found a rise of 0.237°C.

[Z. Physik. Chem. (Leipzig), 6, 446 (1890)]

8-6 Calculate the expected boiling point of a solution which
contained 0.6072 g of a compound of molecular weight 600 in
26.0520 g of CH_2Cl_2.
  [J. K. Ruff, Inorg. Chem., 4, 1448 (1965)]

8-7 Calculate the expected freezing point of a solution containing
1.00 gallon of ethylene glycol, $C_2H_6O_2$, in 18.0 quarts of water.
The density of ethylene glycol is 1.12 g/ml, and 1 quart contains
946.34 cc.

8-8 The molecular weight of a solid compound was determined by the
freezing-point depression of an aqueous solution to be 81.0. If the
solution contained 15.77 g of the compound in 1000 g water, what
was the freezing-point depression of the solution?
  [A. W. Laubengayer and G. F. Condike, J. Am. Chem. Soc.,
   70, 2224 1946]

8-9 Calculate the approximate osmotic pressure of a solution
containing 17.1 g of sucrose, $C_{12}H_{22}O_{11}$, in 500 g of water at 20°C.

8-10 Calculate the molecular weight of a substance from the
vapor-pressure lowering of water at 20.0°, which was 0.2162 mm. The
mass of solute in 1000 g of water was 126.283 g.
  [J. C. W. Frazer, B. F. Lovelace, and T. H. Rogers,
   J. Am. Chem. Soc., 42, 1801 (1920)]

8-11 Calculate the molecular weight of the solute which lowered the
vapor pressure of alcohol 1.084 mm when 5.698 g was dissolved in
78.0 g of the solvent. The vapor pressure of alcohol, C_2H_6O, at the
temperature of the experiment was 42.60 mm.
  [J. Wright, J. Chem. Soc., 115, 1165 (1919)]

8-12 The molecular weight of a compound believed to be $Te_2(CO)_{10}$
was determined by measuring the vapor pressure of a solution

containing 12.5 mg of solute in 300.9 mg of diethyl ether. The
vapor-pressure depression was 3.08 mm at 22° (P° = 464 mm).
Compare the observed molecular weight with that calculated from
the formula.

[J. C. Hileman, D. K. Huggins, and H. D. Kaez, J. Am.
Chem. Soc., 83, 2953 (1961)]

8-13 Calculate the molecular weight of the compound which raised
the boiling point of chloroform 0.475° when 1.0171 g was dissolved
in 60.81 g of solvent.

[E. Beckmann, Z. Physik. Chem. (Leipzig), 6, 446 (1890)]

8-14 A sample of sulfur weighing 0.2096 g was dissolved in 17.79 g
of carbon disulfide. The observed boiling-point elevation was 0.107°.
Calculate the molecular weight of the solute. What is the formula
of the solute?

[A. Helff, Z. Physik. Chem. (Leipzig), 12, 200 (1893)]

8-15 What is the formula of phosphorus in carbon disulfide solution
if 0.1430 g raised the boiling point of 16.87 g of carbon disulfide
0.163°?

[A. Helff, Z. Physik. Chem. (Leipzig), 38, 76 (1901)]

8-16 An aqueous solution of a sugar contained 3.613 g in 1000 g
water, and the freezing point of the solution was -0.0375°C.
Calculate the molecular weight of the sugar.

[E. H. Loomis, Z. Physik. Chem. (Leipzig), 37, 411 (1901)]

8-17 A sample of a compound weighing 23.300 g was dissolved in
1000 g water. The freezing point of the solution was -0.150°. What
was the molecular weight of the compound?

[E. H. Loomis, Z. Physik. Chem. (Leipzig), 37, 411 (1901)]

8-18 The freezing point of a 4.9% aqueous solution was -3.4°. What
was the molecular weight of the solute?

[O. Maas and O. W. Herzberg, J. Am. Chem. Soc., 42, 2570 (1920)]

8-19 Calculate the freezing point of a 9.96% solution containing the same solute as in Prob. 8-18.

8-20 Calculate the molecular weight of a compound if a solution in anhydrous acetic acid containing 74.1 g in 5904 g of solvent has a freezing point of 15.60°C.

[S. English and W. E. S. Turner, J. Chem. Soc., 107, 775 (1915)]

8-21 Estimate the molecular weight of a polymeric material if a solution containing 2.50 g in 100 g of water had an osmotic pressure of 3.50 x 10^{-4} atm at 25°C.

8-22 The compound S_4N_4 was fluorinated with AgF_2, and a product which contained 49.4% S, 21.5% N, and 29.3% F was obtained. The boiling point of a solution of 0.9990 g of the compound in 42.8765 g benzene had a rise of 0.235°. What is the formula of the compound?

[O. Glemser, H. Schröder, and H. Haesler, Z. Anorg. Allgem. Chem., 279, 28 (1954)]

SUPPLEMENTARY PROBLEMS

8-23 Calculate the vapor pressure of a solution of 1.8411 g of naphthalene, $C_{10}H_8$, in 26.53 g of benzene, C_6H_6, if the vapor pressure of pure benzene at the same temperature was 639.85 mm. The observed vapor pressure of the solution was 614.5 mm.

[M. A. Rosanoff and R. A. Dunphy, J. Am. Chem. Soc., 36, 1416 (1914)]

8-24 Frazer and Lovelace observed a vapor-pressure depression of 0.122 mm at 20.0°C when 0.500 mole of mannitol was dissolved in

1000 g water.   Compare this with the calculated value from
Raoult's law.

> [J. Am. Chem. Soc., 36, 2448 (1914)]

8-25   Calculate the vapor pressure of ethyl alcohol at 15°C from
the observation that the vapor pressure of an alcoholic solution
containing 0.02823 g of benzil, $C_{14}H_{10}O_2$, per gram of solvent
was 31.98 mm.

> [O. F. Tower and A. F. O. Germann, J. Am. Chem. Soc., 36,
> 2449 (1914)]

8-26   Make a plot of the observed vapor pressures of the solutions
of oleic acid, $C_{18}H_{34}O_2$, in carbon disulfide versus the mole fraction
of solvent from the data of F. H. Campbell:

% CS_2	Vapor pressure, mm
3.12	40.9
15.75	164.0
25.85	233.7
34.77	290.5
40.51	316.3
56.14	366.6
100.00	431.9

Source: Trans. Faraday Soc., 11, 94 (1915)]

8-27   The molecular weight of a solute was calculated to be 118 when
the vapor pressure of a solution containing 0.784 g in 48.0 g of
CCl_4 was measured.   The vapor pressure of CCl_4 at 15.5°C was 73.3.
What was the vapor pressure of the solution?

> [J. Wright, J. Chem. Soc., 115, 1165 (1919)]

8-28   The boiling-point elevation of a solution of 0.4344 g of
anthracene in 44.16 g of anhydrous acetic acid was 0.140°.   Calculate

the predicted boiling-point elevation and compare the calculated and
observed values.   The formula of anthracene is $C_{14}H_{10}$.
         [E. Beckmann, Z. Physik. Chem. (Leipzig), 6, 449 (1890)]

8-29   Calculate the expected boiling point of a solution containing
0.246 g of P_2I_4 in 4.859 g of carbon disulfide.
         [M. Baudler and G. Fricke, Z. Anorg. Allgem. Chem.,
         320, 11 (1963)]

8-30   Calculate the expected freezing point of a solution containing
1.00 gallon of methanol, CH_4O, in 18.0 quarts of water.   The density
of methanol is 0.793 g/ml and 1 gallon contains 3.7853 liters.

8-31   What is the freezing point of a 0.021 m aqueous solution of
a nonelectrolyte?
         [E. H. Loomis, Z. Physik. Chem. (Leipzig), 37, 411 (1901)]

8-32   Calculate the freezing point of a 0.0800 M solution if the
specific gravity of the solution was 1.0054.   Assume molecular weight
of solute is 341 g/mole.
         [E. H. Loomis, Z. Physik. Chem. (Leipzig), 37, 411 (1901)]

8-33   Calculate the freezing point of a 3.819 mole percent solution
of formamide, $HCONH_2$, in water.English and Turner reported a
freezing point of -2.7° for such a solution.
         [S. English and W. E. S. Turner, J. Chem. Soc., 107,
         775 (1915)]

8-34   The cryoscopic constant of $SC(SH)_2$ was assigned the value
6.75.   To substantiate this value, the molecular weight of sulfur
in a solution of $SC(SH)_2$ was determined.   A molecular weight of 255
was calculated from the freezing-point depression (1.00°) of a
solution containing 13.7 g of solvent.   Calculate the weight of
sulfur in the solution.

[G. Gattow and B. Krebs, Z. Anorg. Allgem. Chem., 321,
149 (1963)]

8-35 A solution contained 5.00 g of polymer (mol. wt. 250,000) in
200 g of water at 20°C. Calculate the osmotic pressure of the
solution.

8-36 A sample of an organic compound weighing 2.3446 g was dissolved
in 26.53 g of benzene, C_6H_6. The vapor pressure of the solution was
607.4 mm, and the vapor pressure of pure benzene at the same
temperature was 639.85 mm. What is the molecular weight of the
solute?

[M. A. Rosanoff and R. A. Dunphy, J. Am. Chem. Soc., 36,
1416 (1914)]

8-37 The vapor pressure of carbon tetrachloride when pure was
observed to be 87.1 mm. The vapor pressure of a solution containing
1.072 g of a solute in 49.0 g of CCl_4 was 85.4 mm. What was the
molecular weight of the solute?

[J. Wright, J. Chem. Soc., 115, 1165 (1919)]

8-38 Calculate the molecular weight of the compound which lowered
the vapor pressure of methanol, CH_4O, at 15°C 0.332 mm when 3.662 g
was dissolved in 100 g of solvent. The vapor pressure of methanol
at 15° was observed to be 73.6 mm.

[O. F. Tower and A. F. O. Germann, J. Am. Chem. Soc., 36,
2449 (1914)]

8-39 Calculate the molecular weight of the compound which depressed
the vapor pressure of isopentane, C_5H_{12}, at 20.3° by 17.40 mm when
0.0697 g was dissolved in 0.891 g of solvent. The vapor pressure
of isopentane at 20.3° was 580.0 mm.

[T. Wartik and H. I. Schlesinger, J. Am. Chem. Soc., 75,
839 (1953)]

8-40 A sample of a hydrocarbon weighing 1.4020 g was dissolved
in 86.52 g of carbon disulfide. The boiling point of the solution
was 46.35°C. Calculate the molecular weight of the solute.
          [E. Beckmann, Z. Physik. Chem. (Leipzig), 6, 447 (1890)]

8-41 A sample of an iodide containing 10.86% phosphorus and
weighing 0.246 g was dissolved in 6.428 g of carbon disulfide. The
boiling-point elevation was 0.161°. What is the formula of the
compound?
          [M. Baudler and G. Fricke, Z. Anorg. Allgem. Chem., 320,
          11 (1963)]

8-42 What is the formula of phosphorus in carbon disulfide
solution if 0.2170 g of solute raised the boiling point of 19.18 g
of CS_2 0.104°?
          [A. Helff, Z. Physik. Chem. (Leipzig), 12, 200 (1893)]

8-43 A sample of phosphorus weighing 0.3635 g was dissolved in
16.87 g of carbon disulfide. The boiling-point elevation of the
solution was 0.363°. Calculate the molecular weight of the
phosphorus. What is the structure of the phosphorus molecules?
          [A. Helff, Z. Physik. Chem. (Leipzig), 38, 76 (1901)]

8-44 An aqueous solution of a sugar containing 68.44 g/liter had
a freezing point of -0.39°C. The density of the solution was
1.0256 g/ml. Calculate the molecular weight of the sugar.
          [E. H. Loomis, Z. Physik. Chem. (Leipzig), 37, 411 (1901)]

8-45 Calculate the molecular weight of the compound if a solution
containing 18.21 g/liter had a freezing point of -0.188°. The
density of the solution was 1.0047 g/ml.
          [E. H. Loomis, Z. Physik. Chem. (Leipzig), 37, 411 (1901)]

8-46 Calculate the molecular weight of sulfur in a solution containing 0.100 g in 9.80 g of SC(SH)$_2$ if the freezing-point depression of the solution was 0.268°C. The cryoscopic constant of SC(SH)$_2$ was determined as 6.75.

[G. Gatton and B. Krebs, Z. Anorg. Allgem. Chem., 321, 149 (1963)]

8-47 An aqueous solution of a polymeric material had an osmotic pressure of 0.40 mm when its concentration was 5.00 mg/g at 25°C. Estimate the molecular weight of the polymer.

Chapter 9

THERMOCHEMISTRY

9-1 HEAT CAPACITY

The concept of specific heat was introduced in Chap. 1 as an example of an intensive property. When heat is withdrawn from the surroundings and added to a system, the temperature of the system rises. The quantity of heat required to raise the temperature of a system 1°C is called the heat capacity of the system: the specific heat of the substance is the heat capacity per gram. When the system is 1 mole of a pure substance, the quantity of heat required to raise the temperature 1° is called the molar heat capacity C. The symbols C_V and C_p are used for heat capacity at constant volume and constant pressure, respectively. The basic unit of energy in the International System of Units is the joule (J). The joule is the energy associated with a force of one newton acting through a distance of one meter (N-m or $kg-m^2-s^{-2}$). Other units commonly used are the erg (cgs) and the calorie (cal). The relationships between these units are:

1 joule (J) = 10^7 ergs
1 calorie (cal) = 4.184 J

These latter two units do not belong to the International System of Units and may someday be abandoned. The calorie, presently, is the more commonly used unit in chemistry.

185

TABLE 9-1   Specific Heats of Some Common Substances

Substance	Temp. range, °C	Sp. heat, cal/g,°C
Water	0-100	1.000
Ice	-20-0	0.505
Steam	100-200	0.475
Mercury	20-100	0.0330
Aluminum	17-100	0.217
Lead	20-100	0.0305
Alcohol	25	0.581
Cadmium	20	0.0549
Selenium (amorphous)	18-38	0.095
Copper	20	0.0912
Gallium	12-23	0.079
Sulfur (rhombic)	0-54	0.1728
Zinc	20	0.0924
Tellurium	15-100	0.0483
Ammonia	-60	1.047
Bismuth	18	0.0303

Example 9-1

The specific heat of ammonia is 1.047 cal/g at -60.0°. Calculate the heat required to raise the temperature of 2.10 moles from -60.0 to -47.0°. The weight of 1 mole of NH_3 is 17.01 g.

C = 17.01 x 1.047 = 17.83 cal/mole °C
17.8 x 2.10 x 13.0 = 485 cal

$$\frac{cal}{mole \times °C} \times mole \times °C = cal$$

9-2 HEAT CONTENT, OR ENTHALPY

Thermochemistry is the study of the heat effects which accompany physical and chemical changes. In exothermic processes, the system

Thermochemistry

evolves or gives off heat. In <u>endothermic</u> processes, the system
absorbs heat. The symbol for the measured heat in such changes is
q— the heat absorbed by a system is +q, and the heat lost by a
system to the surroundings is -q. The heat change of a process
measured at constant pressure or volume is q_p or q_v, respectively.
The heat absorbed at constant volume is equal to the change in
energy of a system, ΔE. The heat absorbed at constant pressure is
equal to the change in <u>heat content</u> or <u>enthalpy</u> of the system, ΔH.
The change in energy is related to the change in enthalpy by:

$$\Delta H = \Delta E + \Delta(PV)$$

where P and V are the pressure and volume, respectively.

9-3 HEAT CHANGES ACCOMPANYING CHANGES IN STATE

Changes in the heat content of a system accompany changes in
state. Four such changes, with the names of the heat content changes,
are listed in Table 9-2.

TABLE 9-2

Process	Name	Symbol	Example
Melting	Heat of fusion	ΔH_{fus}	$H_2O(s) \rightarrow H_2O(\ell)$
Boiling	Heat of vaporization	ΔH_{vap}	$H_2O(\ell) \rightarrow H_2O(g)$
Sublimation	Heat of sublimation	ΔH_{subl}	$I_2(s) \rightarrow I_2(g)$
Crystal structure change	Heat of transition	ΔH_{trans}	$S(r) \rightarrow S(m)$

Example 9-2

Calculate the heat required to change 1.12 moles of ice at 0°C to
water at 0°C.

$H_2O(s) \rightarrow H_2O(\ell)$

The heat of fusion is 79.71 cal/g. 1 mole of H_2O weighs 18.0 g.

1.12 x 18.0 x 79.7 = 1600 cal

mole x $\frac{g}{mole}$ x $\frac{cal}{g}$ = cal

Example 9-3

Calculate the heat required to change 2.20 moles of $H_2O(\ell)$ at 80.0°C to $H_2O(g)$ at 100.0°C. The average specific heat of water over this temperature range is 1.0045 cal/g. The heat of vaporization of water is 539.6 cal/g.

Heat to raise temperature of $H_2O(\ell)$
2.20 x 18.0 x 1.00 x 20.0   =   792 cal
Heat to change $H_2O(\ell) \rightarrow H_2O(g)$
2.20 x 18.0 x 540   =   21,400 cal
Total heat required   22,192 cal
    or with the appropriate number of significant figures,
22,200 cal.

Example 9-4

The heat accompanying the change of amorphous boron to crystalline boron ($\Delta H°_{trans}$) is -0.4 kcal/mole.
          [Natl. Bur. Std. (U.S.) Circ. 500]
    Calculate the difference in heat content between 50 g of B(c) and 50 g of B(amorph) at 25°C.

B(c) $\rightarrow$ B(amorph)
$\Delta H°$ = +0.4 kcal/mole
The sign is reversed from that given in the statement of the problem because the direction of the change is reversed.

$\frac{50}{11}$ = 4.5 moles B
4.5 x 0.4 = 1.8 kcal
This means that 1.8 kcal of heat must be added to change 50 g of crystalline boron at 25°C to 50 g of amorphous boron at 25°C.

9-4 HEAT OF REACTION

The difference in heat content between the products of a chemical
reaction and the reactants is called the heat of reaction or enthalpy
of reaction.

$\Delta H = \Delta H(\text{products}) - \Delta H(\text{reactants})$

In the reaction $2H_2(g) + O_2(g) \rightarrow 2H_2O(\ell)$, $\Delta H = -136.64$ kcal;
In the reaction $3UO_2(s) + O_2(g) \rightarrow U_3O_8(s)$, $\Delta H = -76.01$ kcal.
These are typical enthalpies of reaction.

Absolute values of the heat content H of substances cannot be
determined. The differences in heat content ΔH can be obtained,
however, from measured heats of reaction.

Consistent with the convention that heat absorbed by a system
is positive, ΔH_{react} is negative for an exothermic reaction.

9-5 HEAT OF COMBUSTION

Of the many types of reactions for which heats of reaction have
been determined, those involving reactions of oxygen are among the
most important. These heats of reaction, called heats of combustion,
are determined by oxygen bomb calorimetry. Examples are the heats
of combustion of methane, -212.798 kcal/mole; ethane, -372.820
kcal/mole; and benzene, -780.98 kcal/mole.

9-6 STANDARD ENTHALPY OF FORMATION

The heat involved in the formation of 1 mole of a compound from
the elements at 25°C, each substance being in its standard state,
is the standard enthalpy of formation of the compound. The standard
enthalpy of formation of the elements in their standard states at
25°C has been assigned the value of zero.

9-7 HESS' LAW

The enthalpy (heat content) of a substance is a function of the
physical state of a substance. Changes in enthalpy are functions
of the initial and final states only and are independent of the path.
One illustration of this fundamental principle is Hess' law;

The amount of heat involved in a chemical reaction
is the same whether reaction takes place in one
step or in several steps.

Example 9-5

Calculate the enthalpy of formation of cyanamide, CH_2N_2, from its
enthalpy of combustion, -177.20 kcal/mole.

[D. J. Solley and J. B. Gray, J. Am. Chem. Soc., 70,
2650 (1948)]

From the definition of enthalpy of formation we have

$$C(s) + H_2(g) + N_2(g) \rightarrow CH_2N_2(s) \qquad \Delta H^\circ_f = ?$$

The combustion of cyanamide is given by

$$CH_2N_2(s) + \tfrac{3}{2}O_2(g) \rightarrow$$
$$CO_2(g) + H_2O(\ell) + N_2(g) \qquad \Delta H^\circ = -177.20 \text{ kcal/mole}$$

The other data required (from Table 9-3) are

$$C(s) + O_2(g) \rightarrow CO_2(g) \qquad \Delta H^\circ_f = -93.97 \text{ kcal/mole}$$

$$H_2(g) + \tfrac{1}{2}O_2(g) \rightarrow H_2O(\ell) \qquad \Delta H^\circ_f = -68.32 \text{ kcal/mole}$$

Applying Hess' law, changing the sign of ΔH when the reaction is
reversed, we have

$$N_2(g) + H_2O(\ell) + CO_2(g) \rightarrow$$
$$CH_2N_2(s) + \tfrac{3}{2}O_2(g) \qquad \Delta H^\circ = +177.20 \text{ kcal/mole}$$

$$C(s) + O_2(g) \rightarrow CO_2(g) \qquad \Delta H^\circ_f = -93.97 \text{ kcal/mole}$$

$$H_2(g) + \tfrac{1}{2}O_2(g) \rightarrow H_2O(\ell) \qquad \Delta H^\circ_f = -68.32 \text{ kcal/mole}$$

$$\overline{C(s) + H_2(g) + N_2(g) \rightarrow CH_2N_2(s) \qquad \Delta H^\circ_f = 14.91 \text{ kcal/mole}}$$

Example 9-6

Calculate the standard enthalpy of formation of boron carbide from

the enthalpy of combustion, -683.3 kcal/mole, as determined by
Smith, Dworkin, and Van Artsdalen.

[J. Am. Chem. Soc., 77, 2655 (1955)]

The formation of B_4C from the elements is given by:

$$4B(c) + C(s) \rightarrow B_4C(s) \qquad \Delta H^\circ_f = ?$$

The combustion of B_4C is given by:

$$B_4C(s) + 4O_2 \rightarrow 2B_2O_3(c) + CO_2(g) \qquad \Delta H^\circ_f = -683.3 \text{ kcal/mole}$$

Applying Hess' law:

$$2B_2O_3(s) + CO_2(g) \rightarrow B_4C(s) + 4O_2(g) \qquad \Delta H^\circ_f = +683.3 \text{ kcal/mole}$$

$$C(s) + O_2(g) \rightarrow CO_2(g) \qquad \Delta H^\circ_f = -93.97 \text{ kcal/mole}$$

$$4B(c) + 3O_2(g) \rightarrow 2B_2O_3(c) \qquad \Delta H^\circ_f = 2(-304.1)\text{kcal/mole}$$

$$\overline{C(s) + 4B(c) \rightarrow B_4C(s) \qquad \Delta H^\circ_f = -18.9 \text{ kcal/mole}}$$

The standard enthalpies of formation of a large number of
compounds have been determined, and they are listed in National
Bureau of Standards Circular 500. A revised version of Circular 500
is available as Technical Notes 270-3, 270-4, 270-5, and 270-6 from
the national Bureau of Standards titled, Selected Values of Chemical
Thermodynamic Properties. Most of the values in Table 9-3 are from
Lewis and Randall's Thermodynamics, second ed., revised by K. S.
Pitzer and L. Brewer, McGraw-Hill, New York, 1961: and Handbook
of Chemistry and Physics, 52nd ed., the Chemical Rubber Co.,
Cleveland, 1971.

9-8 CALORIMETRY

Measurements of heats of reaction are often made in a calorimeter.
The weights of the reactants and products are determined as accurately
as possible. The quantity of heat evolved is equal to the heat
absorbed by a known quantity of water surrounding the reaction vessel
(bomb) and the other parts of the calorimeter. The water equivalent
of a calorimeter is the quantity of water which will absorb
the same quantity of heat as the calorimeter with all of its parts.

TABLE 9-3   Some Standard Enthalpies of Formation

Substance	ΔH°_f, kcal/mole at 298° K	Substance	ΔH°_f, kcal/mole at 298° K
$Au(OH)_3$	-100.0	HCl	- 22.019
B_2H_6	7.5	NF_3	- 30.4
BF_3	-265.4	N_2H_4	12.05
B_2O_3	-305.3	NH_3	- 11.04
CH_4	- 15.99	Nb_2O_5	-455.2
CF_4	-216.6	$PrCl_3$	-257.8
CO_2	- 93.969	PH_3	5.5
Cr_2O_3	-272.7	TaN	- 58.2
$H_2O(l)$	- 68.32	Ta_2O_5	-488.8
$H_2O(g)$	- 57.11	$Tl(OH)_3$	-122.6
HBr	- 8.66	WO_3	-201.46
HF	- 64.2	ZrC	- 44.1
HF(aq)	- 78.66	ZrO_2	-261.5
H_2S	- 4.815		

The energy equivalent of a calorimeter is the quantity of heat equivalent to a 1° rise in temperature of the calorimeter assembly.

Example 9-7
The combustion of 2.650 g of indium raised the temperature of a calorimeter 1.055°C. The energy equivalent of the calorimeter was 2412.81 cal/°C. Calculate the molar energy of combustion of indium.

> [C. E. Holly, Jr., E. J. Huber, Jr., and E. H. Meierkord, J. Am. Chem. Soc., 74, 1084 (1952)]

2413 cal/°C x 1.055 °C = 2546 cal

$$\frac{2546 \text{ cal}}{2.650 \text{ g}} \times 114.8 \frac{g}{mole} = 110,300 \text{ cal/mole}$$

Reactions in a calorimeter occur at constant volume. The heat evolved or absorbed at constant volume is equal to the change in energy of the reaction ΔE. To obtain the value of the change in enthalpy of the reaction ΔH, a correction must be made.

$$\Delta H = \Delta E + \Delta(PV)$$

Since V is constant in a calorimeter measurement

$$\Delta H = \Delta E + \Delta P(V)$$
$$\Delta H = \Delta E + (P_2 - P_1)V$$

From $PV = nRT$

$$P_2 = n_2 \frac{RT}{V}$$

$$P_1 = n_1 \frac{RT}{V}$$

$$\Delta H = \Delta E + (n_2 - n_1)\frac{RT}{V} V$$
$$= \Delta E + \Delta nRT$$

$$R = 1.9872 \text{ cal/K-mole}$$

Example 9-8

The energy of combustion of ZrC as measured in a bomb calorimeter was -310,863 cal/mole at 25°C. Calculate the enthalpy of combustion.

[A. D. Mah and B. J. Boyle, J. Am. Chem. Soc., 77,

6513 (1955)]

$$ZrC(s) + 2O_2(g) \rightarrow ZrO_2(s) + CO_2(g)$$

$$\Delta H = \Delta E + \Delta nRT$$

Since the change in volume of the solids may be neglected

$$\Delta n = \text{moles } CO_2 - \text{moles } O_2 = 1 - 2 = -1$$
$$\Delta H = -310,863 + -1 \times 1.9872 \times 298.15$$
$$= -310,863 - 592 = -311,455 \text{ cal}$$
$$= -311.5 \text{ kcal/mole}$$

Example 9-9

The energy of combustion of NbN, as measured in a bomb calorimeter, was -170.38 kcal/mole. Calculate the enthalpy of combustion.

[A. D. Mah and N. L. Gellert, J. Am. Chem. Soc., 78,

3261 (1956)]

$NbN(s) + \frac{5}{4}O_2(g) \rightarrow \frac{1}{2}Nb_2O_5(s) + \frac{1}{2}N_2(g)$

$\Delta H = \Delta E + \Delta nRT$

$\Delta n = \frac{1}{2} - \frac{5}{4} = -\frac{3}{4}$

$\Delta H = -170,380 + (-\frac{3}{4} \times 1.9872 \times 298.15)$

$\quad = -170,380 - 444 = -170,824$

$\quad = -170.82$ kcal/mole NbN

9-9 BOND ENERGIES

The standard bond-dissociation energy (DH°) [S. W. Benson,
J. Chem. Educ., 42, 502 (1965)] is defined as the change in enthalpy
in the breaking of 1 mole of a given bond, reactants and products
being in their standard states of hypothetical ideal gas at 1 atm
and 25°C. The bond dissociation energy of a diatomic substance
is the energy required to cause the dissociation of 1 mole of the
substance, in the gas phase, into its constituent atoms, also in the
gas phase. The bond-dissociation energy of hydrogen, for example,
is 104.2 kcal/mole for the reaction $H_2(g) \rightarrow 2H(g)$.

Example 9-10
Calculate the bond-dissociation energy of O_2.

$O_2 \rightarrow 2O$

$\Delta H°_f(O_2) = 0$

$\Delta H°_f(O) = 59.6$ kcal/mole

$\Delta H°_{react} = 2\Delta H°_f(O) - \Delta H°_f(O_2)$

$\Delta H°_{react} = 2 \times 59.6 - 0 = 119.2$ kcal

$DH°(O—O) = 119.2$ kcal/mole

The bond energies of polyatomic substances may be determined
and recorded as average bond-dissociation energies. The average

bond-dissociation energy $DH°$ of CH_4 is one-fourth of the total
enthalpy needed to break all of the bonds in CH_4.

$$CH_4(g) \rightarrow C(g) + 4H(g)$$

$$DH°(C—H) = \tfrac{1}{4}H°_{react}$$

Average bond-dissociation energies have limited use, especially
from the point of view of reactivity. The average bond-dissociation
energy of the O—H bond in water is 110.5 kcal. However, the energy
required to break the first O—H bond in H—O—H is 118 kcal,
whereas the energy required to break the second is 103 kcal.
This is a significant difference.

Example 9-11

Calculate the average C—H bond energy in CH_4 from the standard
enthalpy of formation of CH_4 and the atomic species resulting from
the dissociation.

$$CH_4(g) \rightarrow C(g) + 4H(g)$$

$$\Delta H°_f(CH_4) = -17.89 \text{ kcal/mole}$$

$$\Delta H°_f(C) = 170.9 \text{ cal/mole}$$

$$\Delta H°_f(H) = 52.10 \text{ cal/mole}$$

$$\Delta H°_{react} = \Delta H°_f(C) + 4\Delta H°_f(H) - \Delta H°_f(CH_4)$$

$$\Delta H°_{react} = 170.9 + 4(52.10) - (-17.89) = 397.2 \text{ kcal}$$

$$DH°(CH_4) = \tfrac{1}{4}\Delta H°_{react} = \frac{397.2}{4} = 99.3 \text{ kcal}$$

The $DH°$ values of the bonds in CH_4 are 104 kcal/mole for $DH°(CH_3—H)$,
106 kcal/mole for $DH°(CH_2—H)$ and $DH°(CH—H)$, and 81 kcal/mole for
$DH°(C—H)$. The average of the four values is 99.3 kcal/mole.

PROBLEMS

In the following problems, those marked with an
asterisk involve calculations using the joule as
a unit of energy.

9-1 Calculate the quantity of heat required to change 4.13×10^{-3}

TABLE   9-4   Standard Enthalpys of Formation
(in kilocalories per gram atom) of Some Monatomic Elements

Element	ΔH°_f	Element	ΔH°_f
H	52.1	Cl	28.9
D	53.0	K	21.3
Be	78.3	Ca	46.0
B	133(±4)	Cu	81.5
C	170.9	Zn	31.2
N	113.0	Ge	78.4
O	59.6	Br	26.7
F	18.9	Ag	69.1
Na	25.8	As	72.9
Mg	35.3	Sn	72
Al	78.0	Sb	60.8
Si	106(±4)	I	25.5
P	79.8	Hg	14.7
S	65.7	Pb	46.7

Source: S. W. Benson, J. Chem. Educ., 42, 517 (1965).

mole of solid aluminum, specific heat 0.274 cal/g, from 600°C
to the liquid state at the melting point, 658°C. The heat of fusion
of aluminum is 76.8 cal/g.

9-2   A sample of sulfur containing 2.00 g-atoms was heated from
20 to 150°C. Calculate the heat absorbed.
   Specific heat: S_r, 0.176 cal/g; S_m, 0.181 cal/g; liquid,
   0.220 cal/g.
   ΔH_{fus} = 13.2 cal/g;
   ΔH_{trans} = 0.096 kcal/g-atom;
   mp = 119°; transition temp. = 95.5°.

9-3   According to J. C. Southard, 4360 cal of heat was required to
change 1 mole of crystalline boron oxide to the glass form at 298 K.
Calculate the difference in heat content between 10.0 g of B_2O_3

(glass) and 10.0 g of B_2O_3(c) at 25°C.

[J. Am. Chem. Soc., 63, 3147 (1941)]

9-4 On a basis of energy evolved per kilogram of fuel, which of
the following substances offers the best possibility as a rocket
fuel? Assume that oxygen is the oxidizer.

    a. Diborane   b. Methane   c. Hydrazine

9-5 If in Prob. 9-4, fluorine is used as an oxidizer in place of
oxygen, what will the heat evolved, per kilogram of fuel, be in each
case if the same substances are used as fuels?

9-6 A sample of tungsten weighing 9.5886 g was burned in a bomb
calorimeter and 43.549 J of heat was evolved. Calculate the
energy of combustion of tungsten (at the temperature of the experiment) in kilocalories per gram atom.

    [G. Huff, E. Squitieri, and P. E. Snyder, J. Am. Chem.

    Soc., 70, 3380 (1948)]

9-7 The standard heat of formation of WO_3 was found to be -201.46
kcal/mole. The heat of combustion (corrected to a constant-pressure
process) of WO_2 was -60.52 kcal/mole. What is the standard enthalpy
of formation of WO_2?

    [A. D. Mah, J. Am. Chem. Soc., 81, 1583 (1959)]

9-8 What is the standard enthalpy of formation of chromium
hexacarbonyl, $Cr(CO)_6$, if the standard enthalpy of combustion was
found to be -443.09 kcal/mole.

    [F. A. Cotton, A. K. Fischer, and G. Wilkinson, J. Am.

    Chem. Soc., 78, 5169 (1956)]

9-9 Humphrey and O'Brien determined the standard enthalpy of
combustion of tin to tin(IV) oxide to be -138,820 cal/mole. The
heat of combustion of SnO(s) at 25°C was found to be -70,470 cal/mole.

Calculate the standard heat of formation of tin(II) oxide.

[J. Am. Chem. Soc., 75, 2805 (1953)]

9-10   The energy of combustion (ΔE°_c) for solid cubane (C_8H_8) was found to be -1156.0 kcal/mole. What is ΔH°_f of cubane?

[Kybett et al., J. Am. Chem. Soc., 88, 626 (1966)]

9-11   Walker et al. recently determined the enthalpy of formation of CF_3CN by two separate measurements. The heat of reaction of

$$2CF_3CN(g) \rightarrow C_2F_6(g) + C_2N_2(g)$$

was -10.54 kcal/mole at 25°C, while the heat of reaction of

$$CF_3CN(g) + \tfrac{5}{3}NF_3(g) \rightarrow 2CF_4(g) + \tfrac{4}{3}N_2(g)$$

was -274.8 kcal/mole at 25°C. Given the following data, calculate the enthalpy of formation of CF_3CN from each reaction.

Compound	ΔH°_f kcal/mole
$CF_4(g)$	-223.05
$NF_3(g)$	- 31.43
$C_2F_6(g)$	-321.2
$C_2N_2(g)$	+ 73.87

[Walker, Sinke, Perette, Jong, J. Am. Chem. Soc., 92, 4525 (1970)]

9-12   Calculate the heat of combustion at constant pressure of TaN. Assume that nitrogen gas and tantalum(V) oxide are the products. The measured value was -184.38 kcal/mole according to Mah and Geller

[A. D. Mah and N. L. Gellert, J. Am. Chem. Soc., 78, 3261 (1956)]

9-13   The heat of combustion of TaN was found to be -184.38 kcal/mol after correction for change in pressure.

$$TaN(s) + \tfrac{5}{4} O_2(g) = \tfrac{1}{2}Ta_2O_5(s) + \tfrac{1}{2}N_2(g)$$

Calculate the standard heat of formation of TaN(s).

[A. D. Mah and N. L. Gellert, J. Am. Chem. Soc., 78, 3261 (1956)]

9-14   Calculate the heat of combustion of ZrC from the data in Table 9-3 and the equation

$ZrC(s) + 2O_2(g) \rightarrow ZrO_2(s) + CO_2(g)$.

[A. D. Mah and B. J. Boyle, J. Am. Chem. Soc., 77, 6513 (1955)]

9-15   Calculate the standard heat of formation of ONF from the following data:

$NO + \frac{1}{2}F = ONF \quad \Delta H° = -37.4 \text{ kcal/mole}$

$\frac{1}{2}N_2 + \frac{1}{2}O_2 = NO \quad \Delta H°_f = +21.6 \text{ kcal/mole}$

[H. S. Johnston and H. J. Bertin, Jr., J. Am. Chem. Soc., 81, 6404 (1959)]

9-16   The heat of combustion of B_4C, as measured in a bomb calorimeter was -682.0 kcal/mole at 298K. Calculate the standard heat of formation of B_4C. (This paper is recommended for student reading.)

[D. Smith, A. S. Dworkin, and E. R. Van Artsdalen, J. Am. Chem. Soc., 77, 2655 (1955)]

9-17   A compound prepared by the action of bromine trifluoride on gold had a weight ratio of compound to gold of 1.290. A sample of the compound was hydrolyzed, and the heat of the reaction with water was -39.4 kcal/mole. Calculate the standard enthalpy of formation of the compound using HF(aq) in the equation.

[A. A. Woolf, J. Chem. Soc., 1954, 4694]

9-18   The heats of combustion of hafnium metal and hafnium nitride, HfN, at 25°C were -265,463 cal/mole and -177,516 cal/mole, respectively, as measured in a bomb calorimeter. The other product of the combustion of HfN was nitrogen gas. What is the standard enthalpy of formation of HfN as calculated from these data?

        [G. L. Humphrey, J. Am. Chem. Soc., 75, 2806 (1953)]

9-19 The energy of formation of arsenic pentafluoride, as measured
by direct combination of the elements in a bomb calorimeter, was
-3933.4 cal/g of arsenic at 25°. Calculate the standard enthalpy
of formation of arsenic pentafluoride and the average bond-dissociation energy of the As—F bond in the compound.

        [P. A. G. O'Hare and W. N. Hubbard, J. Phys. Chem., 69,
        4360 (1965)]

9-20 The value of $\Delta E°_f$ of ruthenium pentafluoride, as measured in a
fluorine bomb calorimeter by direct combination, was -2096.8 cal/g.
Calculate the standard enthalpy of formation of RuF_5.

        [H. A. Porter, E. Greenberg, and W. N. Hubbard, J. Phys.
        Chem., 69, 2308 (1965)]

9-21 The heat of solution of 0.3086 g of $ThCl_4$ was 36.49 cal. What
was the heat of solution, in kilocalories per mole?

        [L. Eyring and E. F. Westrum, Jr., J. Am. Chem. Soc., 72,
        5555 (1950)]

9-22 Calculate the heat of formation of $B_3N_3H_3Cl_3(s)$ from the heat
of solution, -113.8 kcal, as represented by the equation

    $B_3N_3H_3Cl_3(s) + 9H_2O(\ell) + aq \rightarrow 3H_3BO_3(aq) + 3NH_4Cl(aq)$

The standard heats of formation of $NH_4Cl(aq)$ and $H_3BO_3(aq)$ are
-71.76 kcal/mole and -255.2 kcal/mole, respectively.

        [E. R. Van Artsdalen and A. S. Dworkin, J. Am. Chem. Soc.,
        74, 3402 (1952)]

*9-23 The combustion of 2.724 g of indium raised the temperature of
a calorimeter 1.083°C. The energy equivalent of the calorimeter
was 10,095.2 J/°C. Calculate the molar energy of combustion of
indium in kilojoules and kilocalories.

[C. E. Hally, E. J. Huber, Jr., and E. H. Meierkord,
J. Am. Chem. Soc., 74, 1084 (1952)]

*9-24 In one of a series of measurements to calibrate a calorimeter,
the combustion of 1.08907 g of benzoic acid (heat of combustion
26.4284 kJ/g) led to a temperature rise of 2.15266°C. Heat from
other sources amounted to 21.6 cal. Calculate the energy equivalent
of the calorimeter, in calories per degree Celsius.

[D. J. Solley and J. B. Gray, J. Am. Chem. Soc., 70,
2651 (1948)]

9-25 Calculate the bond-dissociation energy of fluorine from the
value of $\Delta H°_f$ of F, which is 18.9 kcal/mole.

9-26 Calculate the average P—H bond-dissociation energy in PH_3.

9-27 Calculate the N—H bond-dissociation energy in NH_2 and compare
it with the average N—H bond-dissociation energy in ammonia. The
standard enthalpy of formation of $NH_2(g)$ is 40 kcal/mole.

9-28 What is the bond-dissociation energy of the bonds in hydrogen
bromide?

SUPPLEMENTARY PROBLEMS

9-29 Calculate the heat evolved if 8.60 moles of Bi is cooled from
300 to 200°C. The heat of fusion of bismuth is 12.64 cal/g; its
melting point is 268°. The specific heat of the liquid is 0.0363
cal/g, and that of the solid is 0.0338 cal/g.

9-30 The standard enthalpy of formation of amorphous boron oxide
was determined as -299.74 kcal/mole, and that of crystalline boron
oxide was -304.1 kcal/mole. Calculate the standard enthalpy of

transition from amorphous boron oxide to crystalline boron oxide.

    [W. D. Good and M. Mansson, J. Phys. Chem., 70, 97 (1966)]

9-31 The standard heat of formation of U_3O_8 is -853.5 kcal/mole.
The standard heat of the reaction $3UO_2 + O_2 \rightarrow U_3O_8$ is -76.01 kcal
according to E. J. Huber, Jr., C. E. Hally, and E. H. Meierkord.
Calculate the standard heat of formation of UO_2.

    [J. Am. Chem. Soc., 74, 3406 (1952)]

9-32 Calculate the standard enthalpy of formation of PrOCl from
the heat of hydrolysis of $PrCl_3$, 22.98 kcal. The equation is:

    $PrCl_3(s) + H_2O(g) \rightarrow PrOCl(s) + 2HCl(g)$

        [C. W. Koch and B. B. Cunningham, J. Am. Chem. Soc., 76,
        1473 (1954)]

9-33 Calculate the standard heat of formation of melamine, $C_3H_6N_6$,
from the heat of combustion at constant pressure, -471.76 kcal/mole.

        [D. J. Solley and J. B. Gray, J. Am. Chem. Soc., 70,
        2650 (1948)]

9-34 What is the standard enthalpy of formation of NbN if the
heat of combustion at 25° (corrected for change in pressure) was
-170.83 kcal/mole?

        [A. D. Mah and N. L. Gellert, J. Am. Chem. Soc., 78,
        3261 (1956)]

9-35 Calculate the heat of formation of WC(s) from the heat of
reaction with oxygen to yield WO_3 and CO_2 (-285.80 kcal/mole) and
the heat of formation of CO_2 from graphite.

        [G. Huff, E. Squitieri, and P. E. Snyder, J. Am. Chem. Soc.,
        70, 3381 (1948)]

9-36 The energy evolved during the combustion in a bomb calorimeter
of 1 g of dicyandiamide, $C_2H_4N_4$, was 3943.1 cal. Calculate the heat

of combustion at constant pressure and the standard enthalpy of
formation of dicyandiamide.
  [D. J. Solley and J. B. Gray, J. Am. Chem. Soc., 70,
  2650 (1948)]

9-37 The energy evolved during combustion of indium to In_2O_3 in a
bomb calorimeter was 959.8 cal/g at 25°C. Calculate the standard
enthalpy of formation of In_2O_3.
  [C. E. Hally, E. J. Huber, Jr., and E. H. Meierkord,
  J. Am. Chem. Soc., 74, 1084 (1952)]

9-38 Calculate the standard enthalpy of formation of cadmium oxide
from the heat of combustion of cadmium at 25°C obtained in a bomb
calorimeter, 540.2 cal/g.
  [A. D. Mah, J. Am. Chem. Soc., 76, 3363 (1954)]

9-39 Calculate the standard enthalpy of formation of TlF_3 from the
heat of hydrolysis, -8.4 kcal. The reaction was
  $TlF_3 + 3H_2O \rightarrow Tl(OH)_3 + 3HF(aq)$
  [A. A. Woolf, J. Chem. Soc., 1954 4694]

9-40 The heat of combustion of ZrB_2 at 25°C was found to be -486.1
kcal/mole. Calculate the standard enthalpy of formation of ZrB_2.
  [E. J. Huber, E. L. Head, and C. E. Halley, J. Phys.
  Chem., 68, 3040 (1964)]

9-41 The standard enthalpy of formation of boron nitride was
calculated from its measured heat of reaction with fluorine, which
was -211.68 kcal/mole (after correction for pressure change). Make
the calculation, assuming that the products were boron trifluoride
and nitrogen.
  [S. S. Wise, J. L. Margrave, H. M. Feder, and W. N.
  Hubbard, J. Phys. Chem., 70, 7(1966)]

9-42 Calculate the average bond-dissociation energy of the B—F
bonds in boron trifluoride from the fact that 25,083 cal/g of B was
evolved during combination of the elements in a bomb calorimeter.
> [G. K. Johnson, H. M. Feder, and W. N. Hubbard, J. Phys.
> Chem., 70, 1 (1966)]

9-43 From the heat of formation of SiF_4 (-385.98 kcal/mole) and the
heat of reaction of the fluorination of α-quartz, calculate the
standard heat of formation of α-quartz.

$SiO_2(\alpha$-quartz$)$ + $2F_2(g)$ → $SiF_4(g)$ + $O_2(g)$
$\Delta H°$ = -168.26 kcal/mole
> [S. S. Wise, J. L. Margrave, H. M. Feder, and W. N. Hubbard,
> J. Phys. Chem., 67, 819 (1963)]

9-44 Calculate the heat of formation of hydrogen fluoride from the
following data and any other required data from Table 9-3.

$SiO_2(\alpha$-quartz$)$ + $2F_2(g)$ →
$\qquad\qquad$ $SiF_4(g)$ + O_2 $\qquad$ $\Delta H°$ = -168.26 kcal/mole
SiO_2(crystobalite) → $SiO_2(\alpha$-quartz$)$ $\Delta H°$ = -0.35 kcal/mole
SiO_2(crystobalite) + $4HF(g)$ →
$\qquad\qquad$ $SiF_4(g)$ + $2H_2O(g)$ $\qquad$ $\Delta H°$ = -24.53 kcal/mole
> [H. M. Feder, W. N. Hubbard, S. S. Wise, and J. L. Margrave,
> J. Phys. Chem., 67, 1148 (1963)]

9-45 The energy of formation of yttrium trifluoride was measured by
direct combination of the elements in a fluorine bomb calorimeter.
The value at 25°C was -409.8 kcal/mole. Calculate the standard
enthalpy of formation of $YF_3(s)$ at 25°C.
> [E. Rudzitis, H.M. Feder, and W. N. Hubbard, J. Phys. Chem.,
> 69, 2305 (1965)]

9-46 Calculate the heat of formation of lithium hydride from its
heat of solution (-31.76 kcal/mole) and the heat of reaction of
lithium metal with a large excess of water (-53.10 kcal/mole).

Thermochemistry   205

[C. E. Messer, L. G. Fasolino, and C. E. Thalmayer,

J. Am. Chem. Soc., 77, 4525 (1955)]

9-47 Calculate the heat of solution of potassium hydride KH(s) +
$H_2O(\ell)$ + aq → KOH(aq) + $H_2(g)$. The heat of reaction of potassium
metal with a large excess of water is -47.05 kcal/mole, and the heat
of formation of KH(s) is -15.16 kcal/mole.

[C. E. Messer, L. G. Fasolino, and C. E. Thalmayer,

J. Am. Chem. Soc., 77, 4525 (1955)]

9-48 The combustion of 2.863 g of indium raised the temperature of
a calorimeter 1.140°C. The energy equivalent of the calorimeter
was 2412.81 cal/°C. Calculate the molar energy of combustion of
indium.

[C. E. Halley, E. J. Huber,Jr., and E. H. Meierkord,

J. Am. Chem. Soc., 74, 1084 (1952)]

9-49 The heat of combustion of liquid bicyclo[2,2,2]octene (C_8H_{12})
was determined recently by Wong and Westrum, Jr. ΔE°_c was -1193.42
kcal/mole. What is the enthalpy of formation of this compound?

[J. Am. Chem. Soc., 93, 5317 (1971)]

9-50 Calculate the enthalpy of formation of In_2O_3 from the following
calorimetric data:

weight of In, 3.116 g

temperature increase, 1.241°C

energy equivalent of calorimeter, 2412.81 cal/°C

[C. E. Hally, E. J. Huber, Jr., and E. H. Meierkord,

J. Am. Chem. Soc., 74, 1084 (1952)]

9-51 The energy equivalent of a calorimeter was determined from
heats of combustion of benzoic acid as 3207.1 cal/°C. Calculate
the water equivalent of the calorimeter. The specific heat of water
is 0.99828 cal/g°C.

[D. J. Solley and J. B. Gray, J. Am. Chem. Soc., 70,
2651 (1948)]

9-52 The combustion of 1.000 g of salicylic acid, $C_7H_6O_3$, in a
calorimeter whose water equivalent was 2,343.8 g produced a temperature rise of 2.240°C. Calculate the molar heat of combustion of
salicylic acid. (The specific heat of water is 0.99828 cal/g°C.)
    [F. A. Cotton, A. K. Fischer, and G. Wilkinson, J. Am.
    Chem. Soc., 78, 5168 (1956)]

9-53 Calculate the H—Cl bond-dissociation energy.

9-54 Compare the average N—F bond-dissociation energy in NF_3,
calculated from the data in Tables 9-3 and 9-4, with the F_2N—F
bond-dissociation energy. The standard enthalpy of formation of
$NF_2(g)$ is 10.0 kcal/mole.

9-55 Calculate the average bond-dissociation energy in hydrogen
sulfide.

9-56 Calculate the average bond-dissociation energy of the bonds
in carbon tetrachloride. $\Delta H°_f(CCl_4)_g$ =   -25.50 kcal/mole.

9-57 Calculate the average bond-dissociation energy of Kr—F.
The standard enthalpy of formation of $KrF_2(g)$ is +13.7 kcal/mole.
    [S. Gunn, J. Am. Chem. Soc., 88, 5924 (1966)]

Chapter 10

SOLUTIONS OF ELECTROLYTES

Electrolytes are substances which conduct electricity in aqueous solutions. Typical electrolytes, such as sodium chloride, potassium sulfate, and sodium hydroxide, are ionic in the solid state. When they dissolve, the solution contains hydrated cations and anions which were present as unhydrated ions in the crystals. Other electrolytes, such as hydrogen chloride, are molecular in nature until they dissolve in water. The molecules react with water to form ions.

10-1 COLLIGATIVE PROPERTIES

Whether the electrolyte is ionic or molecular in nature, the number of solute particles is greater than if no ions were present, and hence the colligative properties are greater.

Example 10-1

$$NaCl \rightarrow Na^+(aq) + Cl^-(aq)$$

Example 10-2

$$K_2SO_4 \rightarrow 2K^+(aq) + SO_4^{2-}(aq)$$

Example 10-3

  $NaOH \rightarrow Na^+(aq) + OH^-(aq)$

Example 10-4

  $HCl(g) + H_2O \rightarrow H_3O^+ + Cl^-(aq)$

  In many cases an equilibrium between un-ionized molecules and
ions is established. Acetic acid and ammonia solutions are typical
examples.

Example 10-5

  $CH_3COOH(aq) + H_2O \rightleftarrows H_3O^+ + CH_3COO^-(aq)$

Example 10-6

  $NH_3(aq) + H_2O \rightleftarrows NH_4^+(aq) + OH^-(aq)$

The concentration of ions is small here because the equilibrium lies
far to the left.

Example 10-7

The vapor pressures of a series of aqueous solutions of potassium
chloride at 20.0°C were measured by B. F. Lovelace, J. C. W. Frazer,
and E. Miller. In one experiment 0.20 mole of KCl lowered the vapor
pressure of 1000 g of water 0.110 mm.   If KCl were molecular and
did not dissociate in solution, the vapor-pressure lowering
(depression) expected from Raoult's law could be calculated.

            [J. Am. Chem. Soc., 38, 515 (1916)]

  $P = P°X$

  $P° = 17.535$ mm at 20.0°C (from table)

  $X = \dfrac{\text{moles of solvent}}{\text{moles of solvent + moles of solute}}$

  $H_2O = 18.0$ g/mole

  Moles of solvent $= \dfrac{1000}{18.0} = 55.56$

$$X = \frac{55.56}{55.56 + 0.20}$$

$$P = 17.54 \frac{55.56}{55.76} = 17.48 \text{ mm}$$

17.54

$\underline{-17.48}$

0.06 mm calculated depression

0.110 mm observed depression

The observed depression is nearly twice as great, indicating that KCl exists as K^+ and Cl^- ions in the solution.

Example 10-8

A solution containing 1.195 g of NaCl in 100 g of H_2O exhibited a freezing-point depression of 0.693°. Compare this result with the values expected if no dissociation occurred, and if complete dissociation occurred.

[L. Kahlenberg, J. Phys. Chem., 5, 353 (1901)]

$\Delta t_f = K_b m$

m = moles NaCl/1000 g H_2O

$1.195 \times \frac{1000}{100} = 11.95$ g NaCl/1000 g H_2O

NaCl   22.99

$\underline{35.45}$

58.44 g/mole

$\frac{11.95}{58.44} = 0.2045$ mole

$\frac{g}{g/mole} = $ mole

$\Delta t_f = 1.86 \times 0.204 = 0.379°$

This is the freezing-point depression expected if sodium chloride were not dissociated. A depression of 2 x 0.379° = 0.758° would be expected if NaCl were completely dissociated and no other complicating effects were present. The observed value indicates that NaCl is indeed highly dissociated, but other effects must be involved.

Example 10-9

The boiling-point elevation of a 0.0500 m solution of sodium
chloride was observed to be 0.0479° by R. P. Smith. Compare this
result with the values predicted assuming no dissociation, and
assuming complete dissociation.

[J. Am. Chem. Soc., 61, 500 (1939)]

$$\Delta t_b = K_b m$$
$$\Delta t_b = 0.52 \times 0.050 = 0.0260°$$

This is the elevation expected if no dissociation occurred. If
complete dissociation occurred and there were no other complicating
effects, the elevation would be 0.052°.

10-2 CONCENTRATION OF IONS

It is very common practice to describe a solution by designating
the concentration of one of the ions present. Chemical reactions
involve ions, and the particular ion in question may be the par-
ticipant in a chemical reaction.

Example 10-10

A 1.60 M solution of sodium perchlorate was used in a kinetics
investigation. The sodium ion concentration and the perchlorate ion
concentration were each 1.60 M as well.

[R. C. Thompson and G. Gordon, Inorg. Chem., 5, 563 (1966)]

$$NaClO_4 \rightarrow Na^+(aq) + ClO_4^-(aq)$$

Example 10-11

In the same study as in Example 10-10, a 1.69 M zinc perchlorate
solution was used. In this case the zinc ion concentration was 1.69 M,
but the perchlorate ion concentration was 3.38 M.

$$Zn(ClO_4)_2 \rightarrow Zn^{2+}(aq) + 2ClO_4^-(aq)$$

Example 10-12

Calculate the quantity of sodium bromide required to make 100 ml of a 0.033 M Br^- solution.

$NaBr \rightarrow Na^+(aq) + Br^-(aq)$

Since 1 mole of NaBr yields 1 mole of $Br^-(aq)$, 0.033 mole of NaBr is required to yield 0.033 mole of $Br^-(aq)$. A 0.033 M solution contains 0.033 mole of solute in 1000 ml of solution. Since 100 ml of solution is required here, only 0.0033 mole of NaBr is needed.

NaBr   23.0

$$\frac{79.9}{102.9} \text{ g/mole}$$

$3.3 \times 10^{-3} \times 1.03 \times 10^2 = 3.40 \times 10^{-1} = 0.340$ g NaBr

mole x g/mole = g

Example 10-13

If the 0.0033 M Br^- solution was prepared from magnesium bromide, how many grams of the salt were required?

$MgBr_2 \rightarrow Mg^{2+}(aq) + 2Br^-(aq)$

Since 1 mole of $Br^-(aq)$ requires 0.5 mole $MgBr_2$, 0.033 mole $Br^-(aq)$ will require only 0.0165 mole $MgBr_2$. For 100 ml of a 0.0165 M solution, 0.00165 mole $MgBr_2$ is required.

$MgBr_2$ 24.3

$$2 \times 79.9 \quad \frac{159.8}{184.1} \text{ g/mole}$$

$1.65 \times 10^{-3} \times 1.84 \times 10^2 = 3.04 \times 10^{-1} = 0.304$ g $MgBr_2$

mole x g/mole = g

Example 10-14

A 3.0 M solution of perchloric acid will yield 3.0 M H_3O^+ and 3.0 M ClO_4^-.

$HClO_4 + H_2O \rightarrow H_3O^+ + ClO_4^-(aq)$

10-3   ANALYTICAL CONCENTRATION

The <u>analytical</u> <u>concentration</u> of a substance is the total number
of moles of a pure substance dissolved in one liter of solution.
The analytical concentration of electrolytes is different from the
concentration of the species actually present in the solution. A
solution prepared by dissolving 7.45 g of KCl in sufficient water
to make a liter of solution has an analytical concentration of KCl
of 0.1 M. For all practical purposes the solution contains no KCl,
only hydrated K^+ and Cl^-. A 0.01 M solution of acetic acid has
an analytical concentration of 0.0100. The actual concentration
of acetic acid molecules present in the solution is 0.00987 M. The
remaining acetic acid is present as hydronium ions and acetate ions.
The symbol C is often used to designate the analytical concentration.
The analytical concentration of $NaClO_4$ in Example 10-10 is 1.60 M.
The analytical concentration of $HClO_4$ in Example 10-14 is 3.0 M.

10-4   pH AND pOH

Very dilute solutions of electrolytes are in common use in the
laboratory. Solutions of acids in which the hydrogen ion concentration (or, more accurately, the concentration of the hydronium ion:
$[H_3O^+]$) has values of 1.0×10^{-4} to 1.0×10^{-10} moles/liter are
frequently encountered. Such concentrations are often expressed on
the pH scale. The pH of a solution is defined by

$$pH = -\log[H_3O^+]$$

Example 10-15
Express the hydrogen ion concentration of a 0.0217 M H_3O^+ solution
on the pH scale.

$[H_3O^+] = 0.0217$ M

$pH = -\log 0.0217 = -(-2 + 0.337)$

$pH = 1.66$

Remember, a pH is a <u>logarithm</u> and is the sum of a characteristic and a mantissa. The characteristic denotes the location of the decimal point. Since the pH scale is limited to dilute solutions, the characteristic is always negative. The mantissa is unique to a particular combination of digits and is always positive. Here the characteristic is -2 and the mantissa from Table D-2 is 0.337.

Example 10-16

Express the hydrogen ion concentration of the following solution on the pH scale.

Method 1.

$$[H_3O^+] = 6.02 \times 10^{-4} \text{ M}$$
$$pH = -\log 6.02 \times 10^{-4}$$
$$= -(-4 + .780) = 3.22$$

In the example, -4 is the characteristic and 0.780 is the mantissa obtained from Table D-2.

Method 2.

$$pH = -\log 6.02 \times 10^{-4} = -(\log 6.02 + \log 10^{-4})$$

The logarithm of a product (6.02×10^{-4}) is the sum of the logarithms of the terms.

$$pH = -[0.780 + (-4.000)] = 3.22$$

Example 10-17

$[H_3O^+]$ moles/liter	Logarithm	pH
(a) 3.71×10^{-8}	pH = - log 3.71×10^{-8}	7.43
(b) 0.00765	pH = - log 0.00765	2.12
(c) 1.00×10^{-5}	pH = - log 1.00×10^{-5}	5.00
(d) 21.2×10^{-9}	pH = - log 21.2×10^{-9}	7.67

Note in (c) that the pH is equal to the absolute magnitude of the exponent 5 because the logarithm of 1.00 is zero. In (d) it is necessary to express the concentration as 2.12×10^{-8} in order to use the exponent as the characteristic of the logarithm in Method 1. Method 2 will lead directly to the answer, however. When the hydrogen ion concentration is expressed as a power of 10, for example, 3.21×10^{-8}, the pH is always less than the absolute value of the exponent but greater than the absolute value of the exponent minus 1.

Example 10-18

If the pH of a solution was 5.34 what was the hydrogen ion concentration in moles per liter?

$$pH = 5.34$$
$$-\log[H_3O^+] = 5.34$$
$$\log[H_3O^+] = -5.34$$

Because this logarithm is <u>negative</u>, it may be expressed as a negative characteristic and a positive mantissa.

$$-6.00 + 0.66 = -5.34$$
$$\log[H_3O^+] = -6.00 + 0.66$$
$$[H_3O^+] = 4.57 \times 10^{-6} \text{ mole/liter}$$

Example 10-19

pH		Logarithm	$[H_3O^+]$ moles/liter
(a)	5.00	$-\log[H_3O^+] = 5.00$	1.00×10^{-5}
(b)	4.38	$-\log[H_3O^+] = 4.38$	4.17×10^{-5}
(c)	8.09	$-\log[H_3O^+] = 8.09$	8.13×10^{-9}
(d)	11.18	$-\log[H_3O^+] = 11.18$	6.61×10^{-12}

In (a) the exponent is equal to the characteristic of the logarithm, since the mantissa is 0. In (b) to (d) the exponent is equal to the characteristic plus 1.

The hydroxide ion concentration of aqueous solutions may be expressed as pOH, which is analogous to pH.

$$pOH = -\log[OH^-]$$

Example 10-20

The pOH of a 0.00200 M solution of sodium hydroxide may be calculated as follows.

Sodium hydroxide is a strong electrolyte, i.e., completely ionized.

$$NaOH \rightarrow Na^+(aq) + OH^-(aq)$$

The [OH^-] is, therefore, 0.00200 mole/liter, since one mole of OH^- is formed for each mole of NaOH dissolved.

$$pOH = -\log 0.00200 = -\log 2.00 \times 10^{-3}$$
$$pOH = -[0.301 + (-3.000)] = 2.70$$

PROBLEMS

10-1 What is the pH of the following solutions whose hydronium ion concentrations are listed?

    a. 6.91×10^{-8} M   b. 8.71×10^{-10} M

    c. 1.21×10^{-3} M   d. 0.0111 M

    e. 3.71×10^{-4} M

10-2 What is the pH of the solutions of strong acids which are listed below?

    a. 3.81×10^{-3} M $HClO_4$ b. 6.72×10^{-4} M HBr

    c. 0.000179 M HCl   d. 5.42×10^{-4} M HNO_3

    e. 9.81×10^{-3} M HCl

10-3 What is the hydrogen ion concentration of the solutions whose pHs are listed?

    a. 9.95   b. 1.97

    c. 5.75   d. 10.08

    e. 2.42

10-4 What is the pH of each of the solutions which has been prepared by diluting the concentrated strong acid?

    a. 1.53 ml of 12.0 M HCl diluted to 1.00 liter

    b. 2.12 ml of 5.43 M HBr diluted to 1.00 liter

    c. 0.862 ml of 3.17 M $HClO_4$ diluted to 2.50 liters

10-5 What was the pH of a solution of nitric acid prepared by

diluting 1.00 ml of a 70.0% solution to 3.25 liters? The specific
gravity of the concentrated acid was 1.42.

10-6 What was the pH of a solution made by diluting 8.50 ml of
38.0% HCl to 1.50 liters? The density of the concentrated acid was
1.19 g/ml.

10-7 What volume of 70.0% perchloric acid (sp. gr. 1.172) would be
required to make 1.00 liter of a solution whose pH was 4.11?

10-8 What volume of 36.5% hydrochloric acid (sp. gr. 1.18) would
be required to make 1.00 liter of a solution whose pH was 3.00?

10-9 Calculate the potassium ion concentration, in moles per liter,
of a solution prepared by dissolving 0.217 g of potassium nitrate
in sufficient water to make 100 ml of solution.

10-10 What is the concentration, in moles per liter, of sodium
ions in a solution containing 17.3 g of sodium sulfate in 1.00 liter
of solution?

10-11 A solution containing 0.200 mole of sodium ions in 100 ml
of solution was required for an experiment. What mass, in grams,
of sodium perchlorate was required to prepare the solution?

10-12 How many grams of potassium perchlorate would be required to
prepare 100 ml of a solution containing 0.200 mole of potassium ion?
After you have made this calculation, consult a handbook for the
solubility of potassium perchlorate in water.

10-13 Is it possible to prepare a solution of lithium fluoride in
which the lithium ion concentration is 0.0100 M? If it is possible,

Solutions of Electrolytes   217

how many grams of lithium fluoride would be required to make 50.0 ml
of such a solution?

10-14 Compare the observed vapor-pressure lowering of 0.217 mm
for a 0.40 m aqueous solution of KCl at 20.0°C with that calculated
from Raoult's Law.
        [B. F. Lovelace, J. C. W. Frazer, and E. Miller, J. Am.
        Chem. Soc., 38, 515 (1916)]

10-15 Calculate the boiling-point elevation expected for a 0.079 m
aqueous solution of sodium chloride assuming complete dissociation.
Compare your answer with the observed elevation of R. P. Smith:
0.0751°.
        [J. Am. Chem. Soc., 61, 500 (1939)]

10-16 Calculate the freezing-point depression expected for a
0.00443 m solution of potassium nitrate in water.
        [L. D. Adams, J. Am. Chem. Soc., 37, 494 (1915)]

10-17 Calculate the predicted boiling-point elevation of a solution
of 1.9501 g of cadmium iodide in 44.69 g of water.
        [E. Beckmann, Z. Physik. Chem. (Leipzig), 6, 460 (1890)]

10-18 Compare the results of a freezing-point depression experiment,
in which 0.699 g of magnesium sulfate in 100 g of water had a
freezing point of -0.154°C, with the value calculated from theory.
        [L. Kahlenberg, J. Phys. Chem., 5, 353 (1901)]

10-19 Plot the vapor pressures calculated for each of the solutions
listed versus the molal concentration. Plot the observed vapor
pressure for each solution on the same graph. Discuss the results.

KCl, m	ΔP, observed*
0.2	0.110
0.4	0.217
0.6	0.329
0.8	0.438
0.9	0.493
1.0	0.547
1.2	0.663
1.5	0.826
2.0	1.102

*Temperature, $20.0^{\circ}C$

10-20 Calculate the molecular weight of cadmium iodide from the fact that 5.6977 g in 44.69 g of water boiled at 100.181°C.

   [E. Beckmann, Z. Physik, Chem. (Leipzig), 6, 460 (1890)]

10-21 Calculate the molecular weight of potassium nitrate from the fact that a 0.00817 m aqueous solution had a freezing point of -0.0295°C. Account for the difference in results in this problem and Prob. 10-20.

   [L. D. Adams, J. Am. Chem. Soc., 37, 494 (1915)]

10-22 Calculate the theoretical freezing-point depression of each of the solutions of KCl listed and plot them versus the molal concentration of KCl. On the same graph plot the observed depressions. Discuss the resulting graph.

   [L. D. Adams, J. Am. Chem. Soc., 37, 494 (1915)]

m	Δt
0.00506	0.0184
0.00963	0.0348
0.01648	0.0590
0.03170	0.1122
0.05818	0.2031
0.11679	0.4014

10-23   The analytical concentration of a solution of strontium
iodate was 0.101 M.   What was the concentration of the various
species in the solution?

10-24   What was the analytical concentration of a solution of
magnesium bromide which contained 0.216 mole of bromide ions per
liter of solution?

10-25   What was the pH of a solution whose analytical concentration
of trifluoroacetic acid was 0.00382 M?

Chapter 11

CHEMICAL EQUILIBRIUM, I

11-1 REVERSIBLE REACTIONS--EQUILIBRIUM

In principle, all chemical reactions are reversible. However, in many instances the reaction proceeds in one direction to such an extent that it is practically complete. Thus fluorine reacts with ammonia, but the reverse reaction is negligible.

$$3F_2(g) + 2NH_3(g) \rightarrow N_2(g) + 6HF(g)$$

Such a reaction is said to be quantitative.

In other reactions, a state of measurable equilibrium is established. The concentration of each reactant decreases until a constant minimum concentration is reached. The concentration of each product increases from zero to a constant maximum concentration. At equilibrium the concentration of each reactant and product remains constant.

11-2 THE EQUILIBRIUM CONSTANT

The relationship of the concentrations of products and reactants may be represented by the equilibrium constant K_c. In the generalized equilibrium reaction

$$aA + bB \rightleftharpoons cC + dD$$

221

$$K_c = \frac{[C]^c[D]^d}{[A]^a[B]^b}$$

The brackets represent the concentration of the particular substance in moles per liter, Note that:

1. The substances on the <u>right</u> side of the chemical equation are placed in the <u>numerator</u>, those on the <u>left</u> in the <u>denominator</u>.

2. The value of the concentration of each substance is raised to a <u>power</u> which is the respective coefficient in the chemical equation.

3. The numerical value of the equilibrium constant changes with change in <u>temperature</u>.

Example 11-1

In an appropriate temperature range, dinitrogen tetroxide is in equilibrium with nitrogen dioxide.

$$N_2O_4(g) \rightleftharpoons 2NO_2(g)$$

Write the equilibrium-constant expression for this equilibrium.

$$K_c = \frac{[NO_2]^2}{[N_2O_4]}$$

Example 11-2

Write the equilibrium-constant expression for the reaction

$$4HCl(g) + O_2(g) \rightleftharpoons 2H_2O(g) + 2Cl_2(g)$$

$$K_c = \frac{[H_2O]^2[Cl_2]^2}{[HCl]^4[O_2]}$$

The form of the equilibrium-constant expression must represent the chemical equation for the equilibrium as it is written. The chemical equation in Example 11-1 may be written

$$\tfrac{1}{2}N_2O_4(g) \rightleftharpoons NO_2(g)$$

The equilibrium-constant expression would be

$$K_c = \frac{[NO_2]}{[N_2O_4]^{\frac{1}{2}}}$$

The numerical value of K_c will be quite different in this case.

11-3 THE DYNAMIC NATURE OF CHEMICAL EQUILIBRIA

The fact that the concentrations of the substances at equilibrium are not changing does not mean that changes are not occurring. At equilibrium two opposing reactions are occurring at the same rate. Since reactions are occurring, the system is said to be in a state of dynamic equilibrium. Despite the fact that the reactions are occurring, the concentrations of substances present remain constant; the substances are formed at the same rate they are consumed.

11-4 HOMOGENEOUS AND HETEROGENEOUS EQUILIBRIA

Equilibria involving substances in the same phase, mixtures of gases or substances in solution, are called homogeneous equilibria. Heterogeneous equilibria involve substances in more than one phase. Examples are precipitates in equilibrium with ions in solution, water vapor in equilibrium with hydrates, and decomposition products in equilibrium with solids. The important consideration here is that the effective concentration of the solid substance is constant regardless of the concentration of the other species present.

Example 11-3

Calcium carbonate decomposes on heating to form carbon dioxide according to the equation

$$CaCO_3(s) \; \rightleftarrows \; CaO(s) + CO_2(g)$$

In a closed container at constant temperature, an equilibrium is established. Write an equilibrium expression for the system.

$$K_c = [CO_2]$$

Both $CaCO_3$ and CaO are solids.

11-5 K_c AND K_p

Equilibrium constants in general are represented by K. The symbol K_c has been used to emphasize the fact that the equilibrium expression contains the <u>concentrations</u> of the substances involved.

When the substances at equilibrium are gases, the <u>partial pressure</u> of each substance may be used in place of the concentration. The equilibrium is then represented by the constant K_p. In the equilibrium $N_2O_4(g) \rightleftarrows 2NO_2(g)$

$$K_p = \frac{P^2_{NO_2}}{P_{N_2O_4}}$$

The decomposition of $CaCO_3$ is represented by $K_p = P_{CO_2}$.

<u>Example 11-4</u>
What is the K_p expression for the equilibrium $4HCl(g) + O_2(g) \rightleftarrows 2H_2O(g) + 2Cl_2(g)$?

$$K_p = \frac{P^2_{H_2O} \cdot P^2_{Cl_2}}{P^4_{HCl} \cdot P_{O_2}}$$

<u>Example 11-5</u>
Write a K_p expression for the equilibrium

$$Na_2SO_3(s) \rightleftarrows Na_2O(s) + SO_2(g)$$

$$K_p = P_{SO_2}$$

11-6 CALCULATION OF EQUILIBRIUM CONSTANTS

The units of K are often omitted because they can easily be determined from the equilibrium expression if necessary. In a more rigorous treatment, the concentrations are referred to standard states and are represented by dimensionless ratios. For these reasons, units will usually not be included here, except for a few illustrations.

Example 11-6
At 8.2°C, 1.00 liter of a chloroform solution was found to contain
0.129 mole of N_2O_4 and 1.17 x 10^{-3} mole of NO_2 at equilibrium.
Calculate K_c for the dissociation of N_2O_4 under these conditions.
    [J. T. Cundell, J. Chem. Soc., 67, 808 (1895)]

$$N_2O_4 \; \overset{\rightarrow}{\leftarrow} \; 2NO_2$$

$$K_c = \frac{[NO_2]^2}{[N_2O_4]} = \frac{(1.17 \times 10^{-3})^2}{1.29 \times 10^{-1}} = 1.07 \times 10^{-5}$$

If the units of K_c are to be required, the units of concentration are substituted in the expression for K_c.

$$K_c = \frac{(mole/liter)^2}{mole/liter} = mole/liter$$

In this example, the units of K_c are moles per liter.

Example 11-7
The decomposition of hydrogen iodide was investigated at 457.6°C.
The concentrations of the species present at equilibrium were
5.62 x 10^{-3} mole/liter of H_2, 5.94 x 10^{-4} mole/liter of I_2, and
1.27 x 10^{-2} mole/liter of HI. Calculate K_c.
    [A. H. Taylor and R. H. Crist, J. Am. Chem. Soc., 63,
    1381 (1941)]

$$2HI(g) \; \overset{\rightarrow}{\leftarrow} \; H_2(g) + I_2(g)$$

$$K_c = \frac{[H_2]\,[I_2]}{[HI]^2}$$

$$K_c = \frac{(5.62 \times 10^{-3}) \times (5.94 \times 10^{-4})}{(1.27 \times 10^{-2})^2} = 2.06 \times 10^{-2}$$

Units

$$\frac{(\text{mole/liter}) \times (\text{mole/liter})}{(\text{mole/liter})^2}$$

Example 11-8

K. V. von Falckenstein investigated the decomposition of HBr at elevated temperatures. In one experiment at 1108°, the partial pressures of the components of the mixture were HBr, 0.998 atm; H_2, 3.82×10^{-3} atm; Br_2, 3.82×10^{-3} atm. Calculate the K_p for the equilibrium.

[Z. Physik. Chem. (Leipzig), 68, 270 (1910)]

$$2HBr(g) \rightleftarrows H_2(g) + Br_2(g)$$

$$K_p = \frac{P_{H_2} \cdot P_{Br_2}}{P^2_{HBr}}$$

$$K_p = \frac{(3.82 \times 10^{-3})^2}{0.998^2} = 1.47 \times 10^{-5}$$

$$\frac{atm^2}{atm^2}$$

Example 11-9

Hydrogen iodide was found to be 22.3% dissociated at 730.8 K. Calculate K_c.

[A. H. Taylor and R. H. Crist, J. Am. Chem. Soc., 63, 1377 (1941)]

$$2HI(g) \rightleftarrows H_2(g) + I_2(g)$$

Assume that the initial concentration of HI was 1 mole/liter.

1.000 x 0.223 = 0.223 mole/liter HI dissociated

1.000 - 0.223 = 0.777 mole/liter HI remaining at equilibrium

$\frac{0.223}{2}$ = 0.112 mole/liter H_2 and I_2 formed.

From the equation, for each mole of HI dissociated, 1/2 mole H_2 and 1/2 mole of I_2 are formed.

$$K_c = \frac{[H_2][I_2]}{[HI]^2}$$

$$K_c = \frac{0.112 \times 0.112}{0.777^2} = \frac{0.0125}{0.604} = 2.07 \times 10^{-2}$$

Example 11-10

At 160°C and 1 atm, $PCl_5(g)$ is 13.5% dissociated in to $PCl_3(g)$ and $Cl_2(g)$. Calculate K_c for the equilibrium.

[W. Fischer and O. Jubermann, Z. Anorg. Allgem. Chem., 235, 337 (1938)]

$$PCl_5(g) \rightleftarrows PCl_3(g) + Cl_2(g)$$

$$K_c = \frac{[PCl_3][Cl_2]}{[PCl_5]}$$

Assume that the initial concentration of $PCl_5(g)$ was 1.00 mole/liter. The number of moles per liter of PCl_5 which dissociated was 1.00 x 0.135 = 0.135. According to the chemical equation, 1.00 mole of PCl_3 and 1.00 mole of Cl_2 are formed for each mole of PCl_5 which dissociates. Therefore, 0.135 mole/liter of PCl_3 and 0.135 mole/liter of Cl_2 were formed. The concentration of PCl_5 remaining is 1.000 - 0.135 = 0.865 mole/liter.

$$K_c = \frac{0.135 \times 0.135}{0.865} = 0.0211$$

11-7 CALCULATION OF CONCENTRATIONS FROM EQUILIBRIUM CONSTANTS

The calculation of the concentration or partial pressure of a
species from an equilibrium constant requires careful consideration
of the relationships among the components of the equilibrium system.

Example 11-11

Calculate the concentration of NO_2 present at equilibrium in a
chloroform solution which contained 0.129 mole/liter of N_2O_4.
K_c for the dissociation of N_2O_4 = 1.07 x 10^{-5}.

$$N_2O_4 \rightleftharpoons 2NO_2$$

$$K_c = \frac{[NO_2]^2}{[N_2O_4]}$$

let X = $[NO_2]$
   $[N_2O_4]$ = 0.129 mole/liter

$$\frac{X^2}{0.129} = 1.07 \times 10^{-5}$$

$$X^2 = 0.129 \times (1.07 \times 10^{-5}) = 1.38 \times 10^{-6}$$

$$X = \sqrt{1.38 \times 10^{-6}} = 1.17 \times 10^{-3} \text{ mole/liter}$$

Example 11-12

Calculate the equilibrium partial pressure of hydrogen formed by the
decomposition of hydrogen bromide if the partial pressure of the
remaining HBr was 0.998 atm. The K_p for the decomposition of HBr
is 3.83 x 10^{-3} at the temperature of the experiment.

$$2HBr \rightleftharpoons H_2 + Br_2$$

$$K_p = \frac{P_{H_2} \cdot P_{Br_2}}{P_{HBr}^2} = 3.83 \times 10^{-3}$$

Let X = partial pressure of H_2, P_{H_2}; then X = P_{Br_2}, since one mole of Br_2 is formed for each mole of H_2 formed according to the equation.

$$\frac{X^2}{0.998^2} = 3.83 \times 10^{-3}$$

$$X^2 = 0.996 \times 3.83 \times 10^{-3} \text{ atm}$$

$$X = 6.17 \times 10^{-2} \text{ atm}$$

Example 11-13

Calculate the percent decomposition of PCl_5 at 160° from the equilibrium constant K_c.

$$PCl_5 \rightleftharpoons PCl_3 + Cl_2$$

From Example 11-10

$$K_c = \frac{[PCl_3][Cl_2]}{[PCl_5]} = 0.0211$$

Assume initial concentration of PCl_5 = 1.00 mole/liter. Let X = $[PCl_3]$ at equilibrium; then $[Cl_2]$ = X. Since one mole of Cl_2 is formed for each mole of PCl_3 formed, X moles of Cl_2 are formed for each X moles of PCl_3 formed. Also $[PCl_5]$ = 1.00 - X, since one mole PCl_5 must dissociate to produce one mole of PCl_3.

$$\frac{X^2}{1.00 - X} = 0.0211$$

$$X^2 = 0.0211 - 0.0211X$$

$$X^2 + 0.0211X - 0.0211 = 0$$

$$X = 0.135 \text{ mole/liter}$$

Example 11-14

Calculate the iodine concentration in an equilibrium mixture of hydrogen iodide and its constituents. The hydrogen concentration was 5.62×10^{-3} mole/liter, and the HI concentration was 1.27×10^{-2} mole/liter. $K_c = 2.06 \times 10^{-2}$

[A. H. Taylor and R. H. Crist, J. Am. Chem. Soc., 63, 1377 (1941)]

$$2HI(g) \rightleftharpoons H_2(g) + I_2(g)$$

$$K_c = \frac{[H_2][I_2]}{[HI]^2} = 2.06 \times 10^{-2}$$

$$\frac{(5.62 \times 10^{-3})[I_2]}{(1.27 \times 10^{-2})^2} = 2.06 \times 10^{-2}$$

$$[I_2] = \frac{(1.61 \times 10^{-4}) \times (2.06 \times 10^{-2})}{5.62 \times 10^{-3}}$$

$$= 5.90 \times 10^{-4} \text{ mole/liter}$$

Example 11-15

A 1.00-mole sample of HI, contained in a 1.00-liter flask, was heated to 730.8 K. Calculate the concentration of each component of the mixture at equilibrium. $K_c = 2.06 \times 10^{-2}$.

$$2HI \rightleftharpoons H_2 + I_2$$

$$K_c = \frac{[H_2][I_2]}{[HI]^2} = 2.06 \times 10^{-2}$$

Let X = moles HI which decomposed; then [HI] = (1.00 - X).

$$[H_2] = [I_2] = \frac{X}{2}$$

since 2 moles of HI gives 1 mole of H_2, and 1 mole of I_2 in the equation

$$\frac{X/2 \times X/2}{(1 - X)^2} = 2.06 \times 10^{-2}$$

Take the square root of both sides.

$$\frac{X/2}{1 - X} = \frac{X}{2 - 2X} = 0.144$$

X = 0.288 - 0.288X

1.288X = 0.288

$$X = \frac{0.288}{1.288} = 0.223 \text{ mole/liter}$$

$[HI] = 1.00 - 0.223 = 0.78 \text{ mole/liter}$

$[H_2] = [I_2] = \frac{0.223}{2} = 0.112 \text{ mole/liter}$

PROBLEMS

11-1 Write concentration equilibrium-constant expressions for the following equilibria. Balance the equations when necessary.

a. $2NO(g) \rightleftarrows N_2(g) + O_2(g)$

b. $N_2(g) + 3H_2(g) \rightleftarrows 2NH_3(g)$

c. $NO_2(g) + H_2(g) \rightleftarrows NH_3(g) + H_2O(g)$

d. $C(s) + O_2(g) \rightleftarrows CO(g)$

e. $H_2(g) + Fe_3O_4(s) \rightleftarrows Fe(s) + H_2O(g)$

11-2 Write K_p expressions for the equilibria in Prob. 11-1.

11-3 Calculate the value of the equilibrium constant for the dissociation of N_2O_4 in chloroform at 8.2°C if its concentration was 0.324 mole/liter and the concentration of NO_2 was 1.85×10^{-3} moles/liter.

[J. T. Cundell, J. Chem. Soc., 67, 808 (1895)]

11-4 In another experiment, J. T. Cundell found that at 8.2°C, a chloroform solution contained 0.778 mole/liter of N_2O_4 and 2.84×10^{-3} mole/liter of NO_2. Calculate K_c for the reaction $2NO_2 \rightleftarrows N_2O_4$.

[J. T. Cundell, J. Chem. Soc., 67, 808 (1895)]

11-5 Calculate the equilibrium constant for the formation of hydrogen iodide from the elements at 763.8 K. At equilibrium, the concentration of iodine was 0.406 mole/liter and the hydrogen and hydrogen iodide concentrations were 0.172 and 1.78 mole/liter, respectively.

[A. H. Taylor and R. H. Crist, J. Am. Chem. Soc., 63,
1381 (1941)]

11-6 The equilibrium constant for the decomposition of hydrogen
iodide at 763.8 K was also determined by A. H. Taylor and R. H. Cris
In one experiment the concentrations at equilibrium were iodine,
0.242; hydrogen, 0.242; and hydrogen iodide, 1.64 mole/liter.
Calculate the value of K_c for the decomposition of HI.
        [J. Am. Chem. Soc., 63, 1381 (1941)]

11-7 The partial pressure of $PCl_3(g)$ and $Cl_2(g)$ in equilibrium with
$PCl_5(g)$ at 422.6 K was 6.08 x 10^{-2} atm. The partial pressure of
$PCl_5(g)$ was 0.453 atm. What was the value of K_p for the system?
        [W. Fischer and O. Jubemann, Z. Anorg. Allgem. Chem.,
        235, 337 (1938)]

11-8 Would the numerical value of K_p in Prob. 11-7 be the same if
the data had been given in mm of Hg, as recorded in the original
paper? $PCl_5(g)$, 344.4 mm; $PCl_3(g)$, 46.2 mm; $Cl_2(g)$, 46.2 mm.

11-9 Determine K_p for the equilibrium represented by:
    $1/6B_2H_6(g) + 2/3BF_3(g) \rightleftarrows HBF_2(g)$
The partial pressures of the components of the equilibrium mixture
were B_2H_6, 2.1 x 10^{-2}; BF_3, 1.4 x 10^{-2}; HBF_2, 1.5 x 10^{-2} atm
at 296 K.
        [R. F. Porter and S. K. Wason, J. Phys. Chem., 69,
        2211 (1955)]

11-10 The hydrolysis of $SmCl_3(s)$ at 893 K was investigated by
C. W. Koch and B. B. Cunningham, who found that the partial
pressures of HCl and H_2O at equilibrium were 418.8 mm and 1.60 mm i
one experiment. The other product of the reaction was SmOCl(s).
They recorded an equilibrium constant of 144 for this experiment.
Calculate the numerical value of K and give the appropriate units.
        [J. Am. Chem. Soc., 75, 796 (1953)]

11-11 Calculate K_p for the equilibrium:

$LaCl_3(s) + H_2O(g) \rightleftharpoons LaOCl(s) + 2HCl(g)$ at 892 K

The partial pressure of HCl was 80.9 mm, and the partial pressure
of H_2O was 2.92 mm.

[C. W. Koch, A. Broido, and B. B. Cunningham, J. Am.

Chem. Soc., 74, 2349 (1952)]

11-12 Sudborough and Miller found that NOCl was 8.22% dissociated
at 796°C into NO and Cl_2. Calculate the equilibrium constant for
the dissociation.

[J. Chem. Soc., 59, 271 (1891)]

11-13 A sample of N_2O_4 gas, initial concentration 0.02968 mole/liter,
was allowed to reach equilibrium at 25°C. The total pressure at
equilibrium was 0.862 atm. Calculate the equilibrium constant for
the dissociation of N_2O_4.

[F. H. Verhoek and F. Daniels, J. Am. Chem. Soc., 53,

1250 (1931)]

11-14 A solution containing 1.00 mole/liter of acetic acid and
0.18 mole/liter of ethyl alcohol at 100°C had produced 0.171 mole
of ester when equilibrium was established. Determine the equilibrium
constant for the esterification.

[M. Bertholet and P. St. Giles, Ann. Chim. Phys., 65,

(3) 385 (1862)]

11-15 When Bertholet and St. Giles mixed 1.00 mole/liter of acetic
acid and 0.33 mole/liter of alcohol, 0.293 mole/liter of ester was
observed. Compare their observation with the quantity of ester
expected from the equilibrium constant calculated in Prob. 11-14.

11-16 Calculate the equilibrium constant for the equilibrium
$2NO(g) + Br_2(g) = 2NOBr(g)$ from the observation that the total
pressure at 296.9 K was 110.5 mm at equilibrium. The initial partial
pressures of NO and Br_2 were 98.4 mm and 41.3 mm, respectively.

[C. M. Blair, Jr., P. D. Brass, and D. M. Yost, J. Am. Chem. Soc., 56, 1916 (1934)]

11-17   The pressure of bromine in equilibrium with $FeBr_3$ at 138°C was 748 mm. What are the values of K_p (in atm) for the two equilibria?

$FeBr_3(s) \rightleftharpoons FeBr_2(s) + \frac{1}{2}Br_2(g)$

$2FeBr_3(s) \rightleftharpoons 2FeBr_2(s) + Br_2(g)$

[N. W. Gregory and B. A. Thackrey, J. Am. Chem. Soc., 72, 3176 (1950)]

11-18   Calculate the pressure of bromine, in torr, in equilibrium with $FeBr_3$ at 108°C. K_p for the reaction $FeBr_3(s) \rightleftharpoons$ $FeBr_2(s) + \frac{1}{2}Br_2(g)$ is 0.456 atm.

11-19   The density of CF_3COOH vapor at 118.1°C and 468.5 torr was 2.784 g/liter. Calculate K_c for the association of the acid $2CF_3COOH \rightleftharpoons (CF_3COOH)_2$.

[R. E. Lundin, F. E. Harris, and L. K. Nash, J. Am. Chem. Soc., 74, 4654 (1952)]

11-20   Calculate the equilibrium partial pressures of the monomer of acetic acid at 121.1°C. K_p (in torr) for the dissociation is 270, and the partial pressure of the dimer was 26.74 torr.

[M. D. Taylor, J. Am. Chem. Soc., 73, 315 (1951)]

11-21   Acetone and iodine react at 371°C to form the compound iodoacetone according to the following reaction

$CH_3COCH_3(g) + I_2(g) \rightleftharpoons CH_3COCH_2I(g) + HI(g)$

A mixture of acetone, originally at 276.5 torr, and iodine at 4.52 torr, had pressures of 275.9 and 3.97 torr, respectively, at equilibrium. What is K_p for this reaction?

[Solly, Golden, and Benson, J. Am. Chem. Soc., 92, 4653 (1970)]

11-22 When a mixture of tetrafluorohydrazine (N_2F_4) and Cl_2 is
exposed to uv light, chlorodifluoramine (NF_2Cl) is formed. In one
experiment, chlorine at 406 mm pressure and chlorodifluoramine at
100 mm pressure were mixed at 49.9°C. At equilibrium the pressure
of each gas was as follows: N_2F_4, 45.1 mm; Cl_2, 451.1 mm; and
NF_2Cl, 9.86 mm. What is K_p for this reaction?
         [R. Petry, J. Am. Chem. Soc., 89, 4600 (1967)]

11-23 In a series of experiments, the equilibrium constant for the
reaction of iodine with propene according to the following equation
was determined.

$$I_2(g) + C_3H_6(g) \rightleftarrows C_3H_5I(g) + HI(g)$$

Some of the results obtained at 545 K were as follows:

Initial pressures			Equilibrium pressures	
P_0 (mm) I_2	P_0 (mm) C_3H_6	P_0 (mm) HI	P_{eq} (mm) HI	P_{eq} (mm) C_3H_5I
23.9	505.8	0	1.80	1.80
11.9	512.2	0	1.25	1.25
24.0	81.6	0	0.756	0.756
16.1	355.3	1.62	2.27	0.645

Determine K_p for each of the above results and the average
value of K_p.
         [A. Rodgers, D. Golden, and S. Benson, J. Am. Chem. Soc.
         88, 3194 (1966)]

11-24 Find the concentration of monomeric dichloroacetic acid in a
carbon tetrachloride solution which contains 0.0129 g of the acid
in 100 ml of solution. The dissociation constant of the dimeric acid
is 9.3×10^{-4}.
         [J. T. Harris, Jr., and M. E. Hobbs, J. Am. Chem. Soc.,
         76, 1419 (1954)]

11-25   If the equilibrium composition of the system $CO_2(g) + H_2(g) \rightleftharpoons$
$CO(g) + H_2O(g)$ was CO_2, 7.15; H_2, 46.93; and $CO = H_2O$, 22.96 mole%,
what was the equilibrium constant?   W. Nernst and F. Haber calculated
a value of 1.57 for K_p.   Is this value consistent with your
calculated K?

[F. Haber and F. Richardt, Z. Anorg. Chem., 38, 5 (1904)]

11-26   The equilibrium constant for the formation of ammonia from the
elements is 0.0266 atm^{-1} at 350°.   Calculate K_c, $1/2N_2(g) + 3/2N_2(g) \rightleftharpoons$
$NH_3(g)$.

[A. T. Larson and R. L. Dodge, J. Am. Chem. Soc., 45,
2918 (1923)]

11-27   Hydrogen iodide gas (1.69 x 10^{-2} mole/liter) was in
equilibrium with its constituents at 457.6°C.   Calculate the hydrogen
concentration if the iodine concentration was 1.52 x 10^{-3} mole/liter.
K_c = 2.06 x 10^{-2}.

[A. H. Taylor and R. H. Crist, J. Am. Chem. Soc., 63,
1377 (1941)]

11-28   A chloroform solution of dinitrogen tetroxide was allowed to
come to equilibrium with nitrogen dioxide at 8.2°C.   Calculate the
concentration of nitrogen dioxide present if the concentration of
dinitrogen tetroxide at equilibrium was 1.29 x 10^{-1} mole/liter.
K_c = 1.07 x 10^{-5}.

[J. T. Cundell, J. Chem. Soc., 67, 808 (1895)]

11-29   Find the equilibrium concentrations of the gases in the
equilibrium $CO_2 + H_2 \rightleftharpoons CO + H_2O$ at 1000°C assuming that 1.00 mole
of CO_2 and 1 mole of H_2 were heated to that temperature. K_c = 1.7
at 1000°C.

11-30   Calculate the equilibrium concentration of the monomer of
CF_3COOH at 103.3°C if 1.00 mole of the dimer was present.   The
equilibrium constant for the dimerization at that temperature is 2.29.

[R. E. Lundin, F. E. Harris, and L. K. Nash, <u>J. Am. Chem.</u>
<u>Soc.</u>, <u>74</u>, 4654 (1952)]

SUPPLEMENTARY PROBLEMS

<u>11-31</u>   Write K_p and K_c expressions for the following equilibria.

a. $2HCl(g) + 1/2O_2(g) \rightleftarrows H_2O(g) + Cl_2(g)$

b. $1/2N_2(g) + 3/2H_2(g) \rightleftarrows NH_3(g)$

<u>11-32</u>   Calculate the value of K_p for the dissociation of HBr at
1024° from the following data: HBr, 0.997 atm; H_2, 3.81×10^{-3} atm;
Br_2, 3.81×10^{-3} atm.

      [K. V. Falckenstein, <u>Z. Physik. Chem. (Leipzig)</u>, <u>68</u>,
      270 (1910)]

<u>11-33</u>   The data for the experiment described in Example 11-8
were given in millimeters of Hg in the original paper, HBr, 758.9 mm;
H_2, 2.90 mm; Br_2, 2.90 mm.   What is the numerical value of K_p if
these data are used directly in the calculations?

<u>11-34</u> Use the data in Prob. 11-32 to calculate K_p for the <u>formation</u>
of HBr from the elements.

<u>11-35</u>   Calculate the value of K_p (include units) for the dissociation
of PCl_5 at 464.5 K from the data of W. Fischer and O. Jubermann:
$PCl_5(g)$, 274.6 mm; $Cl_2(g)$, 153.3 mm.

      [<u>Z. Anorg. Allgem. Chem.</u>, <u>235</u>, 337 (1938)]

<u>11-36</u>   Calculate K_p for the formation of PCl_5 from PCl_3.   At equilib-
rium the partial pressure of $PCl_3(g)$ was 37.4 mm and the partial
pressure of $PCl_5(g)$ was 52.5 mm at 445.3 K.

      [W. Fischer and O. Jubermann, <u>Z. Anorg. Allgem. Chem.</u>, <u>235</u>,
      337 (1938)]

PROBLEMS IN CHEMISTRY

11-37 A gaseous mixture at equilibrium contained HBr (757.96)mm),
H_2 (1.81 mm), and Br_2 (1.81 mm). Calculate K_p for the dissociation
of HBr in this experiment.

> [K. V. von Falckenstein, Z. Physik. Chem. (Leipzig),
> 68, 270 (1910)]

11-38 In another HBr decomposition experiment, the partial pressure
of H_2 was 2.01 mm and the total pressure of the system was 761.12 mm.
What was the value of K_p for the dissociation of HBr in this
experiment?

> [K. V. von Falckenstein, Z. Physik. Chem. (Leipzig), 68,
> 270 (1910)]

11-39 At 1024°, hydrogen bromide was found to be 0.467% dissociated.
Determine the equilibrium constant for the dissociation.

> [K. V. von Falckenstein, Z. Physik. Chem. (Leipzig), 68,
> 270 (1910)]

11-40 Calculate the equilibrium constant for the combination of
deuterium and iodine at 730.8 K. The equilibrium mixture contained
0.10945 x 10^{-5} mole/liter of I_2, 0.44150 x 10^{-5} mole/liter of D_2,
and 1.4160 mole/liter DI.

> [A. H. Taylor, Jr., and R. H. Crist, J. Am. Chem. Soc., 63,
> 1382 (1941)]

11-41 Determine the equilibrium constant for the dissociation of
N_2O_3 at 25°. At a total pressure of 1 atm the gas mixture contained
10.5% N_2O_3.

> [M. Bodenstein et al., Z. Physik. Chem. (Leipzig), 100,
> 75 (1922)]

11-42 The equilibrium partial pressures of PCl_5 and PCl_3 were found
to be 312.5 mm and 99.3 mm, respectively, at 446.4 K. Find K_p for
the dissociation of PCl_5.

Chemical Equilibrium, I   239

[W. Fischer and O. Jubermann, Z. Anorg. Chem., 235,
337 (1938)]

11-43 Calculate the value of K_p for the equilibrium $CO_2(g) + H_2(g) \rightleftharpoons$
$CO(g) + H_2O(g)$ from the data of F. Haber and F. Richardt. The
equilibrium composition observed was CO_2, 0.69; H_2, 80.52; CO, 9.40;
H_2O, 9.40 mole%.
[Z. Anorg. Chem., 38, 5 (1904)]

11-44 Iron(III) bromide was heated to 128°C. The pressure due
to the decomposition, with the formation of iron(II) bromide and
bromine vapor, was 454 mm. Find K_p in atmospheres.
[N. W. Gregory and B. A. Thackrey, J. Am. Chem. Soc., 72,
3176 (1950)]

11-45 Determine the equilibrium constant for the reaction
$CO_2(g) + C(s) \rightleftharpoons 2CO(g)$ at 700°. The gas mixture contains 42.3%
CO_2 and 57.7% CO.

11-46 Calculate the value of K_p for the reaction $2CO_2 \rightleftharpoons 2CO + O_2$
at 1205°C. $K_c = 1.41 \times 10^{-13}$.

11-47 Calculate the equilibrium constant for the association of
CH_3COOH vapor at 81.4°C. The density was 2.444 g/liter at a pressure
of 310.2 torr.
[R. E. Lundin, F. E. Harris, and L. K. Nash, J. Am. Chem.
Soc., 74, 4654 (1952)]

11-48 At 1345.7°, bromine was found to be 4.77% dissociated.
Calculate K_p.
[M. Bodenstein and F. Cramer, Z. Elektrochem., 22, 336 (1916)]

11-49 Calculate K_p for the reaction $CS_2(g) + 4H_2(g) \rightleftharpoons$
$CH_4(g) + 2H_2S(g)$ at 650°C. The partial pressures of CS_2, H_2, CH_4,

and H_2S were 407.0, 270.6, 67.9, and 14.4 mm, respectively.

[E. Terres and E. Sasse, Angew. Chem., 47, 241 (1934)]

11-50   A mixture of hydrogen and iodine was allowed to come to
equilibrium with HI at 448°C. The initial volumes of hydrogen and
iodine (at STP) were 22.13 and 16.18 cc, respectively. At equilibrium
the volume of HI (at STP) was 25.72 cc. Calculate K for the
equilibrium.

[M. Bodenstein, Z. Physik. Chem. (Leipzig), 22, 1 (1897)]

11-51   Calculate the equilibrium constant K_c for the reaction
$CS_2(g) + 4H_2(g) \rightleftharpoons CH_4(g) + 2H_2S(g)$ at 900°. $K_p = 3.0 \times 10^{-5}$.

[E. Terres amd E. Sasse, Angew. Chem., 47, 241 (1934)]

11-52   The vapor-phase dissociation of dimeric acetic acid was
studied by M. D. Taylor. At 100.6°C the pressure of 359.8 ml of the
vapor was 40.68 mm. The pressure which would have been observed if
no dissociation had occurred is 25.33 mm. Find K_p for the
dissociation.

[J. Am. Chem. Soc., 73, 315 (1951)]

11-53   Determine the partial pressure of $N_2O_4(g)$ in equilibrium
with $NO_2(g)$ if the total pressure was 0.157 atm. $K_p = 0.142$
for the dissociation.

[F. H. Verhoeck and F. Daniels, J. Am. Chem. Soc., 53,
1250 (1931)]

11-54   Calculate the equilibrium partial pressures of the components
of the reaction $CS_2 + 4H_2 \rightleftharpoons CH_4 + 2H_2S$ if the total pressure was
2.00 atm at 900°. $K_p = 3.0 \times 10^{-5}$.

[E. Terres and E. Sasse, Angew. Chem., 47, 238 (1934)]

11-55   The equilibrium constant for the formation of ammonia from
the elements $1/2N_2(g) + 3/2H_2(g) \rightleftharpoons NH_3(g)$ at a given temperature
is 6.59×10^{-3} atm^{-1}. Calculate the partial pressure of each gas

at equilibrium if the initial partial pressure of N_2 and H_2 was 5.00 atm each.

11-56 A sample of iodine vapor was observed to exert a pressure of 0.11216 atm at 1274 K. Calculate the value of the pressure due to iodine atoms if the equilibrium constant for the decomposition was 0.1692 atm.

    [M. L. Periman and G. K. Rollefson, J. Chem. Phys., 9, 362 (1941)]

11-57 Determine the partial pressure of F_2 molecules at equilibrium with F atoms at 1115 K if K_p for the dissociation was 7.55 x 10^{-2}. The total pressure was 0.831 atm.

    [R. N. Doescher, J. Chem. Phys., 20, (1952)]

11-58 Use the equilibrium constant calculated in Prob. 11-3 to calculate the concentration of N_2O_4 in equilibrium with 2.13 x 10^{-3}M NO_2 in chloroform.

    [J. T. Cundell, J. Chem. Soc., 67, 808 (1895)]

11-59 Determine the percent monomeric form of benzoic acid present in equilibrium with the dimeric form in CCl_4 solution. K_{diss} = 7 x 10^{-5} at 25°.

    [J. T. Harris, Jr., and M. E. Hobbs, J. Am. Chem. Soc., 76, 1419 (1954)]

11-60 Calculate the percent monomeric acetic acid in equilibrium with the dimeric form at 71.3°C. K_p for the dissociation is 16.4 mm.

    [M. D. Taylor, J. Am. Chem. Soc., 73, 315 (1951)]

11-61 Use the equilibrium constant calculated in Prob. 11-43 to calculate the mole percent of CO_2 present at equilibrium when the equilibrium mixture contained 22.85 mole% H_2 and 27.86 mole% each of CO and H_2O.

242   PROBLEMS IN CHEMISTRY

11-62 In another experiment K_c was found to be 2.01×10^{-2} for
the decomposition of hydrogen iodide. Determine the hydrogen and
iodine concentrations if the hydrogen iodide concentration was
1.00×10^{-2} mole/liter after equilibrium was established.

     [A. H. Taylor and R. H. Crist, J. Am. Chem. Soc., 63,
     1377 (1941)]

11-63 A solution was prepared by dissolving N_2O_4 in chloroform
until the concentration was 0.325 mole/liter. Calculate the NO_2
concentration of the solution after equilibrium was established
at 8.2°C. K_c = 1.07×10^{-5}.

     [J. T. Cundell, J. Chem. Soc., 67, 808 (1895)]

11-64 Find the equilibrium concentrations of the monomer and dimer
of chloroacetic acid in carbon tetrachloride solution if the
analytical concentration of $CH_2ClCOOH$ was 1.00 mole/liter. K_c for
the dissociation of the dimeric form is 6.2×10^{-4}.

     [J. T. Harris and M. E. Hobbs, J. Am. Chem. Soc., 76,
     1419 (1954)]

Chapter 12

CHEMICAL EQUILIBRIUM, II

In Chap. 10, problems connected with solutions of strong electrolytes were discussed. In this chapter, problems concerned with solutions of electrolytes which are slightly ionized are considered.

12-1 WEAK ACIDS AND BASES

Some acids which react with water to form hydronium ions establish an equilibrium which lies far to the left. Hydrocyanic acid is such an acid.

$$HCN + H_2O \rightleftharpoons H_3O^+ + CN^-$$

This equilibrium can be represented by

$$K = \frac{[H_3O^+][CN^-]}{[HCN][H_2O]}$$

In dilute solutions, the quantity of water present is practically constant and can be included in the constant.

$$K \times [H_2O] = \frac{[H_3O^+][CN^-]}{[HCN]}$$

Such an equilibrium constant may be called the ionization constant, K_{ion} or K_a, or the dissociation constant.

$$K_a = \frac{[H_3O^+][CN^-]}{[HCN]}$$

A similar equilibrium involves the base NH_3, and the constant is K_b.

$$NH_3 + H_2O \;\rightleftarrows\; NH_4^+ + OH^-$$

$$K_b = \frac{[NH_4^+][OH^-]}{[NH_3]}$$

Always write the balanced equation for the equilibrium and the dissociation-constant expression before attempting to calculate a numerical answer to a problem involving ionic equilibria. These constants are often expressed as pK_a and pK_b.

$$pK_a = -\log K_a, \text{ and } pK_b = -\log K_b.$$

12-2 CALCULATION OF DISSOCIATION CONSTANTS

Dissociation constants of slightly ionized acids and bases have been calculated from the results of conductance measurements, potentiometric titrations, and spectrometry.

Example 12-1

By conductance measurements, a 0.0500 M solution of acetic acid was observed to be 1.90% dissociated. Calculate K_a for acetic acid.

> [D. A. MacInnes and T. Shedlovsky, J. Am. Chem. Soc., 54, 1435 (1932)]

$$CH_3COOH + H_2O \;\rightleftarrows\; H_3O^+ + CH_3COO^-$$

$$K_a = \frac{[H_3O^+][CH_3COO^-]}{[CH_3COOH]}$$

$0.0500 \times 0.0190 = 0.000950$ mole of CH_3COOH ionized

$[CH_3COOH] = 0.0500 - 0.000950 = 0.0490$ mole CH_3COOH un-ionized

$[CH_3COO^-] = [H_3O^+] = 0.000950$ mole

since 1 mole of CH_3COO^- and 1 mole of H_3O^+ are formed for each mole of CH_3COOH which ionizes.

$$K_a = \frac{(9.50 \times 10^{-4})^2}{4.90 \times 10^{-2}} = \frac{9.02 \times 10^{-7}}{4.90 \times 10^{-2}} = 1.84 \times 10^{-5}$$

12-3 CALCULATIONS FROM DISSOCIATION CONSTANTS

Experimentally determined dissociation constants, Table 12-1, may be used to calculate the approximate concentrations of species in solution.

Example 12-2

Calculate the hydronium ion concentration in a 0.200 M acetic acid solution.

$$CH_3COOH + H_2O \rightleftarrows H_3O^+ + CH_3COO^-$$

$$K_{diss} = \frac{[H_3O^+][CH_3COO^-]}{[CH_3COOH]}$$

At equilibrium, let $X = [H_3O^+]$; then

$X = [CH_3COO^-]$

(1 mole of CH_3COO^- is formed for each mole of H_3O^+.)

$0.200 - X = [CH_3COOH]$

(1 mole of CH_3COOH is consumed for each mole of H_3O^+ formed.)

$$\frac{X^2}{0.200 - X} = 1.75 \times 10^{-5}$$

$$X^2 = 3.50 \times 10^{-6} - 1.75 \times 10^{-5} X$$

$$X^2 + 1.75 \times 10^{-5} X - 3.50 \times 10^{-6} = 0$$

$$X = \frac{-1.75 \times 10^{-5} \pm \sqrt{3.06 \times 10^{-10} + 14.0 \times 10^{-6}}}{2}$$

$$X = \frac{-1.75 \times 10^{-5} \pm \sqrt{3.74 \times 10^{-3}}}{2}$$

TABLE   12-1   Ionization Constants of Weak Electrolytes

	Formula	K_a	pK_a
Acid			
Acetic acid	CH_3COOH	1.75×10^{-5}	4.76
Benzoic acid	C_6H_5COOH	6.30×10^{-5}	4.20
Butyric acid	C_3H_7OOH	1.52×10^{-5}	4.82
Carbonic acid	H_2CO_3	(1) 4.5×10^{-7}	6.35
		(2) 6×10^{-11}	10.22
Formic acid	HCOOH	1.76×10^{-4}	3.75
Hydrocyanic acid	HCN	7.2×10^{-10}	9.14
Nitrous acid	HNO_2	4.5×10^{-4}	3.35
Phosphoric acid	H_3PO_4	(1) 7.5×10^{-3}	2.12
		(2) 2×10^{-7}	6.70
		(3) 1×10^{-12}	12.0
Propionic acid	C_2H_5COOH	1.34×10^{-5}	4.87
Malonic acid	$CH_2(COOH)_2$	(1) 1.12×10^{-3}	2.95
		(2) 3.89×10^{-6}	5.41
o-Phthalic acid	$C_6H_4(COOH)_2$	(1) 1.49×10^{-3}	2.83
		(2) 2.03×10^{-6}	5.69
Salicylic acid	$C_6H_4(OH)(COOH)$	(1) 1.0×10^{-3}	3.00
		(2) 3.98×10^{-15}	14.40
		K_b	pK_b
Base			
Ammonia	NH_3	1.75×10^{-5}	4.76
Pyridine	C_5H_5N	5.62×10^{-6}	5.25
Trimethylamine	$(CH_3)_3N$	4.4×10^{-4}	3.36

$X = 1.86 \times 10^{-3}$ mole/liter

The quantity of H_3O^+ ions is very small compared to the quantity of undissolved acid CH_3COOH present. A simplified, approximate calculation may be made.

$$\frac{X^2}{0.20 - X} = 1.75 \times 10^{-5}$$

Assume that X is very small compared to 0.20 and neglect it in the denominator.

$$\frac{X^2}{0.20} = 1.75 \times 10^{-5}$$

$$X^2 = 3.50 \times 10^{-6}$$

$$X = 1.87 \times 10^{-3} \text{ mole/liter}$$

The result obtained by this shortcut method is usually sufficiently accurate, especially when the value of K_a is less than 10^{-3} and the analytical concentration of the acid is 0.100 or less.

Example 12-3

Calculate the percent dissociation of a 0.10 M HCN solution.

$$HCN + H_2O \rightleftarrows H_3O^+ + CN^-$$

$$K_{diss} = \frac{[H_3O^+][CN^-]}{[HCN]}$$

let $X = [H_3O^+] = [CN^-]$

then $0.10 - X = [HCN]$

$$\frac{X^2}{0.10 - X} = 7.2 \times 10^{-10}$$

Assume that X is very small compared to 0.10 and neglect it in the denominator.

$$X^2 = 0.10 \times 7.2 \times 10^{-10} = 7.2 \times 10^{-11}$$

$$X = 8.4 \times 10^{-6} \text{ mole/liter}$$

$$\text{Percent diss} = \frac{\text{moles HCN dissociated}}{\text{total moles HCN}} \times 100$$

$$\text{Percent diss} = \frac{8.4 \times 10^{-6}}{0.10} \times 100 = 8.4 \times 10^{-3}\%, \text{ or } 0.0084\%$$

Example 12-4

What is the pH of a 0.200 M formic acid solution?

$$HCOOH + H_2O \rightleftarrows H_3O^+ + HCOO^-$$

$$K = \frac{[H_3O^+][HCOO^-]}{[HCOOH]} = 1.76 \times 10^{-4}$$

$$\text{let } X = [H_3O^+] = [HCOO^-]$$
$$\text{then } 0.200 - X = [HCOOH]$$

$$\frac{X^2}{0.200 - X} = 1.76 \times 10^{-4}$$

Assume X is very small compared to 0.2 and neglect it in the denominator.

$$X^2 = 3.52 \times 10^{-5}$$
$$X = 5.93 \times 10^{-3} \text{ mole/liter}$$
$$pH = -\log[H_3O^+]$$
$$= -\log 5.93 \times 10^{-3} = -(-3 + 0.773)$$
$$= 2.23$$

Example 12-5

What is the approximate OH^- concentration of a 0.150 M NH_3 solution?

$$NH_3 + H_2O \rightleftarrows NH_4^+ + OH^-$$

$$K = \frac{[NH_4^+][OH^-]}{[NH_3]}$$

$$\text{let } X = [OH^-] = [NH_4^+]$$
$$\text{then } 0.150 - X = [NH_3]$$

$$\frac{X^2}{0.150 - X} = 1.75 \times 10^{-5}$$

Neglecting X in the denominator

$$x^2 = 0.150 \times 1.75 \times 10^{-5} = 2.62 \times 10^{-6}$$
$$X = 1.62 \times 10^{-3} \text{ mole/liter}$$

12-4   THE ION PRODUCT OF WATER

Water is the most common self-ionizing solvent

$$2H_2O \rightleftharpoons H_3O^+ + OH^-$$

The equilibrium concentrations of the ions are related by the ion product of water K_w.

$$K_w = [H_3O^+][OH^-] = 1.0 \times 10^{-14}$$

In pure water, $[H_3O^+] = [OH^-]$

Example 12-6

Calculate the hydronium ion concentration and hydroxide ion concentration in pure water at 25°C.

$$2H_2O \rightleftharpoons H_3O^+ + OH^-$$
$$K_w = [H_3O^+][OH^-] = 1.0 \times 10^{-14}$$
$$\text{let } X = [H_3O^+] = [OH^-]$$
$$X^2 = 1.0 \times 10^{-14}$$
$$X = 1.0 \times 10^{-7} \text{ mole/liter}$$

12-5 pK_w

In Chap. 10 the pH notation for the concentrations of H_3O^+ was described.

$$pH = - \log[H_3O^+]$$

and similarly

$$pOH = - \log[OH^-]$$

Using the same notation, we may define pK_w.

$$pK_w = - \log K_w$$

since $K_w = [H_3O^+][OH^-] = 1 \times 10^{-14}$

$$\log K_w = \log[H_3O^+] + \log[OH^-]$$

$$- \log K_w = - \log[H_3O^+] + (- \log[OH^-])$$

or

$$pK_w = pH + pOH = 14.0$$

Example 12-7

What is the pOH of a solution of pH 6.0?

$$pH + pOH = 14.0$$

$$pOH = 14.0 - 6.0 = 8.0$$

12-6 THE COMMON-ION EFFECT

The addition of a salt containing an ion, which is also present in a solution of a slightly ionized acid or base, will repress the dissociation of the acid or base. This is expected from Le Chatelier's principle. The addition of sodium acetate to a solution of acetic acid will increase the concentration of undisso-ciated acid and decrease the concentration of hydronium ions.

$$CH_3COOH + H_2O \rightleftarrows H_3O^+ + CH_3COO^-$$

Since sodium acetate is completely ionized, the addition of this salt increases the CH_3COO^- ion concentration and shifts the equilib-rium to the left. From the point of view of the dissociation-constant expression

$$K_{diss} = \frac{[H_3O^+][CH_3COO^-]}{[CH_3COOH]}$$

an increase in $[CH_3COO^-]$ requires that $[CH_3COOH]$ increase and $[H_3O^+]$ decrease if K_{diss} is to remain constant.

$[NH_3]$ = 0.200 - X

(analytical concentration of NH_3 less that ionized)

$$K = \frac{(0.10 + X) \times X}{0.20 - X} = 1.75 \times 10^{-5}$$

Since X is very small compared to 0.1 and 0.2, it may be neglected.

$$\frac{0.10X}{0.20} = 1.75 \times 10^{-5}$$

$$X = 3.52 \times 10^{-5} \text{ mole/liter}$$

12-7 BUFFER SOLUTIONS

One of the practical applications of the common-ion effect is the buffer solution, which resists changes in pH upon the addition of acid or base. A buffer solution contains a weak acid and one of its salts (common anion) or a weak base and one of its salts (common cation), or using Brønsted acid base nomenclature, a buffer solution contains a weak acid and its conjugate base, or a weak base and its conjugate acid.

A solution of acetic acid and sodium acetate will resist change in hydronium ion concentration (pH) when a small quantity of strong acid is added. The hydronium ions added will combine with the acetate ions present to form un-ionized acetic acid molecules. When a strong base is added, the hydroxide ions will react with un-ionized acetic acid.

Example 12-10

Calculate the pH of a buffer solution containing 0.100 mole/liter of sodium acetate in 0.100 M acetic acid.

$$CH_3COOH + H_2O \rightleftarrows H_3O^+ + CH_3COO^-$$

$$CH_3COONa \rightarrow CH_3COO^- + Na^+$$

$$K = \frac{[H_3O^+][CH_3COO^-]}{[CH_3COOH]}$$

let $X = [H_3O^+]$

then $0.100 + X = [CH_3COO^-]$

and

$0.100 - X = [CH_3COOH]$

$$K = \frac{X \times (0.100 + X)}{0.100 - X} = 1.75 \times 10^{-5}$$

Since X is negligible compared to 0.100

$$K = \frac{0.100X}{0.100} = 1.75 \times 10^{-5} = X$$

$$
\begin{aligned}
pH &= -\log[H_3O^+] \\
&= -\log 1.75 \times 10^{-5} = -(-5 + 0.2430) \\
&= 4.76
\end{aligned}
$$

12-8 THE HENDERSON-HASSELBALCH EQUATION

A limiting law that is useful in making buffer calculations is the Henderson-Hasselbalch equation. It is used quite often by biochemists. This equation is given as

$$pH = pK_a + \log \frac{[salt]}{[acid]}$$

for an acid buffer, or in general as

$$pH = pK_a + \log \frac{[conjugate\ base]}{[conjugate\ acid]}$$

Example 12-11

Calculate the pH of a buffer solution containing 0.100 mole/liter of sodium acetate in 0.100 M acetic acid. (Compare this to Example 12-10.)

[conjugate base] = 0.1 M acetate ion (from 0.1 M sodium acetate)

[conjugate acid] = 0.1 M acetic acid

$$pH = 4.76 + \log \frac{[0.1]}{[0.1]}$$

$$pH = 4.76$$

Example 12-12

What is the pH of a 0.1 M ammonium hydroxide solution containing 1 mole/liter of ammonium chloride?

[conjugate base] = 0.1 M NH_3 in water

[conjugate acid] = 1.0 M NH_4^+

pK_b = 4.76

For a basic buffer, the pK_a taken is that of the conjugate acid of the base. This is given by

$$pK_a + pK_b = 14.0$$

so

$$pK_a(NH_4^+) = 9.24$$

$$pH = 9.24 + \log \frac{[0.1]}{[1.0]}$$

$$pH = 9.24 - 1.00$$

$$pH = 8.24$$

12-9 POLYPROTIC ACIDS

Polyprotic acids ionize in two or more steps. Each ionization step reaches an equilibrium.

$$H_2S + H_2O \rightleftarrows H_3O^+ + HS^-$$
$$HS^- + H_2O \rightleftarrows H_3O^+ + S^{2-}$$

Each ionization step is represented by an equilibrium constant.

$$K_1 = \frac{[H_3O^+][HS^-]}{[H_2S]} = 9.1 \times 10^{-8}$$

$$K_2 = \frac{[H_3O^+][S^{2-}]}{[HS^-]} = 1.2 \times 10^{-15}$$

Hydrogen sulfide is chosen as the example of a polyprotic acid because of its importance in the development of wet chemical analyses. A saturated solution of H_2S is approximately 0.10 M. The $[H_3O^+]$ and $[S^{2-}]$ of this solution are of interest.

Example   12-13

Calculate the $[S^{2-}]$ of 0.10 M H_2S solution.

$$H_2S + H_2O \rightleftarrows H_3O^+ + HS^-$$

$$K_1 = \frac{[H_3O^+][HS^-]}{[H_2S]} = 9.1 \times 10^{-8}$$

$$HS^- + H_2O \rightleftarrows H_3O^+ + S^{2-}$$

$$K_2 = \frac{[H_3O^+][S^{2-}]}{[HS^-]} = 1.2 \times 10^{-15}$$

Because $K_1 >> K_2$ practically all of the H_3O^+ comes from the dissociation of H_2S and $[H_3O^+] \approx [HS^-]$ ($\approx$ means approximately equal to.)

Let X = $[H_3O^+]$

then X = $[HS^-]$

$$K_2 = \frac{X[S^{2-}]}{X} = 1.2 \times 10^{-15}$$

Sometimes the overall ionization constant is useful in a calculation.

$$H_2S + 2H_2O \rightleftarrows 2H_3O^+ + S^{2-}$$

$$K_{12} = \frac{[H_3O^+]^2 [S^{2-}]}{[H_2S]} = K_1 \times K_2$$

$$K_{12} = 9.1 \times 10^{-3} \times 1.2 \times 10^{-15} = 1.09 \times 10^{-22}$$

Example 12-14

What is the sulfide ion concentration in a saturated H_2S solution, if the pH is adjusted to 1.0?

pH $= 1.0$

$[H_3O^+] = 1 \times 10^{-1}$

$[H_2S] = 0.1$ M

$$K_{12} = 1.09 \times 10^{-22} = \frac{[H_3O^+]^2[S^{2-}]}{[H_2S]}$$

$$1.09 \times 10^{-22} = \frac{[10^{-1}]^2[S^{2-}]}{[0.1]}$$

$[S^{2-}] = 1.09 \times 10^{-21}$ moles/liter

12-10 ADDITIONAL TYPES OF CALCULATIONS INVOLVING DISSOCIATION CONSTANTS

Example 12-15

Calculate the pH of a solution prepared by mixing 40.0 ml of 0.100 M acetic acid with 60.0 ml of 1.00 M sodium acetate.

$CH_3COOH + H_2O \rightleftarrows H_3O^+ + CH_3COO^-$

$CH_3COONa \rightarrow CH_3COO^- + Na^+$

$$\frac{[H_3O^+][CH_3COO^-]}{[CH_3COOH]} = 1.75 \times 10^{-5}$$

M x V = millimoles solute (Sec. 7-4)

0.100 x 40.0 = 4.00 mM acetic acid.

$$\frac{mM}{ml} \times ml = mM$$

1.00 x 60.0 = 60 mM CH_3COONa

The concentration of acetic acid is 4.00 mM in 100 ml of solution, which is equivalent to 40.0 mM per liter.

$$\frac{40.0}{1000} = 0.0400 \text{ mole/liter}$$

$$\frac{mM}{mM/mole} = mM \times \frac{mole}{mM} = mole$$

The concentration of sodium acetate is 60.0 mM in 100 ml of solution, which is equivalent to 600 mM/liter.

$$\frac{600}{1000} = 0.600 \text{ mole/liter}$$

Let $X = [H_3O^+]$

$[CH_3COO^-] = 0.600 + X$

$[CH_3COOH] = 0.400 - X$

X is negligible compared to 0.600

$$\frac{X \times 0.600}{0.0400} = 1.75 \times 10^{-5}$$

$$X = \frac{(1.75 \times 10^{-5}) \times (4.00 \times 10^{-2})}{6.00 \times 10^{-1}} = 1.17 \times 10^{-6} \text{ mole/liter}$$

$$pH = -\log(1.17 \times 10^{-6}) = 5.93$$

Example 12-16

How many grams of sodium cyanide must be added to 200 ml of a 0.100 M solution of HCN to give a solution whose pH is 6.22?

$$HCN + H_2O \rightleftarrows H_3O^+ + CN^-$$

$$NaCN \rightarrow Na^+ + CN^-$$

$$\frac{[H_3O^+][CN^-]}{[HCN]} = 7.2 \times 10^{-10}$$

First calculate mass of NaCN on the basis of 1 liter of solution.

$$pH = 6.22 = -\log[H_3O^+]$$

$$[H_3O^+] = 6.03 \times 10^{-7} \text{ mole/liter}$$

Let $[CN^-] = X$

$$\frac{(6.03 \times 10^{-7})X}{0.100} = 7.2 \times 10^{-10}$$

$$X = \frac{7.2 \times 10^{-11}}{6.0 \times 10^{-7}}$$

$$= 1.2 \times 10^{-4} \text{ mole/liter}$$

$$1.2 \times 10^{-4} \text{ mole/liter} \times \frac{200}{1000} = 2.4 \times 10^{-5} \text{ mole/200 ml}$$

CH_3COONa 2 x 12.0   =   24.0

                3 x 1.0   =   3.0

                2 x 16.0   =   32.0

                1 x 23.0   =   23.0

                           82.0 g/mole

$2.4 \times 10^{-5} \times 82.0 = 200 \times 10^{-5} = 0.0020$ g

PROBLEMS

12-1   Calculate the value of the dissociation constant of 0.0200 M acetic acid which was observed to be 2.99% dissociated.

     [D. A. MacInnes and T. Shedlovsky, J. Am. Chem. Soc., 54, 1435 (1932)]

12-2   A 0.01173 M solution of n-butyric acid was found to be 3.53% ionized by conductance measurements. Calculate the ionization constant of the acid, C_3H_7COOH.

     [J. F. P. Dippy, J. Chem. Soc., 1938, 1226]

12-3   What is the value of K_a for benzoic acid, C_6H_5COOH, based on the observation that a 0.00823 M solution was 8.65% ionized?

     [D. J. G. Ives, J. Chem. Soc., 1933, 734]

12-4   Calculate the hydronium concentration of 0.025 M acetic acid.

12-5   Calculate the pOH of a 0.200 M solution of ammonia.

12-6   Calculate the concentration of a benzoic acid solution whose pH is 2.12.

12-7   Determine the hydronium ion concentration of a 0.0500 M formic acid solution.

12-8   A. Klemenc and E. Hayek determined the dissociation constants
of nitrous acid at 0, 12.5, and 30°C as 3.2, 4.6, and 6.0 x 10^{-4},
respectively.   Find the hydronium concentrations of a 0.100 M HNO_2
solution at each of the three temperatures.
        [Monatsh. Chem., 54, 407 (1929)]

12-9   What is the cyanide ion concentration of a 0.0120 M HCN
solution?

12-10   Calculate the pOH of a 0.0040 M butyric acid solution.

12-11   Calculate the pH of a 1.02 x 10^{-3} M cyanoacetic acid solution.
K = 3.34 x 10^{-3}.
        [B. Saxton and L. S. Dacken, J. Am. Chem. Soc., 62,
        848 (1962)]

12-12   Find the dissociation constant of an acid if a 0.100 M
solution of the acid has a pH of 3.41.

12-13   Calculate the dissociation constant of an acid if a 0.125 M
solution has a pH of 4.671.

12-14   Determine the pH of a solution of formic acid which contains
10.02 g/liter.

12-15   What is the pH of a 2.00% solution of acetic acid?   The
specific gravity of the solution is 1.0012.

12-16   What is the pH of a 1.00% solution of ammonia, sp. gr. 0.9939?

12-17   Assuming all the [H^+] comes from the first ionization of
H_3PO_4, what is the pH of a 0.1 M H_3PO_4 solution?

12-18 Calculate the hydronium concentration of a 0.120 M acetic acid solution which also contains 0.200 mole of sodium acetate per liter.

12-19 What is the pH of a 0.225 M formic acid solution which also contains 0.300 mole of potassium formate per liter?

12-20 How many grams of sodium acetate must be added to 200 ml of 0.100 M acetic acid to give a solution of pH 4.20?

12-21 Under normal conditions the pH of blood is 7.4. Assuming blood is buffered by the carbonic acid bicarbonate equilibrium, what is the ratio of bicarbonate ion to carbonic acid in blood? (Assume all the CO_2 is present as H_2CO_3.)

12-22 What is the pH of a buffer solution containing 0.750 M acetic acid and 0.45 M sodium acetate?

12-23 How many moles of NaH_2PO_4 would you use to make 100 ml of a buffer of pH 3.0 containing 0.1 M H_3PO_4?

12-24 Use the Henderson-Hasselbalch equation to calculate the ratio of the number of moles of salt to acid necessary to make each of the following buffers:
   a.   sodium acetate, acetic acid, pH = 3.5
   b.   sodium formate, formic acid, pH = 5.0
   c.   sodium dihydrogen phosphate, disodium hydrogen phosphate, pH = 7.0

12-25 A 0.433 M solution of an acid HX also contained 0.623 mole/liter of the salt KX. Find the dissociation constant of the acid if the pH of the solution was 4.882.

12-26 A student in a Freshman lab was given 0.01 mole of a weak organic acid and told to determine the K_a of the acid. He prepared 100 ml of an aqueous solution containing the sample. Fifty ml of this solution was then titrated with NaOH to the equivalence point. The titrated solution was then mixed with the other 50 ml of solution and the pH determined. A value of 4.80 was obtained for the pH of the mixed solutions. What is the approximate K_a value of the acid?

12-27 Determine the $[S^{2-}]$ of a 0.100 M H_2S solution which is also 0.300 M in HCl.

12-28 Calculate the sulfide ion concentration of a 0.10 M H_2S solution which is also 1.00 M in acetic acid.

12-29 Determine the pH of a solution prepared by mixing 30 ml of 0.015 M propionic acid with 20 ml of 0.10 M sodium propionate.

12-30 Calculate the pH of a solution prepared by mixing 150 ml of 0.020 M aqueous ammonia with 80.0 ml of 0.50 M ammonium chloride solution.

12-31 Calculate the pH of a solution made by mixing 500 ml of 0.100 M acetic acid with 500 ml of 0.200 M sodium hydroxide.

12-32 What is the pH of 750 ml of 0.0200 M acetic acid solution to which 250 ml of 0.50 M calcium acetate solution has been added?

SUPPLEMENTARY PROBLEMS

12-33 J. F. P. Dippy and F. R. Williams found that a 0.002907 M solution of benzoic acid was 13.8% dissociated. Calculate pK_a for the acid.

[J. Chem. Soc., 1934 1888]

<u>12-34</u> In an early determination of the ionization constant of acetic acid, J. Kendall found that a 0.03685 M solution was 2.22% ionized. Calculate K_a and pK_a for acetic acid from this measurement.
[J. Chem. Soc., <u>101</u>, 1279 (1912)]

<u>12-35</u> What is the pH of a 0.0250 M solution of propionic acid?

<u>12-36</u> Calculate the pOH of a 0.0110 M benzoic acid solution.

<u>12-37</u> Calculate the pH of a 0.0300 M benzoic acid solution.

<u>12-38</u> What is the pH of a 0.030 M ammonia solution?

<u>12-39</u> What fraction of propionic acid is dissociated in the following solutions?
    a. 0.100 M
    b. 0.0100 M
    c. 0.00100 M

<u>12-40</u> Calculate the pH of a solution prepared by diluting 100 ml of 0.216 M formic acid solution to 2.00 liters.

<u>12-41</u> What is the hydrogen ion concentration of a 0.043 M H_2S solution?

<u>12-42</u> Calculate the pH of a 2.8×10^{-2} M H_2S solution.

<u>12-43</u> What is the pH of 2×10^{-2} M o-phthalic acid solution?

<u>12-44</u> Calculate the pH of a 0.100 M malonic acid solution.

<u>12-45</u> The solubility of salicylic acid in water is 0.16 g/100 g. Calculate the approximate pH of a saturated solution of the acid.

12-46 How many grams of trimethylamine must be present in 250 ml of an aqueous solution to give a pH of 11.91?

12-47 What is the pOH of a 0.0212 M solution of pyridine?

12-48 What concentration of nitrous acid is required to give a solution whose pH is 2.23?

12-49 What is the pH of a 0.208 M phosphoric acid solution?

12-50 Calculate the pH of a 0.125 M hydrogen sulfide solution.

12-51 What is the pH of a 0.309 M sulfurous acid solution? K_1 = 1.54 x 10^{-2}.

12-52 Find the pH of 0.100 M HCN solution.

12-53 What is the pH of a 1.00% solution of formic acid? The density of the solution is 1.0019 g/ml.

12-54 Calculate the pH of a solution of acetic acid which contains 20.02 g/liter.

12-55 What is the hydroxide ion concentration of a 0.125 M ammonia solution?

12-56 What is the concentration of a propionic acid solution whose hydronium concentration is 1.2 x 10^{-3} M?

12-57 What is the ammonium ion concentration of a 0.0200 M solution of ammonia which is also 0.100 M in potassium hydroxide?

12-58 What is the malonate ion concentration of a 0.100 M malonic acid solution which also contains 0.100 M HCl?

12-59 What mass of potassium nitrite must be included in the
solution in Prob. 12-48 to raise the pH from 2.23 to 3.16?

Chapter 13

CHEMICAL EQUILIBRIUM, III

13-1 THE SOLUBILITY PRODUCT

The equilibrium between a solid ionic salt and its saturated aqueous solution is described by the solubility-product expression. This relationship was first observed by Nernst in 1899. The usual example of this principle is the equilibrium between solid silver chloride and its saturated solution.

$$AgCl(s) \rightleftarrows Ag^+ + Cl^-$$

The solubility-product expression is

$$K_s = [Ag^+][Cl^-]$$

where K_s is the solubility-product constant of silver chloride. It has a value of 1.8×10^{-10}.

The solubility-product constant may also be recorded as pK_s, where p serves the same function as in pH.

$$pK_s = -\log K_s$$

For silver chloride, $pK_s = 9.75$.

Table 13-1 is a compilation of K_s and pK_s values obtained by a critical examination of experimental data. The result of a single experiment in the examples and problems here may differ from the values in Table 13-1.

265

TABLE 13-1   Solubility Products

Substance	pK_s	K_s
AgBr	12.28	5.2×10^{-13}
$AgBrO_3$	4.28	5.2×10^{-5}
AgCN	15.92	1.2×10^{-16}
AgCl	9.75	1.8×10^{-10}
Ag_2CrO_4	11.95	1.1×10^{-12}
AgI	16.08	8.3×10^{-17}
$AgIO_3$	7.52	3.0×10^{-8}
AgSCN	12.00	1.0×10^{-12}
Ag_2SeO_4	7.25	5.6×10^{-8}
Ag_2SO_4	4.80	1.6×10^{-5}
$Ba(BrO_3)_2$	5.50	3.2×10^{-6}
$BaCrO_4$	9.93	1.2×10^{-10}
BaF_2	5.98	1.0×10^{-6}
$BaSO_4$	9.87	1.3×10^{-10}
CaF_2	10.31	4.9×10^{-11}
$CaSO_4$	5.92	1.2×10^{-6}
$Cd(IO_3)_2$	9.50	3.2×10^{-10}
CuI	11.96	1.1×10^{-12}
$Cu(IO_3)_2$	7.13	7.4×10^{-8}
Hg_2Cl_2	17.88	1.3×10^{-18}
$KClO_4$	1.97	1.1×10^{-2}
KIO_3	1.70	5.0×10^{-2}
K_2PtCl_6	4.96	1.1×10^{-5}
$La(IO_3)_2$	11.21	6.2×10^{-12}
MgF_2	8.19	6.5×10^{-9}
$PbBr_2$	4.41	3.9×10^{-5}
$PbCl_2$	4.79	1.6×10^{-5}
$PbCrO_4$	13.75	1.8×10^{-14}
$Pb(IO_3)_2$	12.49	3.2×10^{-13}
$PbSO_4$	7.79	1.6×10^{-8}
$PbSeO_4$	6.84	1.4×10^{-7}

TABLE   13-1   Solubility Products (con't.)

Substance	pK_s	K_s
$SrCrO_4$	4.44	3.6×10^{-5}
$Sr(IO_3)_2$	6.48	3.3×10^{-7}
$SrSO_4$	6.49	3.2×10^{-7}
$TlBr$	5.47	3.4×10^{-6}
$TlBrO_3$	4.07	8.5×10^{-5}
$TlCl$	3.76	1.7×10^{-4}
Tl_2CrO_4	12.01	9.8×10^{-13}
TlI	7.19	6.5×10^{-8}
$TlIO_3$	5.51	3.1×10^{-6}

Source: L. Meites (ed), Handbook of Analytical Chemistry,
pp. 1-13, McGraw-Hill, New York, 1963

13-2 CALCULATION OF K_s FROM SOLUBILITY

The solubility of a slightly soluble salt in water is its
analytical concentration. If we assume that the dissolved salt is
completely ionized, the value of K_s can be calculated from the
solubility.

Example 13-1
The solubility of silver bromide in water was determined by C. Bedell
as 1.86×10^{-6} mole/liter at 20°. Calculate the corresponding
value of K_s.
        [Compt. Rend., 207, 632-634 (1938)]

$AgBr(s) \rightleftharpoons Ag^+ + Br^-$

$K_s = [Ag^+][Br^-]$

$[Ag^+] = [Br^-] = C = 1.86 \times 10^{-6}$

where C is the analytical concentration of AgBr.

$[Ag^+] = [Br^-] = C$, because one Ag^+ and one Br^- are formed for each AgBr which dissolves.

$K_s = (1.86 \times 10^{-6})^2 = 3.46 \times 10^{-12}$

The numerical value of 1.86^2 can be found on a slide rule by placing the slide over 1.86 on the C scale. The square of 1.86 will be under the slide on the A scale.

The value of 1.86^2 can also be found by the relationship $\log 1.86^2 = 2 \log 1.86$.

$\log 1.86 = 0.2695$

$$\frac{\ \ 2}{}$$

$\log 1.86^2 = 0.5390$

$1.86^2 = 3.46$

The value of $(10^{-6})^2 = 10^{-6} \times 10^{-6}$, or $10^{-6 \times 2} = 10^{-12}$

Example 13-2

What is the value of the K_s of $Cu(IO_3)_2$ at 25°? The solubility of $Cu(IO_3)_2$ in water was found to be 3.245×10^{-3} mole/liter.

[R. M. Keefer, J. Am. Chem. Soc., 70, 476-479 (1948)]

$Cu(IO_3)_2(s) \rightleftarrows Cu^{2+} + 2IO_3^-$

$K_s = [Cu^{2+}][IO_3^-]^2$

$[Cu^{2+}] = C = 3.245 \times 10^{-3}$ mole/liter

One $Cu(IO_3)_2$ forms one Cu^{2+}ion.

$[IO_3^-] = 2C = 6.490 \times 10^{-3}$

One $Cu(IO_3)_2$ forms two IO_3^-ions.

$K_s = (3.245 \times 10^{-3}) \times (6.490 \times 10^{-3})^2$

$ = 136.6 \times 10^{-9} = 1.37 \times 10^{-7}$

Using a slide rule, place the slide over 6.49 on the C scale and find 6.49^2 on the A scale. Then multiply 6.49^2 by 3.24.

$10^{-3} \times (10^{-3})^2 = 10^{-9}$

Using log tables, $\log 6.490^2 = 2 \log 6.490$

$$
\begin{array}{r}
0.8122 \\
2 \\
\hline
1.6244 \\
0.5112 \\
\hline
2.1356
\end{array}
$$

$10^{-3} \times (10^{-3})^2 = 10^{-9}$

It is very important to realize that the K_s expression describes an experimental observation. When $[A]^n$ appears in the expression, the total concentration of A, regardless of its source, must be raised to the power n. In Example 13-2 the total concentration of IO_3^- [IO_3^-] was raised to the power of 2 because the K_s expression demanded it. The total concentration of IO_3^- was equal to 2C because the only source of IO_3^- ions was $Cu(IO_3)_2$ and one $Cu(IO_3)_2$ formed two IO_3^- ions.

Example 13-3

The average of several determinations of the solubility of silver chloride in water at 25°C was 0.00194 g/liter. Calculate the pK_s of AgCl by using this value.

> [A. Seidell, Solubilities of Inorganic and Metal Organic
> Compounds, 3d ed., p. 32, Van Nostrand, Princeton,
> New Jersey, 1940]

$$AgCl(s) \rightleftarrows Ag^+ + Cl^-$$
$$K_s = [Ag^+][Cl^-]$$
$$[Ag^+] = [Cl^-] = C$$

AgCl   107.870
         35.453
        ‾‾‾‾‾‾‾
        143.323 g/mole

$$C = \frac{1.94 \times 10^{-3}}{1.43 \times 10^{+2}} = 1.36 \times 10^{-5} \text{ mole/liter}$$

$$\frac{g/liter}{g/mole} = \frac{g}{liter} \times \frac{mole}{g} = \text{mole/liter}$$

$$K_s = (1.36 \times 10^{-5})^2 = 1.85 \times 10^{-10}$$
$$pK_s = -\log K_s = -\log(1.85 \times 10^{-10})$$
$$= -[0.27 + (-10.00)] = -(-9.73)$$
$$= 9.73$$

13-3 CALCULATION OF SOLUBILITIES FROM K_s

Example 13-4

Calculate the solubility of thallous chloride in water at 20°C from its K_s, 1.7×10^{-4}. The experimental value obtained by Viktorin and Sirucek was 0.01376 mole/liter.

[Collection Czech. Chem. Comm., 11, 474-493 (1939)]

$TlCl(s) \rightleftarrows Tl^+ + Cl^-$

$K_s = [Tl^+][Cl^-] = 1.7 \times 10^{-4}$

Solubility of TlCl = C

$[Tl^+] = [Cl^-] = C$

$c^2 = 1.7 \times 10^{-4}$

$C = \sqrt{1.7 \times 10^{-4}} = 1.3 \times 10^{-2}$ mole/liter

To find $\sqrt{1.7}$ on the slide rule place the slide over 1.7 on the left side of the A scale. The value of $\sqrt{1.7}$ will be under the slide on the C scale. If you prefer to use logarithms,

$$\log \sqrt{1.7} = \frac{\log 1.7}{2} = \frac{0.2304}{2} = 0.1152$$

$$\sqrt{10^{-4}} = 10^{-4/2} = 10^{-2}$$

$$\sqrt{1.7 \times 10^{-4}} = 1.3 \times 10^{-2} \text{ mole/liter}$$

Example 13-5

Compare the aqueous solubility of lead iodate found by R. M. Keefer and H. G. Reiber, 3.61×10^{-5} mole/liter, with the value calculated from K_s, 3.2×10^{-13}.

[J. Am. Chem. Soc., 63, 689-692 (1941)]

$Pb(IO_3)_2 \rightleftarrows Pb^{2+} + 2IO_3^-$

$K_s = [Pb^{2+}][IO_3^-]^2 = 3.2 \times 10^{-13}$

Solubility of $Pb(IO_3)_2$ = C

$[Pb^{2+}] = C$ [One $Pb(IO_3)_2$ gives <u>one</u> Pb^{2+}.]

$[IO_3^-] = 2C$ [One $Pb(IO_3)_2$ gives <u>two</u> IO_3^-.]

$C \times (2C)^2 = 3.2 \times 10^{-13}$

$4C^3 = 3.2 \times 10^{-13}$

$C^3 = 0.80 \times 10^{-13} = 8.0 \times 10^{-14}$

$C = \sqrt[3]{8.0 \times 10^{-14}} = \sqrt[3]{80 \times 10^{-15}}$

$$= \sqrt[3]{80} \times 10^{-5}$$

$$\log \sqrt[3]{80} = \frac{\log 80}{3} = \frac{1.9041}{3} = 0.6344$$

$$\sqrt[3]{80} = 4.3$$

$$C = 4.3 \times 10^{-5} \text{ mole/liter}$$

Example 13-6

Calculate the solubility of silver iodide in water from the K_s.

$$AgI(s) \rightleftarrows Ag^+ + I^-$$

$$K_s = [Ag^+][I^-] = 8.3 \times 10^{-17}$$

Let C = solubility of AgI; then

$$C = [Ag^+] = [I^-] \text{ (one AgI forms one } Ag^+ \text{ and one } I^-.)$$

$$C^2 = 8.3 \times 10^{-17} = 83 \times 10^{-18}$$

$$C = 9.1 \times 10^{-9} \text{ mole/liter}$$

Example 13-7

Calculate the concentration of barium in a saturated solution of barium iodate at 25°C. The pK_s of barium iodate is 8.82.

[G. Macdougall and C. W. Davies, J. Chem. Soc., 1935, 1416]

$$Ba(IO_3)_2 \rightleftarrows Ba^{2+} + 2IO_3^-$$

$$K_s = [Ba^{2+}][IO_3^-]^2$$

$$- \log K_s = 8.82$$

$$\log K_s = -8.82 = -9. + .18$$

$$K_s = 1.51 \times 10^{-9}$$

Let X = $[Ba^{2+}]$; then

$$2X = [IO_3^-]$$

$$(X)(2X)^2 = 1.51 \times 10^{-9}$$

$$4X^3 = 1.51 \times 10^{-9}$$

$$X^3 = 0.38 \times 10^{-9}$$

$$X = \sqrt[3]{.38 \times 10^{-3}}$$

$$X = 7.24 \times 10^{-4} \text{ mole/liter}$$

13-4 COMMON-ION EFFECT APPLIED TO SOLUBILITY

The common-ion effect applies to equilibria involving slightly soluble salts. The addition of a soluble salt with an ion in common with the slightly soluble salt diminishes the concentration of the other ion. The addition of potassium sulfate to a saturated solution of strontium sulfate will diminish the concentration of strontium ions.

$$SrSO_4(s) \; \rightleftharpoons \; Sr^{2+} + SO_4^{2-}$$

$$K_s \; = \; [Sr^{2+}][SO_4^{2-}] \; = \; 3.2 \times 10^{-7}$$

Example 13-8

Calculate the strontium ion concentration in a 0.200 M K_2SO_4 solution. Let $X \; = \; [Sr^{2+}]$; then

$$0.200 + X \; = \; [SO_4^{2-}]$$

$$X \times (0.200 + X) \; = \; 3.2 \times 10^{-7}$$

X is negligibly small compared to 0.200

$$0.200X \; = \; 3.2 \times 10^{-7}$$

$$X \; = \; 1.6 \times 10^{-6} \; mole/liter$$

In this example the excess sulfate ion was very large. If the excess is small, X cannot be neglected and the quadratic equation must be involved.

Example 13-9

Calculate the solubility of lead iodate in a 0.020 M KIO_3 solution.

$$Pb(IO_3)_2(s) \; \rightleftharpoons \; Pb^{2+} + 2IO_3^-$$

$$K_s \; = \; [Pb^{2+}][IO_3^-]^2 \; = \; 3.2 \times 10^{-13}$$

Let C = solubility of $Pb(IO_3)_2$ (analytical concentration); then

$$[Pb^{2+}] \; = \; C$$

$$[IO_3^-] \; = \; 0.020 + 2C$$

$$C(0.020 + 2C)^2 \; = \; 3.2 \times 10^{-13}$$

2C<<0.020 and can be neglected

$4.0 \times 10^{-4} C = 3.2 \times 10^{-13}$

$C = 0.80 \times 10^{-9} = 8.0 \times 10^{-10}$ mole/liter

Example 13-10

50.0 ml of 2.50×10^{-2} M silver nitrate solution was added to 50.0 ml of 5.00×10^{-2} M sodium chloride solution. Calculate the silver ion concentration of the resulting solution.

$Ag^+ + NO_3^- + Na^+ + Cl^- \rightleftharpoons AgCl(s) + Na^+ + NO_3^-$

$AgCl(s) \rightleftharpoons [Ag^+][Cl^-]$

$50.0 \times 2.50 \times 10^{-2} = 1.25$ mM $AgNO_3$

$50.0 \times 5.00 \times 10^{-2} = 2.50$ mM NaCl

(Twice the number of mM of NaCl required to precipitate the $AgNO_3$ has been added.)

1.25 mM of NaCl remains in 100 ml of solution. The concentration of NaCl is 12.5 mM/liter or 0.0125 mole/liter. Let $X = [Ag^+]$; then

$X + 0.0125 = [Cl^-]$

$(X)(X + 0.0125) = 1.8 \times 10^{-10}$

$1.25 \times 10^{-2} X = 1.8 \times 10^{-10}$

$X \ll 0.0125$ and can be neglected in $(X + 0.0125)$

$X = \dfrac{1.8 \times 10^{-10}}{1.25 \times 10^{-2}} = 1.44 \times 10^{-8}$ mole/liter

13-5 COMPLEX IONS

Complex ions exist in equilibrium with their constituents in solution. The constituents are metal cations and neutral molecules or anions. The neutral molecules and anions are called ligands.

The most familiar example to general-chemistry students is the silver ammine complex $Ag(NH_3)_2^+$. The equilibrium may be represented by

$Ag(NH_3)_2^+ \rightleftharpoons Ag^+ + 2NH_3$

and the equilibrium constant

$$K_{diss} = \frac{[Ag^+][NH_3]^2}{[Ag(NH_3)_2^+]}$$

13-6 DISSOCIATION CONSTANT

K_{diss} is called the <u>dissociation constant</u> of the complex ion
(see Table 13-2). Some tables list formation constants, which are,
of course, reciprocals of the dissociation constants.

TABLE 13-2 Dissociation Constants of Complex Ions

Equilibrium	K_{diss}
$Ag(CN)_2^- \rightleftharpoons Ag^+ + 2CN^-$	1.4×10^{-20}
$Ag(NH_3)_2^+ \rightleftharpoons Ag^+ + 2NH_3$	6.8×10^{-8}
$Ag(S_2O_3)_2^{3-} \rightleftharpoons Ag^+ + 2S_2O_3^{2-}$	6×10^{-14}
$Cd(NH_3)_4^{2+} \rightleftharpoons Cd^{2+} + 4NH_3$	2.8×10^{-7}
$Co(NH_3)_6^{2+} \rightleftharpoons Co^{2+} + 6NH_3$	1.2×10^{-5}
$Co(NH_3)_6^{3+} \rightleftharpoons Co^{3+} + 6NH_3$	2.2×10^{-34}
$Cu(NH_3)_4^{2+} \rightleftharpoons Cu^{2+} + 4NH_3$	4.7×10^{-15}
$Hg(CN)_4^{2-} \rightleftharpoons Hg^{2+} + 4CN^-$	4×10^{-42}
$Ni(NH_3)_6^{2+} \rightleftharpoons Ni^{2+} + 6NH_3$	2×10^{-9}
$Zn(NH_3)_4^{2+} \rightleftharpoons Zn^{2+} + 4NH_3$	3×10^{-10}
$Zn(OH)_4^{2-} \rightleftharpoons Zn^{2+} + 4OH^-$	3×10^{-16}

The dissociation of a complex ion, such as $Ag(NH_3)_2^+$, is a
stepwise process similar to the ionization of polyprotic acids.
Each step constitutes an equilibrium with an equilibrium constant.

$$Ag(NH_3)_2^+ \rightleftharpoons AgNH_3^+ + NH_3$$

$$K_1 = \frac{[AgNH_3^+][NH_3]}{[Ag(NH_3)_2^+]}$$

$$AgNH_3^+ \rightleftharpoons Ag^+ + NH_3$$

$$K_2 = \frac{[Ag^+][NH_3]}{[AgNH_3^+]}$$

The equilibria involved in the stepwise dissociation of complex ions may become extremely complicated. The problems in this treatment will be confined to those in which the ligand concentration is sufficiently high to eliminate the necessity for considering the intermediate steps in the dissociation. Only the overall dissociation constant need be considered.

$$K_{diss} = K_1 \times K_2 = \frac{[Ag(NH_3)^+][NH_3]}{[Ag(NH_3)_2^+]} \times \frac{[Ag^+][NH_3]}{[Ag(NH_3)^+]}$$

Example 13-11

Determine the silver ion concentration in a 0.200 M solution of $Ag(NH_3)_2NO_3$.

$$Ag(NH_3)_2^+ \rightleftharpoons Ag^+ + 2NH_3$$

$$K_{diss} = \frac{[Ag^+][NH_3]^2}{[Ag(NH_3)_2^+]} = 6.8 \times 10^{-8}$$

Let $X = [Ag^+]$; then

$2X = [NH_3]$ and $0.200 - X = [Ag(NH_3)_2^+]$

$$\frac{4X^3}{0.200 - X} = 6.8 \times 10^{-8}$$

$$4X^3 = 1.36 \times 10^{-8}$$

$$X = \sqrt[3]{0.34 \times 10^{-8}} = \sqrt[3]{3.4 \times 10^{-3}}$$

$$X = 1.5 \times 10^{-3} \text{ mole/liter}$$

Example 13-12

How many moles of ammonia must be added to one liter of a 0.100 M silver nitrate solution to reduce the silver ion concentration to 2.00×10^{-7} M?

$$Ag(NH_3)_2^+ \; \rightleftharpoons \; Ag^+ + 2NH_3$$

$$K_{diss} = \frac{[Ag^+][NH_3]^2}{[Ag(NH_3)_2^+]} = 6.8 \times 10^{-8}$$

(1) Sufficient ammonia must be added to combine with practically all of the silver ions present. (2) An additional amount to further reduce the silver ion concentration is also necessary.

1. $Ag^+ + 2NH_3 \rightarrow Ag(NH_3)_2^+$
    0.100 mole + 0.200 mole $\rightarrow$ 0.100 mole

2. $Ag(NH_3)_2^+ \; \rightleftharpoons \; Ag^+ + 2NH_3$

$$K_{diss} = \frac{[Ag^+][NH_3]^2}{[Ag(NH_3)_2^+]}$$

Let $[NH_3]$ = X
$[Ag^+]$ = 2.0×10^{-7}
$[Ag(NH_3)_2^+]$ = $0.100 - 2 \times 10^{-7}$

$$\frac{2.0 \times 10^{-7}x^2}{0.100} = 6.8 \times 10^{-8}$$

$$X^2 = \frac{6.8 \times 10^{-9}}{2.0 \times 10^{-7}} = 3.4 \times 10^{-2}$$

X = 0.184 mole/liter
0.200 + 0.184 = 0.384 mole/liter

13-7 HYDROLYSIS

In Chap. 10, solutions of salts whose ions did not react with the solvent water were considered. Since anions are bases and some hydrated cations are acids, reactions with water may be expected.

Reactions of this type are called hydrolytic reactions. The
phenomenon is called hydrolysis.

Hydrolysis occurs when salts of strong bases and weak acids,
weak bases and strong acids, and weak bases and weak acids are
dissolved in water. Examples are sodium acetate, ammonium chloride,
and ammonium acetate.

13-8 SODIUM ACETATE

This will serve as an illustration of the hydrolysis of a salt
of a strong base and a weak acid. When CH_3COONa dissolves in water,
the ions dissociate, $CH_3COO^- + Na^+$. The solution is basic to
indicators. The following reversible reactions are involved.

$$CH_3COO^- + H_3O^+ \rightleftarrows CH_3COOH + H_2O$$

$$2H_2O \rightleftarrows H_3O^+ + OH^-$$

They may be combined:

$$CH_3COO^- + H_2O \rightleftarrows CH_3COOH + OH^-$$

The equilibrium constant expression has the form

$$K_h = \frac{[CH_3COOH][OH^-]}{[CH_3COO^-]}$$

and is called the hydrolysis constant. The numerical value of K_h
may be obtained by making suitable substitutions.

$$[CH_3COOH] = \frac{[H_3O^+][CH_3COO^-]}{K_a}$$

$$[OH^-] = \frac{K_w}{[H_3O^+]}$$

$$K_h = \frac{[H_3O^+][CH_3COO^-]/K_a \times K_w/[H_3O^+]}{[CH_3COO^-]}$$

$$K_h = \frac{[H_3O^+][CH_3COO^-]}{K_a} \times \frac{K_w}{[H_3O^+]} \times \frac{1}{[CH_3COO^-]} = \frac{K_w}{K_a}$$

Example 13-13

Calculate the hydroxide ion concentration of a 0.100 M sodium acetate solution.

$$CH_3COONa \rightarrow CH_3COO^- + Na^+$$

$$CH_3COO^- + H_2O \rightleftarrows CH_3COOH + OH^-$$

$$K_h = \frac{[CH_3COOH][OH^-]}{[CH_3COO^-]} = \frac{K_w}{K_a} = \frac{1.00 \times 10^{-14}}{1.75 \times 10^{-5}}$$

$$K_h = 5.72 \times 10^{-10}$$

Let $X = [OH^-]$; then

$X = [CH_3COOH]$ (1 CH_3COOH is formed for each OH^-)

$0.100 - X = [CH_3COO^-]$ (1 CH_3COO^- is consumed for each

OH^- formed)

$$\frac{X^2}{0.100 - X} = 5.72 \times 10^{-10}$$

X may be neglected in the denominator.

$$\frac{X^2}{1.00 \times 10^{-1}} = 5.72 \times 10^{-10}$$

$$X^2 = 5.72 \times 10^{-11}$$

$$X^2 = 57.2 \times 10^{-12}$$

$$X = 7.56 \times 10^{-6} \text{ mole/liter}$$

13-9 AMMONIUM CHLORIDE

Ammonium chloride is the salt of a weak base and a strong acid. When it dissolves in water, the ions dissociate: $NH_4^+ + Cl^-$. The solution is acidic to indicators. The reactions which occur in the solution are

$$NH_4^+ + OH^- \rightleftharpoons NH_3 + H_2O$$

$$2H_2O \rightleftharpoons H_3O^+ + OH^-$$

The overall equilibrium reaction is

$$NH_4^+ + H_2O \rightleftharpoons H_3O^+ + NH_3$$

$$K_h = \frac{[H_3O^+][NH_3]}{[NH_4^+]}$$

From

$$K_b = \frac{[NH_4^+][OH^-]}{[NH_4]}$$

we get

$$[NH_3] = \frac{[NH_4^+][OH^-]}{K_b}$$

and from

$$K_w = [H_3O^+][OH^-]$$

we get

$$[H_3O^+] = \frac{K_w}{[OH^-]}$$

Substituting

$$K_h = \frac{K_w/[OH^-] \times [NH_4^+][OH^-]/K_b}{[NH_4^+]}$$

$$= \frac{K_w}{[OH^-]} \times \frac{[NH_4^+][OH^-]}{K_b} \times \frac{1}{[NH_4^+]} = \frac{K_w}{K_b}$$

Example 13-14

What is the pH of a 0.100 M NH_4Cl solution?

$$NH_4^+ + H_2O \rightleftharpoons NH_3 + H_3O^+$$

$$K_h = \frac{[NH_3][H_3O^+]}{[NH_4^+]} = \frac{1.00 \times 10^{-14}}{1.75 \times 10^{-5}} = 5.72 \times 10^{-10}$$

Let $X = [H_3O^+]$; then

$X = [NH_3]$

$0.100 - X = [NH_4^+]$

$$\frac{X^2}{0.100 - X} = 5.72 \times 10^{-10}$$

X may be neglected in the denominator

$$\frac{X^2}{1.00 \times 10^{-1}} = 5.72 \times 10^{-10}$$

$X^2 = 5.72 \times 10^{-11} = 57.2 \times 10^{-12}$

$X = 7.56 \times 10^{-6}$ mole/liter

$pH = -\log [H_3O^+] = -\log 7.56 \times 10^{-6}$

$\quad = -(-6. + .878) = 5.12$

13-10 AMMONIUM ACETATE

Ammonium acetate yields NH_4^+ and CH_3COO^- ions in solution. The ammonium ion is an acid and the acetate ion is a base.

$NH_4^+ + H_2O \rightleftarrows NH_3 + H_3O^+$

$CH_3COO^- + H_2O \rightleftarrows CH_3COOH + OH^-$

The H_3O^+ ions formed combine with the OH^- ions formed.

$NH_4^+ + CH_3COO^- \rightleftarrows NH_3 + CH_3COOH$

$$K_h = \frac{[NH_3][CH_3COOH]}{[NH_4^+][CH_3COO^-]}$$

$$K_b = \frac{[NH_4^+][OH^-]}{[NH_3]}$$

$$K_a = \frac{[H_3O^+][CH_3COO^-]}{[CH_3COOH]}$$

$$[NH_3] = \frac{[NH_4^+][OH^-]}{K_b}$$

$$[CH_3COOH] = \frac{[H_3O^+][CH_3COO^-]}{K_a}$$

$$K_b = \frac{[NH_4^+][OH^-]/K_b \times [H_3O^+][CH_3COO^-]/K_a}{[NH_4^+][CH_3COO^-]}$$

$$K_h = \frac{K_w}{K_b \times K_a}$$

PROBLEMS

13-1 What is the value of K_s for $TlIO_3$ based on the solubility
0.001844 mole/liter determined by V. K. La Mer and F. H. Goldman?
        [J. Am. Chem. Soc., 51, 2632 (1929)]

13-2 Calculate the solubility-product constant of silver iodate at
75° from the solubility determined by W. P. Baxter, 0.8403 mM/1000 g
of water. The density of water at 25°C is 0.99707 g/ml.
        [J. Am. Chem. Soc., 48, 615-621 (1926)]

13-3 A. E. Hill and J. P. Simmons found the solubility of $AgIO_3$
to be 0.0503 g/liter at 25°. Calculate the K_s.
        [J. Am. Chem. Soc., 31, 821-839 (1909)]

13-4 Determine a value of the K_s of the Ag_2SeO_4 from the solubility,
0.00242 mole/liter at 25°.
        [R. W. Gelbach and G. B. King, J. Am. Chem. Soc., 64,
        1054 (1942)]

13-5 The solubility of $Ba(BrO_3)_2$ was determined by R. M. Keefer,
H. G. Reiber, and C. S. Bisson as 7.895 g/1000 g H_2O. Calculate
the value of K_s based on this solubility determination.
        [J. Am. Chem. Soc., 62, 2951-2955 (1940)]

13-6 According to R. Dolique, 0.1361 g $SrSeO_4$ dissolves in 100 ml
of water at 10°. Calculate a value of the K_s based on this
solubility.
        [Bull. Soc., Chim. France, 10, 50 (1943)]

13-7 Three determinations of the solubility of silver sulfate in water at 25° gave 0.02677, 0.02675, and 0.02676 mole/1000 g water. Find a value of the K_s of silver sulfate from this data.

[G. Akerlof and H. C. Thomas, J. Am. Chem. Soc., 56, 596 (1934)]

13-8 Calculate the value of K_s for $La(IO_3)_3$ from the solubility in water and compare the solubilities calculated from K_s with those listed below.

Added salt	Concentration	Solubility of $La(IO_3)_3$, moles/liter
None		9.426×10^{-4}
$La_2(SO_4)_3$	M/30	1.830×10^{-4}
$La_2(SO_4)_3$	M/60	1.532×10^{-4}
$La(NO_3)_3$	M/30	9.398×10^{-4}
$LaCl_3$	M/30	9.233×10^{-4}

Source: V. K. La Mer and F. H. Goldman, J. Am. Chem. Soc., 51, 2632 (1932)

13-9 What is the value of the solubility-product constant of praseodymium iodate, $Pr(IO_3)_3$, if the iodate ion concentration is 3.4×10^{-3} mole/liter?

[F. H. Firsching and T. R. Paul, J. Inorg. Nucl. Chem., 28, 2415 (1966)]

13-10 Determine the solubility of silver cyanide from the K_s for AgCN.

13-11 Calculate the lead ion concentration in a saturated solution of lead selenate.

13-12 Calculate the concentration of calcium ions in a solution prepared by adding 25.0 ml of 0.125 M $Ca(NO_3)_2$ to 25.0 ml of 0.125 M Na_2SO_4.

13-13 Calculate the cadmium ion concentration of a saturated solution of cadmium iodate.

13-14 A solution was prepared by mixing 30.0 ml of 0.250 M silver nitrate solution with 30.0 ml of 0.125 M sodium sulfate solution. What was the silver ion concentration of the solution after mixing?

13-15 Calculate the silver ion concentration in each of the solutions listed. Compare your calculated results with those observed by J. J. Renier and D. S. Martin, Jr.

[J. Am. Chem. Soc., 78, 1835 (1956)]

$[IO_3^-]$	$[Ag^+] \times 10^6$ (observed)
0.000940	90.2
0.00248	35.4
0.0103	9.05
0.0501	1.96
0.0634	1.68
0.492	1.33

13-16 A sample of sodium fluoride weighing 0.420 g was added to 100 ml of a saturated solution of calcium fluoride. Calculate the calcium ion concentration before and after addition of the sodium fluoride.

13-17 What is the solubility of lead sulfate in 0.020 M sodium sulfate solution?

13-18 Calculate the silver ion concentration of a solution prepared by dissolving 1.99 g of $KAg(CN)_2$ in sufficient water to make 100 ml of solution.

13-19 What is the Ag^+ concentration of a solution prepared by dissolving 0.100 mole of silver nitrate in 1 liter of a solution in which the ammonia concentration is 0.0870 moles/liter?

13-20 What is the Cu^{2+} concentration in a solution prepared by dissolving 2.00×10^{-3} mole of $CuSO_4$ in 100 ml of solution containing 0.0850 mole/liter of ammonia?

13-21 A solution which contained 1.00 mole of potassium cyanide in 1.00 liter of a 0.100 M silver nitrate solution was prepared. Determine the dissociation constant of the silver cyanide complex ion, $Ag(CN)_2^-$, if the silver ion concentration of the solution was 6.76×10^{-20} mole/liter.

13-22 What is the pH of a 0.500 M CH_3COONa solution?

13-23 Calculate the hydroxide ion concentration of a 0.0125 M NH_4Cl solution.

13-24 Calculate the pH of a 5.00% potassium acetate solution, sp. gr. 1.0191.

13-25 What is the pH of 8.00% ammonium nitrate, sp. gr. 1.0313?

13-26 The pH of a saturated solution of calcium hydroxide is 12.58. Calculate K_s.

13-27 In a series of experiments the pH's of solutions of KCN which had hydrolyzed were determined colorimetrically. Find a value of pK_{HCN} from one of these experiments in which the KCN was 12.5% hydrolyzed and the pH was 10.34.

13-28 Calculate the pH of a solution prepared by adding 500 ml of 0.100 M sodium hydroxide to 500 ml of 0.100 M acetic acid.

13-29 What is the pH of a solution prepared by adding 250 ml of a 0.200 M solution of hydrochloric acid to 500 ml of 0.100 M ammonia?

13-30 A solution was prepared by mixing 200 ml of 0.125 M propionic acid with 100 ml of 0.250 potassium hydroxide. What was the pH of the solution?

SUPPLEMENTARY PROBLEMS

13-31 The solubility of lanthanum(III) iodate in water at 25°C as determined by V. K. La Mer and F. H. Goldman was 8.90×10^{-4} mole/liter. Find the K_s of $La(IO_3)_3$.
    [J. Am. Chem. Soc., 51, 2632 (1929)]

13-32 Calculate the K_s of a saturated solution of $PbCl_2$. The solubility is 0.03367 mole/liter.
    [A. Seidell and W. F. Linke, Solubilities of Inorganic and Organic Compounds, supplement to 3d ed., p. 504, Van Nostrand Company, Princeton, New Jersey, 1952]

13-33 The solubility of Ag_2CrO_4 is given by H. Schafer as 0.029 g/liter at 25°. Calculate a value of the K_s from this measurement.
    [Z. Anorg. Chem., 45, 310 (1909)]

13-34 What is the K_s of $TlIO_3$ at 25° from its solubility, 0.001844 mole/liter?
    [V. K. La Mer and F. H. Goldman, J. Am. Chem. Soc., 51, 2632 (1929)]

13-35 Calculate K_s for lanthanum iodate, $La(IO_3)_3$, from the iodate
ion concentration of a saturated solution, 3.1×10^{-3} mole/liter.

> [F. H. Firsching and T. R. Paul, J. Inorg. Nucl. Chem.
> 28, 2415 (1966)]

13-36 The solubility of yttrium iodate was found by a determination
of the iodate ion concentration of a saturated solution, 5.8×10^{-3}
mole/liter. Calculate the K_s for $Y(IO_3)_3$.

> [F. H. Firsching and T. R. Paul, J. Inorg. Nucl. Chem.,
> 28, 2415 (1966)]

13-37 The solubility of calcium hydroxide at 0°C is 0.185 g/100
g H_2O. What is the molal concentration? Assume that the specific
gravity is 1.00 and calculate K_s.

13-38 What is the solubility of copper iodate at 25°C?

13-39 What is the silver ion concentration of a saturated solution
of silver chromate?

13-40 What is the solubility of europium iodate $Eu(IO_3)_3$ from the
K_s, 1.1×10^{-11}?

> [F. H. Firsching and T. R. Paul, J. Inorg. Nucl. Chem.,
> 28, 2415 (1966)]

13-41 What is the chloride ion concentration in a saturated solution
of mercurous chloride?

13-42 Calculate the copper(II) ion concentration of a solution
prepared by adding 50.0 ml of 0.100 M sodium iodate to 25.0 ml of
0.100 M copper nitrate.

13-43 What is the thallium(I) ion concentration in a 0.0050 M
potassium chromate solution?

13-44 Determine the mercurous ion concentration in a 0.100 M KCl
solution. What is the mercurous ion concentration in a saturated
solution of KCl? The solubility of KCl in water is 26.34 wt% at
25°. The density of the saturated solution is 1.178 g/ml.

13-45 What is the Gd^{3+} concentration of a saturated solution of
$Gd(IO_3)_3$? K_s = 1.8 x 10^{-11}.
   [F. H. Firsching and T. R. Paul, J. Inorg. Nucl. Chem.,
   28, 2415 (1966)]

13-46 A saturated solution of barium chromate which also contained
0.020 mole/liter of sodium chromate was prepared. What was the
barium ion concentration of the solution?

13-47 Saturated solutions of silver iodide in aqueous potassium
iodide which contained 0.118, 0.134, 0.277, 0.437, 0.584, 0.766,
and 1.06 moles/liter of iodide ion were prepared. Calculate the
silver ion concentration of each of these solutions. Plot the
silver ion concentration versus the iodide ion concentration.
   [E. L. King, H. J. Krall, M. L. Pandow, J. Am. Chem. Soc.,
   74, 3493 (1952)]

13-48 Calculate the solubility of lead iodate in a 0.0200 M
calcium iodate solution.

13-49 Calculate the silver ion concentration in a solution made by
dissolving 0.0200 mole of silver nitrate in 750 ml of 0.200 M
sodium cyanide solution.

13-50 Calculate the cadmium ion concentration in a solution made
by mixing 100 ml of 0.0100 M cadmium nitrate solution with 150 ml
of 0.100 M aqueous ammonia.

13-51 What is the ammonia concentration required to reduce the
silver ion concentration of a solution from 0.120 mole/liter to
1.00 x 10^{-7} mole/liter?

13-52 What concentration of sodium acetate is necessary to prepare
an aqueous solution of pH 9.38?

13-53 Determine the mass in grams of ammonium chloride required
to prepare 100 ml of a solution of pH 4.63.

13-54 A 0.0245 M KCN solution hydrolyzed to form HCN to the extent
of 3.18%. Determine a value of K_{HCN} from this observation.
        [H. T. S. Britton and E. N. Dodd, J. Chem. Soc., 102,
        2333 (1931)]

13-55 The hydrolysis constant of KCN was reported to be 2.55 x 10^{-5}
by H. T. S. Britton and E. N. Dodd. Calculate the percent hydrolysis
of a 0.245 M KCN solution.
        [J. Chem. Soc., 102, 2333 (1931)]

13-56 Calculate the pH of 2.00% $NaNO_2$, sp. gr. 1.0125.

13-57 A volume of 20.0 ml of 0.0992 M sodium hydroxide solution was
run from a buret into 40.0 ml of 0.0496 M acetic acid. Calculate
the pH of the solution.

13-58 Find the pH of a solution resulting from the titration of
32.6 ml of 0.104 M sodium hydroxide with 40.7 ml of 0.0833 M
acetic acid. The solutions were titrated in a beaker that contained
50.0 ml of water.

Chapter 14

OXIDATION-REDUCTION

Oxidation is the process in which an element increases its oxidation number; in reduction there is a decrease in oxidation number. A loss of electrons accompanies oxidation, and a gain of electrons accompanies reduction.

14-1 OXIDATION NUMBERS

Oxidation numbers are numbers assigned to atoms in a rather arbitrary fashion to designate electron transfers in oxidation-reduction reactions. They represent the charges that atoms would have if the electrons were assigned according to an arbitrary set of rules.

14-2 ASSIGNMENT OF OXIDATION NUMBERS

The assignment of oxidation numbers to elements is governed by the following set of rules.

1. The oxidation number of elements in the free (uncombined) state, for example, Li, H_2, P_4, S_8, and Tl, is 0.
2. The oxidation number of fluorine is -1 in all compounds, for example, HF, OF_2, and KF.
3. The oxidation number of oxygen in compounds, for example, H_2O, Li_2O, NO, and Cl_2O, usually is -2. Oxygen has an

289

oxidation number of -1 in peroxides, for example, Na_2O_2, H_2O_2. It is +2 in OF_2.

4. The oxidation number of hydrogen in compounds, for example, H_2O, NH_3, HCl, and PH_3, is usually +1. In metal hydrides, for example, LiH and CaH_2, it is -1.

5. The sum of the oxidation numbers in compounds is equal to 0, for example, in H_2SO_4, $2(+1) + (+6) + 4(-2) = 0$

6. The oxidation number of simple ions is equal to the charge on the ion, for example, Cl^-, F^-, O^{2-}, Na^+, Mg^{2+}.

7. The sum of the oxidation numbers in a complex ion is equal to the charge on the ion, for example, NO_3^-, $+5 + 3(-2) = -1$.

Example 14-1
Assign oxidation numbers to each of the elements in the following substances: H_2Te, N_2, SF_4, Li_2SO_4, K_2O_2 (potassium peroxide).

H_2Te $2(+1) + (-2)$

N_2 0

SF_4 $(+4) + 4(-1)$

Li_2SO_4 $2(+1) + (+6) + 4(-2)$

K_2O_2 $2(+1) + 2(-1)$

Example 14-2
Assign oxidation numbers to the elements in the following ions: I^-, Sc^{3+}, NO_2^-, HSO_3^-, BF_4^-.

I^- -1

Sc^{3+} +3

NO_2^- $+3 + 2(-2) = -1$

HSO_3^- $(+1) + (+4) + 3(-2) = -1$

BF_4^- $(+3) + 4(-1) = -1$

Example 14-3
Indicate which elements are oxidized and which are reduced in the reaction:

$HBr(aq) + HSO_4^- \rightarrow Br_2 + SO_2 + 2H_2O$

$$3H^+ + 2Br^- + HSO_4^- \rightarrow Br_2 + SO_2 + 2H_2O$$

+1	-1	(+1)	(+6)	(-2)	0	(+4)(-2)	(+1)(-2)	

reduction ↓ (top), oxidation ↑ (bottom)

14-3 OXIDATION-REDUCTION REACTIONS IN SOLUTION

Reactions which occur in the absence of polar solvents have been described in Sec. 4-3. Many very useful and important oxidation-reduction reactions occur in aqueous solution. A typical example is the reaction of Fe^{2+} ions with MnO_4^-, the basis of a volumetric method of analysis for iron. Such reactions may be separated into two parts: the oxidation part and the reduction part. In the reaction of Fe^{2+} with MnO_4^- the parts are:

$$Fe^{2+} \rightarrow Fe^{3+} \quad \text{(oxidation)}$$
$$8H^+ + MnO_4^- \rightarrow Mn^{2+} + 4H_2O \quad \text{(reduction)}$$

14-4 BALANCING OXIDATION-REDUCTION REACTIONS BY THE ION-ELECTRON METHOD

Method

A. REACTIONS IN ACID SOLUTION
1. Divide the reaction into half-reactions.
2. Balance the atoms in each half-reaction.
   a. Add H^+ or H_2O where necessary to balance H and O.
3. Balance the charges in each half-reaction.
   a. Add electrons to the side which is more positive.
4. Make the loss of electrons equal to the gain of electrons by multiplying each half-reaction by the appropriate integer.
5. Add the two balanced half-reactions.
6. Check the balance of atoms and charges.

Example 14-4

Write a balanced equation for the reaction of Fe^{2+} with MnO_4^-.
The products are Fe^{3+} and Mn^{2+}.

Step 1

$$Fe^{2+} \rightarrow Fe^{3+}$$

$$MnO_4^- \rightarrow Mn^{2+}$$

Step 2

$$Fe^{2+} \rightarrow Fe^{3+}$$

$$MnO_4^- \rightarrow Mn^{2+} + 4H_2O$$

to provide 4O's

$$8H^+ + MnO_4^- \rightarrow Mn^{2+} + 4H_2O$$

to provide 8H's

Step 3

$$Fe^{2+} \rightarrow Fe^{3+} + e^-$$

$$+2 \quad = \quad +3 \quad + \quad (-1)$$

An electron was added to the right side to provide a charge of +2,
equal to the charge on the left side.

$$8H^+ + MnO_4^- + 5e^- \rightarrow Mn^{2+} + 4H_2O$$

$$+7 \quad + \quad (-5) \quad = \quad +2$$

$5e^-$ must be added to the left side to give a charge of +2, equal
to the charge of the right side.

Step 4

$$5(Fe^{2+} \rightarrow Fe^{3+} + e^-)$$

$$5Fe^{2+} \rightarrow 5Fe^{3+} + 5e^-$$

$$8H^+ + MnO_4^- + 5e^- \rightarrow Mn^{2+} + 4H_2O$$

Step 5

$$5Fe^{2+} \rightarrow 5Fe^{3+} + 5e^-$$

$$8H^+ + MnO_4^- + 5e^- \rightarrow Mn^{2+} + 4H_2O$$

$$5Fe^{2+} + 8H^+ + MnO_4^- \rightarrow 5Fe^{3+} + Mn^{2+} + 4H_2O$$

Step 6

$5Fe^{2+}$	$5Fe^{3+}$
$8H^+$	$4 \times 2H$
1 Mn	1 Mn

$$\begin{array}{c} 4\ 0 \qquad 4\ 0 \\ 5Fe^{2+} + 8H^+ + MnO_4^- \rightarrow 5Fe^{3+} + Mn^{2+} + 4H_2O \\ (+10) + (+8) + (-1) = +17 = (+15) + (+2) = +17 \end{array}$$

Example 14-5

Chlorine is produced by the reaction of hydrochloric acid with potassium dichromate. Write a balanced ionic equation for the reaction. Dichromate ions are reduced to chromium(III) ions.

Step 1

$$Cl^- \rightarrow Cl_2$$
$$Cr_2O_7^{2-} \rightarrow Cr^{3+}$$

Step 2

$$2Cl^- \rightarrow Cl_2$$
$$14H^+ + Cr_2O_7^{2-} \rightarrow 2Cr^{3+} + 7H_2O$$

Step 3

$$2Cl^- \rightarrow Cl_2 + 2e^-$$
$$2(-) = 0 + 2(-1)$$
$$14H^+ + Cr_2O_7^{2-} \rightarrow 2Cr^{3+} + 7H_2O$$
$$14(+1) + (-2) = 2(+3)$$
$$+12 \neq +6$$

Six electrons must be added to the left side.

$$14H^+ + Cr_2O_7^{2-} + 6e^- \rightarrow 2Cr^{3+} + 7H_2O$$
$$+14 + (-2) + (-6) = +6 = +6$$

Step 4

$$3[2Cl^- \rightarrow Cl_2 + 2e^-]$$
$$6Cl^- \rightarrow 3Cl_2 + 6e^-$$
$$14H^+ + Cr_2O_7^{2-} + 6e^- \rightarrow 2Cr^{3+} + 7H_2O$$

Step 5

$$\begin{array}{r} 6Cl^- \rightarrow 3Cl_2 + 6e^- \\ 14H^+ + Cr_2O_7^{2-} + 6e^- \rightarrow 2Cr^{3+} + 7H_2O \\ \hline 6Cl^- + 14H^+ + Cr_2O_7^{2-} \rightarrow 3Cl_2 + 2Cr^{3+} + 7H_2O \end{array}$$

Step 6

  6Cl   3 x 2 Cl

  14 H   7 x 2 H

  2 Cr   2 x 2 Cr

  7 O   7 O

$6Cl^- + 14H^+ + Cr_2O_7^{2-} \rightarrow 3Cl_2 + 2Cr^{3+} + 7H_2O$

-6 + (14) + (-2) = +6 = 0 + 2(+3) + 0 = +6

B. REACTIONS IN BASIC SOLUTIONS

    Reactions which occur in basic solution may not contain H^+. H_2O molecules and OH^- ions must be added.

Example 14-6

Balance the equation for the oxidation of AsO_3^{3-} to AsO_4^{3-} by MnO_4^-. In basic solution MnO_4^- is reduced to MnO_2.

Step 1

  $AsO_3^{3-} \rightarrow AsO_4^{3-}$

  $MnO_4^- \rightarrow MnO_2$

Step 2

  $H_2O + AsO_3^{3-} \rightarrow AsO_4^{3-}$

  to provide 1 O

In acid solution we could add H^+ to the right side. Since we cannot do so here because the reaction occurs in basic solution, we must add OH^- in place of H_2O.

  $OH^- + AsO_3^{3-} \rightarrow AsO_4^{3-}$

  to provide 1 O

To balance the H introduced with the O in OH^-, we must add H_2O to the right side.

  $OH^- + AsO_3^{3-} \rightarrow AsO_4^{3-} + H_2O$

Balance by multiplying OH^- by 2

  $2OH^- + AsO_3^{3-} \rightarrow AsO_4^{3-} + H_2O$

  $2H_2O + MnO_4^- \rightarrow MnO_2 + 4OH^-$

Step 3

$$2OH^- + AsO_3^{3-} \rightarrow AsO_4^{3-} + H_2O$$

$$2(-1) + (-3) = -5 \neq -3$$

Add $2e^-$ to the <u>right</u> side.

$$2OH^- + AsO_3^{3-} \rightarrow AsO_4^{3-} + H_2O + 2e^-$$

$$-2 \;\; + (-3) = -5 = -3 \qquad\qquad + (-2) = -5$$

$$2H_2O + MnO_4^- \rightarrow MnO_2 + 4OH^-$$

$$0 \;\; + (-1) \;\; \neq \;\; 0 \;\; + 4(-1)$$

Add $3e^-$ to the <u>left</u> side.

$$2H_2O + MnO_4^- + 3e^- \rightarrow MnO_2 + 4OH^-$$

$$0 + (-1) \;\; + (-3) = -4 = 0 + 4(-1) = -4$$

Step 4

$$3(2OH^- + AsO_3^{3-} \rightarrow AsO_4^{3-} + H_2O)$$

$$2(2H_2O + MnO_4^- \rightarrow MnO_2 + 4OH^-)$$

Step 5

$$6OH^- + 3AsO_3^{3-} + 4H_2O + 2MnO_4^- \rightarrow$$

$$3AsO_4^{3-} + 3H_2O + 2MnO_2 + 8OH^-$$

$$3AsO_3^{3-} + H_2O + 2MnO_4^- \rightarrow 3AsO_4^{3-} + 2MnO_2 + 2OH^-$$

Step 6

18 O	18 O
2 H	2 H
3 As	3 As
2 Mn	2 Mn

$$3AsO_3^{3-} + H_2O + 2MnO_4^- \rightarrow 3AsO_4^{3-} + 2MnO_2 + 2OH^-$$

$$3(-3) \;\; + 0 \;\; + 2(-1) = 3(-3) \;\; + \;\; 0 \;\; + 2(-1)$$

$$-11 = -11$$

<u>Important</u>: Whenever possible when writing half-reactions, use the actual species that exist in solution.

Example 14-7

HNO_3 $H^+ + NO_3^-$ not N^{5+}

Na_2CrO_4 $2Na^- + CrO_4^{2-}$ not Cr^{6+}

KIO_3 $K^+ + IO_3^-$ not I^{5+}

14-5 STOICHIOMETRY OF OXIDATION-REDUCTION REACTIONS

A balanced ionic equation is often all that is required when considering an oxidation-reduction reaction. When the quantities of substances required to prepare the solution must be known, a complete "molecular" equation is necessary. The computation of a yield based on a starting material may also require such an equation.

Example 14-8

Write a stoichiometric equation for the reaction of potassium permanganate with iron(II) sulfate in sulfuric acid solution.

The balanced ionic equation is

$$5Fe^{2+} + 8H^+ + MnO_4^- \rightarrow 5Fe^{3+} + Mn^{2+} + 4H_2O$$

Potassium ions K^+ and sulfate ions SO_4^{2-} were omitted, since no change involved them. K^+ ions are needed to balance the MnO_4^- anions; SO_4^{2-} ions are required to balance the cations Fe^{2+}, H^+, Fe^{3+}, and Mn^{2+}.

$$5FeSO_4 + 4H_2SO_4 + KMnO_4 \rightarrow 5Fe_2(SO_4)_3 + MnSO_4 + 4H_2O$$

This equation is not complete, since potassium is present on the left but not on the right. If it is added, it must be as the ion K^+ balanced by an anion, SO_4^{2-} in this case.

$$5FeSO_4 + 4H_2SO_4 + KMnO_4 \rightarrow 5Fe(SO_4)_3 + MnSO_4 + K_2SO_4 + 4H_2O$$

Check

5Fe   10Fe

1K   2K

$$10FeSO_4 + 4H_2SO_4 + 2KMnO_4 \rightarrow 5Fe_2(SO_4)_3 + 2MnSO_4 + K_2SO_4 + 4H_2O$$

Check

10Fe	10Fe
$14SO_4$	$18SO_4$
8H	8H
2K	2K
2Mn	2Mn
8O	8O

$$10FeSO_4 + 8H_2SO_4 + 2KMnO_4 \rightarrow 5Fe_2(SO_4)_3 + 2MnSO_4 + K_2SO_4 + 8H_2O$$

Check

10Fe	10Fe
$18SO_4$	$18SO_4$
16H	16H
2K	2K
2Mn	2Mn
8O	8O

Example 14-9

Calculate the weight of cadmium sulfide required to make 22.0 g
of $Cd(NO_3)_2$ by the action of nitric acid on cadmium sulfide.

$$CdS(s) + NO_3^- \rightarrow Cd^{2+} + NO + S$$
$$3[CdS \rightarrow Cd^{2+} + S + 2e^-]$$
$$2[4H^+ + NO_3^- + 3e^- \rightarrow NO + 2H_2O]$$

$$\overline{3CdS(s) + 8H^+ + 2NO_3^- \rightarrow 3Cd^{2+} + 3S + 2NO + 4H_2O}$$

$$3CdS(s) + 8HNO_3 \rightarrow 3Cd(NO_3)_2 + 3S + 2NO + 4H_2O$$

3 moles → 3 moles

1 mole → 1 mole

CdS	112.4	$Cd(NO_3)_2$		112.4
	$\underline{32.1}$		$2 \times 14.0 =$	28.0
	$\overline{144.5}$ g/mole		$5 \times 16.0 =$	$\underline{80.0}$
				$\overline{220.4}$ g/mole

$$\frac{22.0}{220} = 0.1000 \text{ mole of } Cd(NO_3)_2 \text{ prepared}$$

0.100 mole CdS required

0.100 x 145 = 14.5 g CdS

PROBLEMS

<u>14-1</u> Indicate which elements are oxidized and which are reduced in the following reactions.

a. Fe_3O_4 + $4H_2$ → $3Fe$ + $4H_2O$

b. MnO_2 + $4HCl$ → $MnCl_2$ + Cl_2 + $2H_2O$

c. $2K_2Cr_2O_7$ + $2H_2O$ + $3S$ → $3SO_2$ + $4KOH$ + $2Cr_2O_3$

d. $2Ca_3(PO_4)_2$ + $5C$ + $6SiO_2$ → $6CaSiO_3$ + P_4 + $5CO_2$

e. $CaCO_3$ → CaO + CO_2

f. $2KClO_3$ → $2KCl$ + $3O_2$

g. H_2SO_4 + $2KOH$ → K_2SO_4 + $2H_2O$

h. NH_4NO_2 → N_2 + $2H_2O$

i. Mg_2Si + $4HCl$ → $2MgCl_2$ + SiH_4

j. $6KOH$ + $3Br_2$ → $5KBr$ + $KBrO_3$ + $3H_2O$

<u>14-2</u> Divide the following reactions into half-reactions. Designate whether oxidation or reduction has occurred in each case.

a. NO_2 + $HClO$ → H^+ + NO_3^- + $2Cl^-$

b. Mg + H^+ + NO_3^- → Mg^{2+} + N_2 + H_2O

c. IO_3^- + H_2S → I_2 + SO_2 + H_2O

d. H^+ + CrO_4^{2-} + HNO_2 → Cr^{3+} + NO_3^- + H_2O

e. H^+ + MnO_2 + Cl^- → Mn^{2+} + Cl_2 + H_2O

14-3 Write balanced ionic equations for the following reactions which occur in neutral or acid solution.

a. $I_2(s) + S_2O_3^{2-} \rightarrow I^- + S_4O_6^{2-}$

b. $Br^- + BrO_3^- \rightarrow Br_2(\ell)$

c. $CuS(s) + NO_3^- \rightarrow Cu^{2+} + S_8(s) + NO(g)$

d. $Cl^- + MnO_4^- \rightarrow Cl_2(g) + Mn^{2+}$

e. $Zn(s) + NO_3^- \rightarrow Zn^{2+} + NH_4^+$

f. $Cu(s) + NO_3^- \rightarrow Cu^{2+} + NO(g)$

g. $Cu(s) + HSO_4^- \rightarrow Cu^{2+} + SO_2(g)$

h. $H_2S(aq) + Cr_2O_7^{2-} \rightarrow Cr^{3+} + S_8(s)$

i. $Cu^{2+} + 2I^- \rightarrow CuI(s) + I_2(s)$

j. $ClO_3^- + SO_2(g) \rightarrow ClO_2(g) + HSO_4^-$

k. $ICl(aq) \rightarrow Cl^- + IO_3^- + I_2$

l. $H_2SO_3(aq) + Cr_2O_7^{2-} \rightarrow HSO_4^- + Cr^{3+}$

m. $Mn^{2+} + S_2O_8^{2-} \rightarrow MnO_4^- + HSO_4^-$

n. $MnO_4^- + H_2O_2 \rightarrow Mn^{2+} + H_2O$

o. $Ag(s) + NO_3^- \rightarrow Ag^+ + NO_2(g)$

14-4 Write balanced ionic equations for the reactions which occur in basic solution.

a. $ClO_2(g) + SbO_2^- \rightarrow ClO_2^- + Sb(OH)_6^-$

b. $Cl_2(g) + IO_3^- \rightarrow Cl^- + IO_4^-$

c. $I_2 \rightarrow I^- + IO_3^-$

d. $Fe_3O_4(s) + MnO_4^- \rightarrow Fe_2O_3(s) + MnO_2(s)$

e. $MnO_2(s) + H_2O_2 \rightarrow MnO_4^-$

14-5 The following problems are a little more difficult than
average to balance. Balance them.

a. $C_2H_5OH + K_2Cr_2O_7 + H_2SO_4 \rightarrow$
$$HC_2H_3O_2 + Cr_2(SO_4)_3 + K_2SO_4 + H_2O$$

b. $As_2S_5 + HNO_3 \rightarrow NO_2 + H_2O + H_3AsO_4 + H_2SO_4$

c. $CrI_3 + Cl_2 + KOH \rightarrow KIO_4 + K_2CrO_4 + KCl + H_2O$

d. $As_2S_3 + HClO_3 + H_2O \rightarrow HCl + H_3AsO_4 + H_2SO_4$

14-6 Calculate the mass of iodine that can form by the reaction
of sulfuric acid with 1.66 g of potassium iodide. The HSO_4^- ions
are reduced to HSO_3^-.

14-7 Determine the volume, in milliliters, of the NO gas formed
by the reaction of 12.7 g of Cu with nitric acid at 27°C and 570 mm.

14-8 Sodium iodate is made by the reaction of iodine with an excess
of sodium chlorate. The chlorate ion is reduced to chloride ion.
How many grams of sodium chlorate would be required to react with
100 g of iodine? How many grams would be required to provide a 20%
excess?
       [H. H. Willard, Inorg. Syn., 1(1939), 169]

14-9 How many grams of sodium iodate would be required to prepare
225 g of $Na_3H_2IO_6$ by oxidation with chlorine gas in basic solution?
The chlorine is reduced to chloride ion in the reaction.
       [H. H. Willard, Inorg. Syn., 1(1939), 169-170]

14-10 K_2ReCl_6 has been made by reduction of ReO_4^- in hydrochloric
acid solution by KI. How many grams of K_2ReCl_6 can be made from
8.00 g $KReO_4$ if the yield is 85%?
          [L. C. Hurd and V. A. Reinders, Inorg. Syn., 1(1939),
          178-179]

14-11 A vanadium sulfate containing 26.13% vanadium and 74.16%
sulfate ion has been prepared by the action of sulfur on V_2O_5 in
sulfuric acid solution. The sulfur was oxidized to SO_2 gas. The
weight of vanadium(V) was 36.4 g, and the weight of sulfur was
9.6 g. Was either of these substances in excess? The vanadium
compound collected weighed 75 g. Calculate the percent yield.

[R. T. Claunch and M. M. Jones, Inorg. Syn., 7(1963),
92-93]

14-12 Cyanogen $(CN)_2$ can be made by the oxidation of aqueous sodium
cyanide by aqueous copper(II) sulfate. Insoluble copper(I) cyanide
is also formed. Sodium cyanide solution is added to the copper(II)
sulfate solution until the copper(II) ion is completely reduced.
How many grams of sodium cyanide will be required to reduce 500 g
of copper(II) sulfate? The yield of cyanogen was 21 g. Calculate
the percent yield. What volume would the cyanogen produced
occupy at 27° and 760 mm?

[G. J. Janz, Inorg. Syn., 5(1957), 43]

14-13 Telluric acid, H_6TeO_6, is formed by the oxidation of TeO_2
with permanganate ions in nitric acid solution. A sample of
H_6TeO_6 weighing 12.5 g was obtained from 100 g of TeO_2. Calculate
the percent yield.

[F. C. Mathers, C. M. Rice, H. Broderick, and R. Forney,
Inorg. Syn., 3(1950), 145]

14-14 Copper(I) iodide can be quantitatively prepared according to
the following unbalanced reaction:

$$CuSO_4 + KI + Na_2S_2O_3 \rightarrow CuI + K_2SO_4 + Na_2S_4O_6 + Na_2SO_4$$

How many grams of CuI can be made starting with 25.0 g of $CuSO_4 \cdot 5H_2O$?

[G. B. Kauffman and R. P. Pinnell, Inorg. Syn., 6(1960), 3]
[G. B. Kauffman, Inorg. Syn., 11(1968), 215]

14-15 Potassium dithioferrate(III) can be prepared by the direct reaction of iron, sulfur, and potassium carbonate at 900°C. Potassium sulfate and carbon dioxide are the other products. Eight grams of Fe powder gave 18.5 g of potassium dithioferrate(III). What is the % yield?

> [J. L. Deutsch and H. B. Jonassen, Inorg. Syn., 6(1960),
> 170]

Chapter 15

ELECTROCHEMISTRY

15-1 ELECTROLYSIS

When an electric current passes through a molten salt, such as
sodium chloride, cations are reduced at the cathode and anions
are oxidized at the anode. The two reactions may be written as
follows.

Cathode reaction:

$$Na^+ + e^- \rightarrow Na(s) \qquad (reduction)$$

Anode reaction:

$$Cl^- \rightarrow \tfrac{1}{2}Cl_2(g) + e^- \qquad (oxidation)$$

Solutions of electrolytes behave in a similar fashion. Since
the solvent water may also react at the electrodes, the situation
may be more complicated.

The quantitative relationships of the reactions at the electrodes
are the same for molten salts and for solutions of electrolytes.
They are described by Faraday's laws.

15-2 FARADAY'S LAWS OF ELECTROLYSIS

1. The mass of a substance produced or consumed at an electrode
   is proportional to the quantity of electricity which passes
   through the solution.

The quantity of electricity is the amount of charge passed and
is measured in coulombs (SI symbol is C). Charge is equal to
current x time. Coulombs = amperes x seconds.

   2. One <u>gram equivalent</u> of a substance is produced or consumed
      at an electrode by the passage of 96,487 C (1 faraday, F)
      of charge.

If one faraday of electric charge is passed through molten
sodium chloride, the quantity of sodium deposited is 23.0 g, 1 gram
equivalent of sodium.

$$Na^+ + e^- \rightarrow Na$$

A gram equivalent of a substance is its atomic or molecular weight
divided by the number of electrons involved in the electrode process
in question. The weight of a gram equivalent of copper is
63.54/2 = 31.77 g

$$Cu^{2+} + 2e^- \rightarrow Cu$$

The coulomb is a measure of electric charge and is equal to 1 ampere
flowing for one second; coulombs = amperes x seconds. One faraday,
96,487 C, is the charge of 1 mole (6.022×10^{23}) of electrons. One
gram atom of sodium ions (23.0 g) reacts with 1 faraday (6.022×10^{23}
electrons) to form 1 gram atom (6.022×10^{23} atoms) of sodium. One
mole of Cu^{2+} ions requires 2 faradays to produce 1 mole of Cu atoms.
One gram equivalent of copper ions ($1/2 \times 6.022 \times 10^{23}$ ions)
requires 1 faraday (6.022×10^{23} electrons).

15-3 DETERMINATION OF THE FARADAY

The recommended value of the faraday is 96,486.7 C/equivalent.
      [Natl. Bur. Std. (U.S.) Tech. News Bull., <u>53</u>, (1),
      January, 1971]
For calculations involving three significant figures, the value
96,500 C may be used.

Electrochemistry

Calculate a value of the faraday from the observation that the
passage of 3,671.3 C resulted in the deposition of 4,828.4 mg
of iodine from a solution containing iodide ions.

      [S. J. Bates and G. W. Vinal, J. Am. Chem. Soc., 36,
      916 (1914)]

$$I^- \rightarrow I + e^-$$

1 gram equivalent of I weighs $\dfrac{126.90}{1}$ g

$$1\ F = \frac{3,671.3}{4.8284} \times 126.90 = 96,489\ C/equiv.$$

$$\frac{C}{equiv.} = \frac{C}{\cancel{g}} \times \frac{\cancel{g}}{equiv.} = C/equiv.$$

Calculate the mass of copper deposited from a solution of copper(II)
sulfate by the passage of a current of 2.21 A for 3 hr 15 min.

$$Cu^{2+} + 2e^- \rightarrow Cu$$

Quantity of charge $= 2.21 \times 3.25 \times 60 \times 60 = 25,900$ C

$$A \times \cancel{hr} \times \frac{\cancel{min}}{\cancel{hr}} \times \frac{sec}{\cancel{min}} = A \times sec = C$$

$1\ F = (96,500\ C)$, will deposit $\dfrac{63.5}{2} = 31.8$ g

$$31.8 \times \frac{25,900}{96,500} = 8.54\ g\ Cu\ deposited$$

$$\frac{g}{\cancel{g}} \times \cancel{g} = g$$

15-4 GALVANIC CELLS

In Chap. 14 it was shown that oxidation-reduction reactions could
be divided into two half-reactions. The overall reaction is the

combination of the two half-reactions, and electrons are transferred
from the reducing agent to the oxidizing agent. It is possible to
construct an apparatus in which the reactants are separated in
such a way that the electrons are transferred through a wire. Such
an apparatus is called a galvanic or Voltaic cell. Practical
galvanic cells, which provide a useful current for a sufficient
length of time, are called batteries. Figure 15-1 represents one
galvanic cell, often called the Daniell cell. In one compartment
the half-reaction $Zn \rightarrow Zn^{2+} + 2e^-$ occurs, and in the other, the
reaction $Cu^{2+} + 2e^- \rightarrow Cu$ takes place. The metal electrodes are
connected by a wire through which electric current flows. The two
cell compartments are separated by a salt bridge or porous diaphragm,
which enables ions to carry the current and complete the circuit.
The electrode at which <u>oxidation</u> occurs is called the <u>anode,</u> and
<u>reduction</u> occurs at the <u>cathode.</u> The two parts of the cell are
called <u>half-cells</u>.

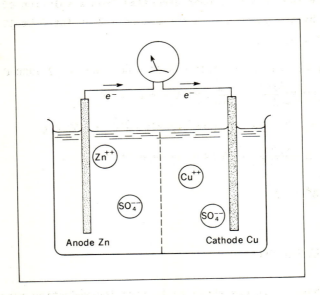

FIGURE 15-1   Schematic diagram of a Daniell cell

In the <u>half-cells</u> of the Daniell cell the chemical changes are
as follows.

Anode reaction:

$$Zn \rightarrow Zn^{2+} + 2e^- \qquad \text{(oxidation)}$$

Cathode reaction:

$$Cu^{2+} + 2e^- \rightarrow Cu \qquad \text{(reduction)}$$

The total reaction is the sum of the two half-reactions

$$Zn + Cu^{2+} \rightarrow Zn^{2+} + Cu$$

The electric current flows through the external circuit by
virtue of a <u>potential</u> which is established between the <u>electrodes</u>
when the circuit is completed. The value of the potential depends
upon the nature of the substances which comprise each <u>half-cell</u>,
the activities of the substances involved, and the temperature.
Many galvanic cells have been investigated; it is possible to
measure the potential developed by a number of these cells.

15-5 STANDARD POTENTIALS: MEASUREMENT

Since each cell may be considered as the sum of two half-cells,
it would be desirable to have a means of measuring the potential
developed by each half-cell. Such a measurement is not possible,
and it is necessary to adopt a table of relative potentials. The
potentials of the half-cells are compared to the <u>normal hydrogen
electrode</u>, to which the arbitrary value 0.000 volt (V) is assigned.
The normal hydrogen electrode may be represented by

$$H^+(a = 1) + e^- \rightleftarrows \tfrac{1}{2}H_2(1 \text{ atm}) \qquad V^\circ = 0.000 \text{ V}$$

The pressure of $H_2(g)$ is 1 atm, the activity of H^+ is 1 m, and the
temperature is 25°C. The standard potential of the $Zn^{2+} + 2e^- = Zn$
electrode may be determined from a cell such as that illustrated in
Fig. 15-1. The complete cell is represented by the notation
$Zn|Zn^{2+}(1 \text{ m})||H^+(1 \text{ m})|H_2(1 \text{ atm}),Pt$. The single line denotes a

junction between an electrode and a solution; the double line
represents a junction between two solutions.

Example 15-3
What is the galvanic cell designation of the overall reaction
$Zn(s) + CuSO_4(aq) \rightarrow Cu(s) + ZnSO_4(aq)$ if the activities are 1 m?

   $Zn | ZnSO_4(1\ m) || CuSO_4(1\ m) | Cu$

Example 15-4
Write equations for the half-cell reactions and the overall
reaction for a cell consisting of a lead electrode immersed in a
$1\ m\ Pb(NO_3)_2$ solution in one compartment and hydrogen gas at 1 atm
bubbled over a platinum electrode immersed in 1 m HCl.

Oxidation	$Pb \rightleftharpoons Pb^{2+} + 2e^-$	(anode)
Reduction	$2H^+ + 2e^- \rightleftharpoons H_2$	(cathode)

$$Pb + 2H^+ \rightleftharpoons H_2 + Pb^{2+}$$

The cell designation is

   $Pb | Pb(NO_3)_2(1\ m) || HCl(1\ m) | Pt, H_2(1\ atm)$

15-6 STANDARD POTENTIALS: TABLES

Table 15-1 is a list of a few standard electrode potentials
required for the solution of problems at the end of this chapter.
The half-reactions are written as reductions. The standard potential
$V°$ is the potential of the electrode in a cell in which the other
electrode is the normal hydrogen electrode. Similar tables are to
be found in many general chemistry textbooks. In some textbooks the
symbol $E°$ may be used to represent the standard potential of an
electrode.

   In Table 15-1 the substances which are more easily oxidized are
at the top, and the substances which are more easily oxidized than

TABLE 15-1 Standard Electrode Potentials

Electrode-reaction	$V°$, volts
$K^+ + e^- = K$	-2.925
$Ca^{2+} + 2e^- = Ca$	-2.866
$Na^+ + e^- = Na$	-2.714
$Mg^{2+} + 2e^- = Mg$	-2.363
$Al^{3+} + 3e^- = Al$	-1.662
$V^{2+} + 2e^- = V$	-1.186
$Zn^{2+} + 2e^- = Zn$	-0.7628
$Fe^{2+} + 2e^- = Fe$	-0.4402
$Cd^{2+} + 2e^- = Cd$	-0.4029
$PbSO_4 + 2e^- = Pb + SO_4^{2-}$	-0.3588
$Cd^{2+} + Hg + 2e^- = Cd(Hg)$	-0.3516
$In^{3+} + 3e^- = In$	-0.343
$Co^{2+} + 2e^- = Co$	-0.277
$V^{3+} + e^- = V^{2+}$	-0.256
$Sn^{2+} + 2e^- = Sn$	-0.136
$Pb^{2+} + 2e^- = Pb$	-0.126
$2H^+ + 2e^- = H_2$	0.0000
$UO_2^{2+} + e^- = UO_2^+$	+0.05
$S + 2H^+ + 2e^- = H_2S\,(aq)$	+0.142
$Cu^{2+} + e^- = Cu^+$	+0.153
$AgCl + e^- = Ag + Cl^-$	+0.2222
$Hg_2Cl_2 + 2e^- = 2Hg + 2Cl^-$	+0.2676
$Cu^{2+} + 2e^- = Cu$	+0.337
$Cu^+ + e^- = Cu$	+0.521

TABLE 15-1   Standard Electrode Potentials (con't)

Electrode-reaction	$V°$, volts
$I_2 + 2e^- = 2I^-$	+0.5355
$Hg_2SO_4 + 2e^- = 2Hg + SO_4^{2-}$	+0.6151
$Fe^{3+} + e^- = Fe^{2+}$	+0.771
$Ag^+ + e^- = Ag$	+0.7991
$NO_3^- + 4H^+ + 3e^- = NO + 2H_2O$	+0.96
$Br_2(\ell) + 2e^- = 2Br^-$	+1.0652
$Br_2(aq) + 2e^- = 2Br^-$	+1.087
$O_2 + 4H^+ + 4e^- = 2H_2O(\ell)$	+1.229
$Tl^{3+} + 2e^- = Tl^+$	+1.25
$Cr_2O_7^{2-} + 14H^+ + 6e^- = 2Cr^{3+} + 7H_2O$	+1.33
$Cl_2 + 2e^- = 2Cl^-$	+1.3595
$MnO_4^- + 8H^+ + 5e^- = Mn^{2+} + 4H_2O$	+1.51
$H_2O_2 + 2H^+ + 2e^- = 2H_2O$	+1.776
$Ag^{2+} + e^- = Ag^+$	+1.980
$F_2(g) + 2e^- = 2F^-$	+2.87

hydrogen have $V°$ values which are (-). Each of these substances is the negative electrode when placed in a battery in which the other electrode is the normal hydrogen electrode. Another widely used compilation of electrode potentials assigns a positive sign to these half-reactions and a negative sign to those below the normal hydrogen electrode. These potentials are known as oxidation potentials. The former assignment of signs has the approval of the International Union of Pure and Applied Chemistry and may be found in many physical chemistry and freshman textbooks. These are also referred to as reduction potentials.

15-7 USE OF STANDARD ELECTRODE POTENTIALS

If two half-reactions are combined and the potential is calculated, a positive value indicates that the reaction will proceed
as written. If the electromotive force (emf) is negative, the
reaction will not proceed spontaneously as written. When standard
potentials are used, the species in solution are at unit activity and
gases are at one atmosphere. The emf is given as

$$E^{\circ}_{cell} = V^{\circ}_{right} - V^{\circ}_{left} = V^{\circ}_{cathode} - V^{\circ}_{anode}$$

where right and left represent the right and left electrodes as
designated in the cell diagram. The right electrode is always the
cathode, the left, the anode. Once a cell is written a given way
in a problem, it must not be changed. To determine the emf of a
cell, both electrodes are written as reduction reactions and then
they are subtracted in the same manner as above (right electrode -
left electrode).

Example 15-5

Will cadmium metal reduce Cu^{2+}?

Half-reactions

Cathode	$Cu^{2+} + 2e^- \rightleftarrows Cu$	V°	=	0.337 V
Anode	$Cd^{2+} + 2e^- \rightleftarrows Cd$	V°	=	-0.403 V
	$Cd + Cu^{2+} \rightleftarrows Cd^{2+} + Cu$	E°_{cell}	=	0.740 V

$$E^{\circ}_{cell} = V^{\circ}_{right} - V^{\circ}_{left} = 0.740 \text{ V}$$

Since E°_{cell} is +, the reaction will occur as written.

Example 15-6

Will an acid solution of potassium permanganate oxidize bromide
ions to bromine?

Cathode

$$MnO_4^- + 8H^+ + 5e^- \rightleftarrows Mn^{2+} + 4H_2O \qquad V^{\circ} = 1.51 \text{ V}$$

Anode

$Br_2 + 2e^- \rightleftharpoons 2Br^-$ $V° = 1.06$ V

$2MnO_4^- + 16H^+ + 10e^- \rightleftharpoons 2Mn^{2+} + 8H_2O$ $V° = 1.51$ V

$5Br_2 + 10e^- \rightleftharpoons 10Br^-$ $V° = 1.06$ V

$\overline{5Br_2 + 2MnO_4^- + 16H^+ \rightleftharpoons 2Mn^{2+} + 8H_2O + 10Br^-}$ $E°_{cell} = +0.45$ V

$E°_{cell} = V°_{right} - V°_{left} = +0.45$ V

> Note: The value of $V°$ for each half-reaction
> is not altered when the equation is multiplied by
> an integer, since $V°$ is an intensive property.
> The reaction proceeds as written, since $E°_{cell}$ is
> positive.

Example 15-7

Calculate the emf of a cell in which an iron electrode is immersed
in 1 m iron(II) in the left compartment and a zinc electrode is
inserted in 1 m zinc nitrate in the right compartment.

$Fe|Fe(NO_3)_2(1$ m$)||Zn(NO_3)_2(1$ m$)|Zn$

By definition, oxidation occurs in the left compartment.

Cathode $Zn^{2+} + 2e^- \rightleftharpoons Zn$ $V° = -0.763$ V

Anode $Fe^{2+} + 2e^- \rightleftharpoons Fe$ $V° = -0.440$ V

$\overline{Fe + Zn^{2+} \rightarrow Fe^{2+} + Zn}$ $E°_{cell} = -0.323$ V

$E°_{cell} = V°_{right} - V°_{left} = -0.323$ V

The reaction will not occur spontaneously as written, since $E°_{cell}$
is negative.

Example 15-8

Calculate the emf of a cell comprising a $UO_2^+ \rightleftharpoons UO_2^{2+}$ and a
$V^{2+} \rightleftharpoons V^{3+}$ electrode.

> Assume that the first half reaction occurs as an oxidation.

Cathode $V^{3+} + e^- \rightleftharpoons V^{2+}$ $V° = -0.256$ V

Anode $UO_2^{2+} + e^- \rightleftharpoons UO_2^+$ $V° = -0.05$ V

$\overline{UO_2^+ + V^{3+} \rightleftharpoons V^{2+} + UO_2^{2+}}$ $E°_{cell} = -0.306$ V

The reaction will not occur spontaneously as written.

15-8 THE EFFECT OF CONCENTRATION ON THE emf OF CELLS

The values of the standard electrode potentials in Table 15-1 are for reactions in which all gases are at one atmosphere and all substances in solution are at unit activity. Pure solid substances are assigned an activity of 1. When the activities are at other values, the potential of an electrode is given by the Nernst equation:

$$V = V° - \frac{0.0592}{n}\log Q$$

where $V°$ is the standard potential and n is the number of electrons involved in the electrode reaction. Q is the reaction quotient.

$$Q = \frac{[right]}{[left]}$$

Example 15-9

Calculate the potential of a zinc-zinc ion electrode in which the zinc ion activity is 0.00100 m.

$$V = V° - \frac{0.0592}{n}\log\frac{[right]}{[left]}$$

$$Zn^{2+} + 2e^- = Zn$$

$$n = 2$$

$$V° = -0.763 \text{ V}$$

The activity of the solid Zn is 1.

$$V = -0.763 - \frac{0.0592}{2}\log\frac{1}{[0.00100]}$$

$$= -0.763 - 0.0296 \times (+3.000)$$

$$= -0.763 - 0.0888 = -0.852 \text{ V}$$

The potential of a cell is given by an equivalent form of the Nernst equation:

$$E = E° - \frac{0.0592}{n}\log Q$$

Example 15-10

Calculate the potential of the cell:

$Cd \mid Cd^{2+}(0.100) \mid \mid H^{+}(0.200) \mid Pt, H_2(0.5 \text{ atm})$

$2H^{+} + 2e^{-} \rightleftharpoons H_2$	$V° =$	0.000 V
$Cd^{2+} + 2e^{-} \rightleftharpoons Cd$	$V° =$	-0.403 V
$Cd + 2H^{+} \rightarrow Cd^{2+} + H_2$	$E° =$	0.403 V

$$E_{cell} = E°_{cell} - \frac{0.0592}{n} \log\frac{[\text{right}]}{[\text{left}]}$$

$$= E°_{cell} - \frac{0.0592}{n} \log\frac{[Cd^{2+}]P_{H_2}}{[H^{+}]^2}$$

$$= 0.403 - 0.0296 \log\frac{0.1 \times 0.5}{(0.2)^2}$$

$$E_{cell} = 0.403 - 0.0296 \log 1.25$$
$$= 0.403 - 0.0296(0.0969) = 0.403 - 0.003$$
$$= 0.400 \text{ V}$$

Example 15-11

Calculate the emf of the following cell:

$Pt, H_2(1 \text{ atm}) \mid HCl(1 \text{ m}) \mid \mid AgCl \mid Ag.$

First calculate the standard potential of the cell.

$AgCl + e^{-} \rightleftharpoons Ag + Cl^{-}$	$V° =$	$+0.2222$ V
$H^{+} + e^{-} \rightleftharpoons \frac{1}{2} H_2$	$V° =$	0.0000 V
$\frac{1}{2}H_2 + AgCl \rightleftharpoons H^{+} + Cl^{-} + Ag$	$E° =$	$+0.2222$ V

$$E = E° - \frac{0.0592}{1} \log\frac{[H^{+}][Cl^{-}]}{P_{H_2}^{\frac{1}{2}}}$$

$[H^{+}] = 1 \text{ m}$

The chloride ion concentration must be calculated from the K_s of AgCl.

$$K_s = [Ag^{+}][Cl^{-}] = 1.8 \times 10^{-10}$$

$$[Cl^{-}] = [Ag^{+}] = x$$

$$x^2 = 1.8 \times 10^{-10}$$

$$x = \sqrt{1.8} \times 10^{-5} \text{ mole/liter}$$

(At such low concentration, the molar concentration is equal to the molal concentration.)

$$E = 0.222 - \frac{0.0592}{1} \log \frac{[1][1.34 \times 10^{-5}]}{1}$$

$$E = 0.222 - 0.0592 \times (-4.89) = 0.222 + 0.289 = 0.511 \text{ V}$$

15-9 CALCULATION OF EQUILIBRIUM CONSTANTS

If current is drawn from a cell, the cell reaction will come to equilibrium and the emf will be zero.

For a cell at equilibrium:

$$E_{cell} = 0$$

$$E = E° - \frac{0.0592}{n} \log Q$$

at equilibrium

$$E = 0 \text{ and } Q = K$$

$$E° - \frac{0.0592}{n} \log K = 0$$

$$\log K = \frac{nE°}{0.0592}$$

Example 15-12

Calculate the equilibrium constant for the reaction

$Zn + Cu^{2+} \rightleftarrows Cu + Zn^{2+}$

Cathode(reduction)	$Cu^{2+} + 2e^- \rightleftarrows Cu$	$V° = +0.337 \text{ V}$
Anode(oxidation)	$Zn^{2+} + 2e^- \rightleftarrows Zn$	$V° = -0.763 \text{ V}$
$E° = V°_{right} - V°_{left}$	$Zn + Cu^{2+} \rightleftarrows Zn^{2+} + Cu$	$E° = 1.100 \text{ V}$

$$\log K = \frac{2 \times 1.10}{0.0592} = 37.2$$

$$K = 1.6 \times 10^{37}$$

PROBLEMS

15-1 In a very carefully executed experiment of S. J. Bates and W.
Vinal, 4,104.69 mg of silver was deposited.

>   [J. Am. Chem. Soc., 36, 916 (1914)]

a. What was the quantity of electricity passed through the
solution?

b. Calculate the weight of iodine which could be deposited by
the passage of the same current through a solution containing iodide
ions.

>   (Bates and Vinal obtained 4,828.62 mg.)

15-2 Calculate the weight of copper to be expected from an
electrolysis in which 2.8197 g of silver was deposited in a
coulometer in series with the electrolysis cell.

>   [T. W. Richards, E. Collins, and G. W. Heimrad, Proc. Am.
>   Acad. Arts Sci., 35, 123 (1899-1900)]

15-3 A solution containing 0.0407 g of ruthenium was prepared by
dissolving the appropriate mass of K_2RuCl_6. A current of 0.050 A
was used to remove all of the ruthenium by electrolysis. What time
was required?

>   [E. F. Smith and H. B. Harris, J. Am. Chem. Soc., 17,
>   652 (1895)]

15-4 Calculate the quantity of cadmium which may be deposited from
a solution of cadmium sulfate by the passage of a 1.02 A current
for 2 hr 20 min.

15-5 A quantity of electricity deposited 0.200 g atom of copper
from an aqueous solution. What weight of silver could be deposited
by the same current?

15-6 Calculate the value of the electronic charge from the value
of the faraday and Avogadro's number.

15-7 What is the volume of hydrogen (at STP) which could be formed by the passage of 7.00 A for 6 hr through a solution of potassium azide in anhydrous hydrazoic acid?

[A. W. Browne and G. E. F. Lundell, J. Am. Chem. Soc., 31, 446 (1909)]

15-8 A current of 300 mA was passed for 120 min through a solution in which a copper anode and a tellurium cathode were suspended. The loss in weight of the copper anode was 0.669 g, and of the tellurium cathode 1.120 g. What was the current efficiency at each electrode?

[A. J. Paulson, Inorg. Chem., 3, 941 (1964)]

15-9 Determine the value of the faraday from the weight of iodine, 4,828.5 mg, released by the passage of 3,671.3 C of electricity.

[S. J. Bates and G. W. Vinal, J. Am. Chem. Soc., 36, 916 (1914)]

15-10 Calculate the weight of silver which should be deposited by the current in Example 15-1; use the value of the faraday calculated from the data in that problem. Bates and Vinal obtained 4,105.23 mg of silver in the experiment.

[S. J. Bates and G. W. Vinal, J. Am. Chem. Soc., 36, 916 (1914)]

15-11 The electrochemical equivalent of an element is the quantity of the element deposited by the passage of 1 C of electricity. Calculate the electrochemical equivalent of hydrogen, oxygen, and nickel in milligrams per coulomb.

15-12

a. Calculate the emf of each of the following cells.
b. Write equations for the cell reactions which occur.

$Zn|Zn^{2+}(1\ m)||Cl^-(1\ m)|Pt,Cl_2$

$Cd|Cd^{2+}(1\ m)||Cl^-(1\ m),\ AgCl(s)|Ag$

$Cu|Cu^{2+}(1 \text{ m})||V^{2+}(1 \text{ m})|V$

$Pb|PbSO_4(s), SO_4^{2-}(1 \text{ m})||H_3O^+(1 \text{ m})|Pt, H_2 (1 \text{ atm})$

$Cl_2,Pt|Cl^-(1 \text{ m})||AgCl(s), Cl^-(1 \text{ m})|Ag$

15-13 Determine the standard potential of the cell

$In|In_2(SO_4)_3(1 \text{ m})||Hg_2SO_4(s), SO_4^{2-}(1 \text{ m})|Hg.$

E. M. Hattox and T. DeVries reported a value of 0.954 V.

[J. Am. Chem. Soc., 58, 2126 (1936)]

15-14 Find the standard potential of the cell

$Cd(Hg)|Cd^{2+}(1 \text{ m})||Cl^-(1 \text{ m})AgCl|Ag.$

H. S. Harned and M. E. Fitzgerald reported 0.57300 V.

[J. Am. Chem. Soc., 58, 2624 (1936)]

15-15 Determine if each of the following reactions will occur spontaneously to an appreciable extent. The equations are not balanced.

   a. $Fe^{2+} + Cr_2O_7^{2-} \rightarrow Fe^{3+} + Cr^{3+}$

   b. $MnO_4^- + Ag + H^+ \rightarrow Mn^{2+} + Ag^+ + H_2O$

   c. $H_2S + NO_3^- \rightarrow S + NO$

   d. $Co + Zn^{2+} \rightarrow Co^{2+} + Zn$

   e. $Ag + Sn^{2+} \rightarrow Ag^+ + An$

15-16 Calculate the emf of the cell:

$Cd|Cd^{2+}(0.200 \text{ m})||Cu^{2+}(0.100 \text{ m})|Cu$

15-17 Calculate the emf of the cell:

$Pt,H_2(1 \text{ atm})|H^+(1 \times 10^{-7} \text{ m})||H^+(1 \text{ m})|H_2(1 \text{ atm}), Pt$

15-18 Calculate the values of the equilibrium constants of the reactions in the cells in Prob. 15-12 as written.

15-19 A sulfur vapor electrode has been developed which can be used
in the temperature range 490-860°C. The cell diagram is:

$Ag|Ag_2S, AgCl|S(vapor)$, graphite

Here AgCl is the solvent and the sulfur vapor exists mainly as S_2.

    a.   Write the cathode reaction.

    b.   Write the anode reaction.

    c.   Write the cell reaction.

    d.   Write the Nernst equation for the emf of the cell.

    e.   What should be the slope of the line formed when E (cell)
is plotted vs $\log(P_{S_2})$ when the cell is saturated with
Ag_2S? (This information was used by the authors to verify
that sulfur is S_2 in the vapor state.)

        [W. T. Thompson and S. W. Flengas, Can. J. Chem., 46,
1611 (1968)]

15-20 Fredericks and Temple have measured the emf and determined
the E° of the following cell at various temperatures.

Pt, $O_2|O^{2-}||AgNO_3(1.0\ m)|Ag$

At 300°C with the O^{2-} concentration fixed at 10^{-5} m, the following
data were obtained as a function of the oxygen pressures

$\log P_{O_2}$	E(V)
-0.20	0.338
-0.40	0.344
-0.60	0.349
-0.80	0.356

    a.   Write the cell reaction.

    b.   Write the Nernst equation for the emf of the cell (remember
$Ag^+ = 1$ m).

    c.   Does this cell obey the Nernst equation at these pressures?

    d.   What is E° for the cell at 300°C?

        [Inorg. Chem., 11, 968 (1972)]

SUPPLEMENTARY PROBLEMS

15-21 Determine the quantity of iodine that could be deposited from
a solution of iodide ions by the passage of 3,671.06 C of electricity.
          [S. J. Bates and W. Vinal, J. Am. Chem. Soc., 36, 916 (1914)]

15-22 What is the quantity of electricity required to deposit
0.636 g of copper by electrolysis?
          [T. W. Richards, E. Collins, and G. W. Heimrad, Proc.
          Am. Acad. Arts Sci., 35, 123 (1899-1900)]

15-23 Compare the value of the faraday calculated from Avogadro's
number and the electronic charge with that calculated from the
data obtained by S. J. Bates and G. W. Vinal: the passage of
3,666 C deposited 4,099 mg of silver.
          [J. Am. Chem. Soc., 36, 916 (1914)]

15-24 What time would be required to deposit 0.1174 g of nickel
from a solution of nickel nitrate by the passage of a current
of 193 mA?

15-25 What volume of hydrogen, measured at 25°C and 735 Torr, could
be collected by the passage of a current of 0.321 A for 4.00 hr
through a solution of barium hydroxide?

15-26 The preparation of lithium hydroxide by the electrolysis of
a 35% solution of lithium chloride using a platinum anode led to
a current efficiency of 95%. What weight of lithium hydroxide was
formed by the passage of 2.68 A for 1 hr?

15-27 Calculate the number of coulombs passed through a solution
containing silver ions from the weight of silver deposited,

4,099.03 mg. The quantity of electricity passed was determined
from emf and resistance measurements to be 3,666.0 C.

> [S. J. Bates and G. W. Vinal, J. Am. Chem. Soc., 36,
> 916 (1914)]

15-28 An electrolysis was conducted with a silver anode, whose
weight loss was 4.18685 g. How many coulombs of electricity passed
through the cell?

> [W. M. Bovard and G. A. Hulett, J. Am. Chem. Soc., 39,
> 1077 (1917)]

15-29 The electrolysis of a solution containing copper(I), nickel,
and zinc as complex cyanides produced a deposit weighing 0.175 g.
The deposit contained 72.8% by weight of copper, 4.3% nickel, and
22.9% zinc. Assume that no other element was released and calculate
the number of coulombs passed through the solution.

> [J. Faust and R. Montillon, Trans. Electrochem. Soc., 65,
> 361 (1934)]

15-30 In the experiment described in Prob. 15-8, a precipitate
weighing 1.746 g was collected. What was the precipitate and what
was the current efficiency of its formation?

> [A. J. Pauson, Inorg. Chem., 3, 941 (1964)]

15-31 Calculate the standard potential of the cell
$$Ag \mid AgCl \mid HCl(1 \text{ m}) \mid Hg_2Cl_2 \mid Hg$$
and compare the result with the emf (+45.5 mV) observed by M. H.
Lietzke and J. V. Vaughan.

> [J. Am. Chem. Soc., 77, 876 (1955)]

15-32 Calculate the standard potential of the cell
$$Ag \mid AgCl(s), Cl^-(1 \text{ m}) \mid\mid Cu^{2+}(1 \text{ m}) \mid Cu$$

15-33 Determine if it is worthwhile to attempt to prepare each of the following compounds:

    a. CuF b. FeI_3

    c. TlI_3 d. AgF_2

15-34 What is the emf of the cell

    $Ag|AgCl, Cl^-(1\ m)||H^+(0.00100\ m)|H_2(0.5\ atm),Pt$

15-35 Determine the emf of the cell

    $H_2(740\ mm), Pt|H^+, Cl^-(1\ m),AgCl|Ag$

(Correct the pressure of hydrogen for the water content.)

15-36 The standard potential of the $Ag|AgSCN(s)$ electrode was found to be 0.08951 V. Calculate the potential of the cell

    $Pt, H_2(1\ atm)|HClO_4\ (m_1), KSCN\ (m_2)|AgSCN(s), Ag$

when m_1 = 1.4138 x 10^{-3} m and m_2 = 1.5270 m.

       [C. E. Vanderzee and W. E. Smith, J. Am. Chem. Soc., 78, 723 (1956)]

15-37 What are the values of the equilibrium constants of the reactions in the cells in Prob. 15-15?

15-38 The emf of the cell

    $Pt, O_2|O^{2-}||AgNO_3\ (1.0\ m)|Ag$

has been measured at various O^{2-} concentrations with the oxygen gas pressure fixed at 1 atm. The following data were obtained at 320°C.

log $[O^{2-}]$	E (V)
-6.50	0.2403
-6.00	0.2694
-5.50	0.2986
-5.00	0.3277

a. Write the cell reaction.

b. Write the Nernst equation for the emf of the cell.

c. Does this cell obey the Nernst Equation? (Plot E vs log $[O^{2-}]$.)

d. How does the slope of the line compare to the theoretical value expected at this temperature?

e. What is E° for this cell at 320°C?

[Fredericks and Temple, Inorg. Chem., 11, 968 (1972)]

Chapter 16

ENTROPY AND FREE ENERGY

16-1   ENTROPY

A fundamental property of a system is the <u>entropy</u> of a system. Entropy is a measure of the disorder of a system.

Like enthalpy, entropy changes are a function of the initial and final states of a system and are independent of the path. For this reason entropy is called a state function. The change in entropy, for a system in going from state 1 to state 2, is equal to the <u>reversible</u> heat absorbed by the system divided by the absolute temperature of the system.

$$\Delta S = S_2 - S_1 = q_{rev}/T$$

The entropy of a system is an extensive property and, in chemistry, is usually measured in units of calories per degree. One calorie per degree is often written as 1 eu (entropy unit). The entropy per gram, or per mole, of material is an intensive property.

16-2   ENTROPY CHANGES ACCOMPANYING CHANGES IN STATE

In Sec. 9-3 it was shown that the enthalpy content of a system changes with a change of state. Since these changes occur reversibly, the entropy changes of these systems can be calculated.

Example 16-1

Calculate the entropy change associated with the boiling of 1 mole of water at 100° C and 1 atm pressure. The heat of vaporization of water is 540 cal g^{-1}.

q_{rev} = 18 x 540 x 1 = 9720 cal

g/mole x cal/g x mole = cal

T = 373 K

ΔS = 9720/373 = 26.1 eu

cal/deg = eu

The entropy of the gaseous phase is greater than that of the liquid phase at the normal boiling point of the liquid, since there is more disorder in the gaseous phase than the liquid phase.

Example 16-2

Calculate the entropy change associated with the melting of 1.12 mole of ice at 0° C to water at 0° C.

   In Example 9-2 the heat absorbed by this system was found to be 1600 cal. Since this is a reversible heat absorbed

ΔS = 1600/273 = 5.86 eu

cal/deg = eu

The entropy of the liquid phase is greater than that of the solid phase at the melting point.

16-3 ENTROPY CHANGES AT CONSTANT TEMPERATURE

   If an ideal gas undergoes a reversible compression or expansion at constant temperature, the entropy change of the gas is given as

ΔS = nR $\ln(V_2/V_1)$ = nR $\ln(P_1/P_2)$

here ln = natural logarithm

R = gas constant = 1.987 cal deg^{-1} $mole^{-1}$

V_2 = final volume of gas

V_1 = initial volume of gas
P_2 = final pressure of gas
P_1 = initial pressure of gas

Example 16-3

Calculate the entropy change for the reversible expansion of
1 mole of an ideal gas from 20 liters to 40 liters at constant
temperature.

$$\Delta S = 1 \times 1.987 \times \ln(40/20) = 1.38 \text{ eu}$$
$$\text{mole} \times \text{cal/deg mole} = \text{cal/deg} = \text{eu}$$

Example 16-4

Calculate the entropy change for the reversible compression of
1.25 moles of an ideal gas from 1 atm to 5 atm at constant temperature.

$$\Delta S = 1.25 \times 1.987 \times \ln(1/5) = -4.00 \text{ eu}$$
$$\text{moles} \times \text{cal/deg mole} = \text{cal/deg}$$

16-4 IRREVERSIBLE CHANGES

Whenever a system undergoes an irreversible change, the heat
absorbed or given off by the system is always less than the heat
change of a similar process under reversible conditions. Here the
entropy change is greater than q_{irr}/T, or

$$\Delta S_{system} > q_{irr}/T$$

where q_{irr} is the heat change under irreversible conditions. To
calculate the entropy change of such a system, a method in going
from the same initial to the same final states (under reversible
conditions) is assumed. Since entropy is a state function, the
entropy change of the reversible process is the same as the entropy
change of the irreversible process.

Example 16-5

Calculate the entropy change of 1 mole of an ideal gas that expands, isothermally (at constant temperature) against a vacuum from 20 liters to 40 liters. Here no heat is given off to or absorbed from the surroundings.

Examples 16-4 and 16-5 both measure the entropy change in going from the same initial state to the same final state. The entropy change is the same in both examples.

$$\Delta S = 1 \times 1.987 \times \ln(40/20) = 1.38 \text{ eu}$$
$$\text{mole} \times \text{cal/deg mole} = \text{cal/deg} = \text{eu}$$

16-5 ENTROPY OF THE UNIVERSE

Whenever a system absorbs heat from its surroundings or gives off heat to the surroundings, both the system and the surroundings undergo a change in entropy. The heat change of the surroundings is the negative of the heat change of the system.

$$q_{system} = - q_{surroundings}$$

The sum of the entropy change of the system plus that of the surroundings is called the entropy change of the universe.

$$\Delta S_{universe} = \Delta S_{system} + \Delta S_{surroundings}$$

For reversible processes

$$\Delta S_{universe} = q_{system}/T + q_{surroundings}/T = 0$$

The entropy change of the universe for a reversible process is zero.

For an irreversible process, the entropy change of the surroundings is given as

$$\Delta S_{surroundings} = - q_{irr}/T$$

The entropy change of the universe is

$$\Delta S_{universe} = \Delta S_{system} - q_{irr}/T > 0$$

For any irreversible process there is always a net increase in the entropy of the universe.

Example 16-6

Calculate the entropy change of the universe in Example 16-5.

Since there was no heat absorbed by the surroundings, the entropy change of the surroundings was zero.

$$\Delta S_{universe} \ = \ 1.38 + 0 \ = \ 1.38 \ eu$$

Example 16-7

If one mole of an ideal gas is allowed to expand, isothermally at 215°C, from 20 liters to 40 liters against a constant pressure of 1 atm, approximately 488 cal will be absorbed from the surroundings. What is the entropy change of the universe in this expansion?

$$q_{irr} \ = \ 488 \ cal \ ; \ T \ = \ 273 + 215 \ = \ 488 \ K$$

Since this is Example 16-4 again, the entropy change of the system is 1.38 eu. The entropy change of the universe is

$$\Delta S_{universe} \ = \ 1.38 - 488/488 \ = \ 0.38 \ eu$$
$$cal/deg \ = \ eu$$

16-6 BOLTZMANN EQUATION

The entropy of a system is related to the probability of that particular state of the system to exist. From statistical theory, the entropy per molecule can be shown to be given by:

$$S \ = \ k \ ln \ W$$

here k = Boltzmann constant = 1.38×10^{-16} erg/deg (this is the gas constant per molecule), and

W = number of distinct ways in which the molecules can be

distributed throughout the system and still end up with the same
state of the system. This equation was first given by L. Boltzmann
in 1896.

Example 16-8

When CO crystallizes in the solid state, there are two possible
orientations of the molecule. One with the O atoms of one molecule
opposite a C atom of another molecule, and the other with the O
atom of a molecule opposite another O atom of another molecule.
Calculate the entropy of CO if both these orientations have the
same probability of occurance.

The entropy per molecule of the system is

$$S = 1.38 \times 10^{-16} \times \ln 2 = 9.56 \times 10^{-17} \text{ erg/deg}$$

The entropy per mole is

$$S = 9.56 \times 10^{-17} \times 6.02 \times 10^{23} \times 1/4.1840 \times 10^{-7} = 1.38 \text{ eu}$$

erg/deg molecule x molecule/mole x cal/erg = cal/deg mole

> [J. Clayton and W. F. Giauque, J. Am. Chem. Soc., 54,
> 2620 (1932)]

16-7 ABSOLUTE ENTROPIES

The Boltzmann equation is the theoretical basis for what is
called the third law of thermodynamics. There is only one distinct
way in which the molecules of a perfect pure substance can be
arranged in the solid state, at absolute zero. W is therefore
equal to 1 and S is equal to 0 for this case.

The third law of thermodynamics states that at absolute zero
(0 K) the entropy of a pure perfect crystalline substance is zero.
By using this as a reference point it is possible to obtain entropy
values of a substance at any temperature. These entropies are called
absolute entropies. (see Table 16-1).

TABLE 16-1   Absolute Entropies*
$S°$ (cal/deg-mole) at 298 K

Substance	Entropy
B (s)	1.56
BF_3 (g)	60.70
C (graphite)	1.36
CO (g)	47.30
CO_2 (g)	51.06
$CaCO_3$ (c)	22.2
CaO (c)	9.5
CH_4 (g)	44.50
Cl_2 (g)	53.29
F_2 (g)	48.6
HBr (g)	47.44
HCl (g)	44.62
H_2 (g)	31.21
H_2O (g)	45.11
H_2O (ℓ)	16.72
NH_3 (g)	46.01
N_2 (g)	45.77
O_2 (g)	49.00
P (s)	10.6
PH_3 (g)	50.2
S (c)	7.62
SO_2 (g)	59.4
SO_3 (g)	61.24

*Source: Roy A. Keller, Basic Tables in Chemistry, McGraw-Hill, New York, 1967, pp. 297-368

Absolute entropies can be used to calculate the entropy changes that accompany chemical reactions, in the same manner that enthalpy changes were calculated:

$$\Delta S° = S°(\text{products}) - S°(\text{reactants})$$

Example 16-9

Commercially, NH_3 is obtained by the fixation of N_2 from the atmosphere by the Haber process. The reaction is

$$N_2(g) + 3H_2(g) = 2NH_3(g).$$

What is the molar entropy change of this reaction?

$$\Delta S° = 2S°(NH_3) - S°(N_2) - 3S°(H_2)$$

$$\Delta S° = 2(46.0) - 45.7 - 3(31.21) = -47.3 \text{ eu}$$

There is a decrease in entropy in the above example, since four gas molecules are combining to form two gas molecules.

Example 16-10

$CaCO_3$ when heated decomposes into CaO and CO_2. What is the entropy change of this reaction?

$$CaCO_3(s) = CaO(s) + CO_2(g)$$

$$\Delta S° = S°(CaO) + S°(CO_2) - S°(CaCO_3)$$

$$\Delta S° = 9.5 + 51.1 - 22.2 = 38.4 \text{ eu}$$

There is an increase in entropy here, due to the formation of the gaseous product. Gases normally have a much higher entropy than solids or liquids.

16-8 GIBBS FREE ENERGY

Both the enthalpy change and the entropy change of a reaction are measures of the tendency of a reaction to occur spontaneously. Neither of these functions, by themselves, can be used to determine whether a reaction will occur spontaneously. If the functions are properly combined, a new state function is formed that tells in which direction a chemical reaction will proceed. This new function is called the Gibbs free energy and is defined as:

$$G = H - TS$$

Absolute values of the free energy, G, cannot be determined. The difference in free energy between products and reactants can be determined from standard free energies.

16-9 STANDARD FREE ENERGY OF FORMATION

The free energy change for one mole of compound formed from the elements at 25°C, each substance being in its standard state, is the standard free energy of formation of the compound (see Table 16-2). As with enthalpy, the standard free energy of formation of all elements in their standard state at 25°C has been assigned a value of zero. The standard free energy change of a reaction $\Delta G°$ can be obtained from the standard enthalpy of formation and from the entropy change of the formation reaction at 25°C.

$$\Delta G°_f = \Delta H°_f - T\Delta S°_f$$

Example 16-11

What is the Gibbs free energy change for the formation reaction of $H_2O(\ell)$ at 25°C?
This reaction is

$$H_2(g) + \tfrac{1}{2}O_2(g) \rightarrow H_2O(\ell)$$

From Table 9-3, $\Delta H°_f = -68.32$ kcal/mol

$$\Delta S°_f = S°(H_2O(\ell)) - S°(H_2(g)) - \tfrac{1}{2}S°(O_2(g))$$

$$= 16.72 - 31.21 - \tfrac{1}{2}(49.00) = -38.99 \text{ cal/mol}$$

$$= -0.0390 \text{ kcal/mol}$$

$$\Delta G°_f = -68.32 - (298)(-0.0390) = -56.70 \text{ kcal/mol}$$

TABLE 16-2 Free Energy of Formation*

$\Delta G°_f$ (kcal/mol) at 25°C

Substance	Free energy
$BF_3(g)$	-261.3
$B_2O_3(s)$	-283.0
$CaCO_3(s)$	-269.78
$CaO(s)$	-406.5
$CH_4(g)$	- 12.14
$C_2H_4(g)$	+ 16.28
$C_2H_6(g)$	- 7.86
$CO(g)$	- 32.81
$CO_2(g)$	- 94.26
$HBr(g)$	- 12.72
$HCl(g)$	- 22.77
$HF(g)$	- 64.7
$H_2O(g)$	- 54.64
$H_2O(\ell)$	- 56.69
$H_2S(g)$	- 7.89
$PH_3(g)$	+ 4.36
$SO_2(g)$	- 71.79
$SO_3(g)$	- 88.52

*Source: Roy A. Keller, Basic Tables in Chemistry, McGraw-Hill, New York, 1967, pp. 297-368.

From tables of $\Delta G°_f$ of compounds, the standard free energy of any reaction can be determined.

Example 16-12

What is the standard free energy change for the combustion of 1 mole of methane?

$$CH_4(g) + O_2(g) \rightarrow CO_2(g) + H_2O\ (\ell)$$

$$\Delta G° = \Delta G°_f(CO_2)_g + \Delta G°_f(H_2O)_\ell - \Delta G°_f(CH_4)_g - \Delta G°_f(O_2)_g$$

$\Delta G°$ = $-94.26 - 56.69 + 12.14 + 0$

= -138.81 kcal/mol

Note that this same result could have been obtained from tables of $\Delta H°$ and $S°$ and the definition of $\Delta G°$.

Example 16-13

Repeat the calculation in Example 16-12 using the values of $\Delta H°$ and $S°$ from the appropriate tables.

$\Delta H°_{react}$ = $\Delta H°_f(CO_2)_g + \Delta H°_f(H_2O)_\ell - \Delta H°_f(CH_4)_g - \Delta H°_f(O_2)_g$

$\Delta H°_{react}$ = $-93.97 - 68.32 - (-15.99) - (0)$

= -146.30 kcal/mol

$\Delta S°_{react}$ = $S°(CO_2)_g + S°(H_2O)_\ell - S°(CH_4)_g - S°(O_2)_g$

$\Delta S°_{react}$ = $51.1 + 16.73 - 44.5 - 49.0$

= -25.7 cal/mol-deg

$\Delta G°$ = $\Delta H°_{react} - T\Delta S°_{react}$

= $-146.30 - (298)(-25.7)(10^{-3})$

= -138.64 kcal/mol

16-10 FREE ENERGY AND THE EQUILIBRIUM CONSTANT

The change in the standard free energy of a reaction is related to the equilibrium constant of the reaction by the equation

$\Delta G°$ = $-RT \ln K$ = -2.303 RT log K

where ln K is the natural logarithm of the equilibrium constant and log K is the logarithm to the base 10. This relationship can be used to either calculate $\Delta G°$ for a reaction, or to determine equilibrium constants from tables of $\Delta G°$.

Example 16-14

K_p for the dissociation of HBr at 1108 K was found to be 1.47×10^{-5}. What is the free energy change for this reaction?

$$\Delta G°_{1108} = -(1.987)(1108) \ln(1.47 \times 10^{-5}) = 24,499 \text{ cal/mol}$$
$$\text{cal/mol-deg} \times \text{deg} = \text{cal/mol}$$

Example 16-15

What is the equilibrium constant for the reaction of sulfur dioxide
with oxygen to form sulfur trioxide?

$$SO_2(g) + \tfrac{1}{2}O_2(g) \rightarrow SO_3(g)$$

$$\Delta G° = \Delta G°_f(SO_3)_g - \tfrac{1}{2}\Delta G°_f(O_2)_g - \Delta G°_f(SO_2)_g$$

$$\Delta G° = -88.52 - 0 - (-71.79)$$

$$= -16.73 \text{ kcal/mol}$$

$$\log K_p = 16.73 \times 10^3/(1.987)(298)(2.303) = 12.268$$

$$K_p = 1.87 \times 10^{12}$$

16-11 FREE ENERGY AND SPONTANEOUS REACTIONS

For a reaction which is not at equilibrium, ΔG is given as

$$\Delta G = \Delta G° + RT \ln Q$$

Here Q is the reaction quotient (see Sec. 15-8). The value of ΔG
tells whether a reaction can occur spontaneously from a thermodynamic
point of view. If ΔG is negative, the reaction, as written, can be
spontaneous. The rate of such a reaction, though, may be too slow
to measure. If ΔG is positive, the reaction will be non-spontaneous.
If ΔG is zero, the system is at equilibrium.

Example 16-16

The equilibrium constant for the reaction of sulfur dioxide with
oxygen to form sulfur trioxide was found in Example 16-15 to be
1.87×10^{12}. In which direction will this reaction tend to go,
if these substances are mixed in a container at the following
partial pressures: SO_3, 1 atm; SO_2, $\tfrac{1}{4}$ atm, O_2, $\tfrac{1}{4}$ atm?

$$SO_2 + \tfrac{1}{2}O_2 \rightleftharpoons SO_3$$

$$\Delta G = \Delta G^\circ + RT \ln Q$$

$$\Delta G = \Delta G^\circ + RT \ln (P_{SO_3})/(P_{SO_2})(P_{O_2}^{\tfrac{1}{2}})$$

$$\Delta G = -16.73 \times 10^3 + (2.303)(1.987)[298 \log (1/(\tfrac{1}{4})(\tfrac{1}{4})^{\tfrac{1}{2}}]$$

$$\Delta G = -15,500 \text{ cal/mol}$$

Since ΔG is negative, the reaction will go to the right, ie, SO_2 will react with O_2 to form more SO_3.

16-12 CALCULATION OF FREE ENERGY CHANGES FROM CELL emf's

In Section 15-9 the equilibrium constant for a reaction was calculated from E° values by use of the equation:

$$nFE^\circ = + RT \ln K$$

since $\Delta G^\circ = -RT \ln K$

$$\Delta G^\circ = -nFE^\circ$$

Example 16-17

Johnson et al, determined E° for the cell

$\quad$ Li(ℓ)|LiX (saturated with LiH(s))|H_2(g), 1 atm, Fe,

to be

$\quad E^\circ = 0.9081 - 7.699 \times 10^{-4} T$

Here LiX was any halide and T may range from 673 to 853 K. The cell reaction is

$\quad$ Li(ℓ) + $\tfrac{1}{2}H_2$(g) $\rightarrow$ LiH(s)

What is the free energy change for this reaction at 800 K?
Here n = 1 and F = 23.06 kcal/V

$$\Delta G^\circ = -(1)(23.060)[0.9081 - 7.699 \times 10^{-4} \times 800]$$

$$\frac{kcal}{mol \nearrow V} \times V^{\diagup} = kcal/mol$$

$$\Delta G^\circ = -6.738 \text{ kcal/mol}$$

[C. E. Johnson, R. R. Heinrich, and C. T. Crouthamel,
J. Phys. Chem., 70, 242 (1966)]

The free energy change of a reaction which is not at equilibrium
can be calculated from:

$\Delta G = -nFE$

Example 16-18

What is the free energy change for the cell reaction in Example
15-10?

The cell reaction is

$$Cd + 2H^+ \rightleftharpoons Cd^{2+} + H_2$$

and E = 0.400 V

$\Delta G = -(2)(96,500)(0.400)VC$

moles x C/mole x V = VC

$\Delta G = -77200$ VC

1 VC = 1 joule

$\Delta G = -77200/4.184 = -18,451$ cal

joule x cal/joule = cal

> Note: An excellent article on entropy and its
> importance to everyday life is that by Henry
> Bent in Chemistry 54, (9), 6 (1971), entitled
> "Haste Makes Waste-Pollution and Entropy."

PROBLEMS

16-1 The heat of fusion of hydrogen fluoride is 46.93 calories
per gram at 189.79 K. What is the entropy change of fusion?
[J. Hu, D. White and H. Johnston, J. Am. Chem. Soc., 75,
1232 (1953)]

16-2 The heat of vaporization of diborane is 3412 cal/mole at
180.32 K. What is the entropy change of vaporization?

     [J. Clarke, E. Rifkin and H. Johnston, J. Am. Chem. Soc.,
     75, 781 (1953)]

16-3 The heat of sublimation of iodine at 298.1 K is 14,877 cal/mole.
The absolute entropy of iodine at 298.1 K is 27.9 eu/mole for the
solid state. What is the entropy of the gaseous state of iodine at
this temperature?

     [W. F. Giauque, J. Am. Chem. Soc., 53, 511 (1931)]

16-4 The absolute entropy of mercury was measured at the melting
point for the solid and the liquid state of mercury. The absolute
entropy of the solid state was 14.217 eu per mole and for the liquid
state, 16.559 eu per mole. The melting point is 234.29 K. What is
the heat of fusion of mercury?

     [R. Busey and W. Giauque, J. Am. Chem. Soc., 75, 806
     (1953)]

16-5 What is the entropy change of the system associated with the
compression of 1.33 moles of an ideal gas from 10 liters to 5 liters
at constant temperature?

16-6 What is the entropy change of the system associated with the
expansion of 0.75 mole of an ideal gas from 1 to 0.75 atm at constant
temperature?

16-7 If in Prob. 16-5, 700 cal were absorbed by the surroundings
at 25°C, what was the entropy change of the universe?

16-8 If in Prob. 16-6, 25 cal were absorbed by the system from the
surroundings at 25°C, what was the entropy change of the universe?

16-9 Nitrous oxide is a linear molecule which can orientate itself
in two possible ways at absolute zero. What is the entropy of this
molecule at absolute zero, according to the Boltzmann equation?
          [R. W. Blue and W. F. Giaque, J. Am. Chem. Soc., 57,
          991 (1935)]

16-10 Linus Pauling developed a structure for the crystalline state
of water which leads to 3/2 distinct ways in which the ice molecules
may be orientated at absolute zero. How does the entropy of this
structure compare to an observed value of 0.87 eu/mole for ice?
          [J. Am. Chem. Soc., 57, 2680 (1935)]

16-11 The absolute entropy of H_2S at its boiling point, 212.77 K,
was found to be 46.38 eu from heat capacity measurements and 46.44 eu
from spectroscopic measurements. Heat capacity measurements assume
the entropy of a substance to be zero at absolute zero, while
spectroscopic measurements do not. How many orientations of the
molecule are predicted at absolute zero from the above data?
          [W. Giauque and R. Blue, J. Am. Chem. Soc., 58, 831 (1936)]

16-12 Calculate the entropy change for each of the following
reactions. Balance the reactions where necessary.
    a. $C(graphite) + O_2(g) = CO_2(g)$
    b. $H_2(g) + Cl_2(g) = HCl(g)$
    c. $CO(g) + O_2(g) = CO_2(g)$
    d. $S(s) + O_2(g) = SO_3(g)$
    e. $SO_2(g) + O_2(g) = SO_3(g)$
    f. $H_2(g) + O_2(g) = H_2O(g)$

16-13 The absolute entropy of LiOH was found to be 10.231 eu per
mole at 298.16 K. The entropy change for the reaction

$$LiOH \cdot H_2O(s) \rightarrow LiOH(s) + H_2O(g)$$

was 38.29 eu per mole at 298.16 K. What is the absolute entropy of LiOH·H$_2$O(s) at this temperature? (Use the value for the entropy of water given in Table 16-1.)

> [T. Bauer, H. Johnston, and E. Keir, J. Am. Chem. Soc., 72, 5174 (1950)]

16-14 Calculate the free energy change for each of the following reactions. Balance the reactions where necessary. Tell whether the reactions are spontaneous or not spontaneous.

a. $B(s) + O_2(g) \rightleftarrows B_2O_3(s)$
b. $C_2H_4(g) + O_2(g) \rightleftarrows CO_2(g) + H_2O(g)$
c. $C_2H_4(g) + O_2(g) \rightleftarrows CO_2(g) + H_2O(\ell)$
d. $C_2H_6(g) + O_2(g) \rightleftarrows CO(g) + H_2O(\ell)$
e. $CaCO_3(s) \rightleftarrows CaO(s) + O_2(g)$
f. $Br_2(\ell) + HF(g) \rightleftarrows HBr(g) + F_2(g)$

16-15 Using the enthalpy of formation of BF$_3$ from Table 9-3 and the absolute entropies of B(s), F$_2$(g) and BF$_3$(g) from Table 16-1, verify the value of the free energy of formation given in Table 16-2.

16-16 Using the value of the enthalpy of formation for HCl(g) from Table 9-3 and the absolute entropies of HCl(g), H$_2$(g) and Cl$_2$(g) from Table 16-1 verify the value for the free energy of formation of HCl(g) listed in Table 16-2.

16-17 The free energy change for the reaction

$$GdCl_3(s) + H_2O(g) = GdOCl(s) + 2HCl(g)$$

was determined by measuring the equilibrium pressures of H$_2$O and HCl. In one experiment at 895 K the pressures obtained for duplicate determinations were: HCl, 610, 636 (mm) and H$_2$O, 2.46, 2.87 (mm). What is $\Delta G°$ for the reaction?

> [C. Koch and R. Cunningham, J. Am. Chem. Soc., 75, 796 (1953)]

16-18   The free energy of formation of Ag_2O was measured at 302°C
by determining the equilibrium pressure of O_2.   A value of 20.5 atm
was obtained at this temperature.   What is $\Delta G°_f$ of Ag_2O at this
temperature?

[G. N. Lewis, J. Am. Chem. Soc., 28, 154 (1906)]

16-19   The E° of the cell

   $Ag(s) | AgCl(\ell) | Cl_2(g, 1 \text{ atm}),$ C rod

was found to be 0.9001 V at 500°C.   What is the cell reaction?   What
is $\Delta G°$ for this reaction?

[E. J. Salstrom, J. Am. Chem. Soc., 56, 1274 (1934)]

16-20   Thompson and Flengas have determined the free energy of
formation of $Ag_2S(s)$ from 500 to 800°C by use of the cell

   $Ag | Ag_2S$ (saturated) AgCl, $KCl | S(g, 1 \text{ stm}),$ C

where AgCl and KCl are the solvents.

E° in millivolts was found to be given by E° = 455.3 - 0.1919 T
in this temperature range, where T is in degrees Kelvin.   What is
$\Delta G°$ for the formation reaction at 500°C?

[J. Electrochem. Soc., 118, 419 (1971)]

16-21   E. J. Salstrom determined the free energy of formation of
$PbBr_2(\ell)$ by measuring E° of the cell

   $Pb(\ell) | PbBr_2(\ell) | Br_2(g, 1 \text{ atm}).$

A tungsten wire was used to make contact with the molten lead and a
graphite rod in contact with the molten $PbBr_2$. Br_2 was passed over
the carbon rod.   At 500°C the researcher got a value of -47.600 kcal
for the formation of 1 mole of $PbBr_2(\ell)$.   What was E° for the cell?

[J. Am. Chem. Soc., 52, 4648 (1930)]

SUPPLEMENTARY PROBLEMS

16-22 The heat of vaporization of hydrogen fluoride at 741.4 mm is
89.45 cal/g. The boiling point of HF at this pressure is 292.61 K.
What is the entropy change of vaporization?

> [J. Hue, D. White, and H. Johnston, J. Am. Chem. Soc., 75,
> 1232 (1953)]

16-23 Diborane has a heat of fusion of 1069.1 cal/mole at 108.30 K.
What is the entropy change of fusion?

> [J. Clarke, E. Rifkin, and H. Johnston, J. Am. Chem. Soc.,
> 75, 781 (1953)]

16-24 The entropy of the liquid state of chlorine trifluoride at
284.91 K, the boiling point of the substance, was found to be
43.66 cal/deg-mole. The heat of vaporization at this temperature
is 6,580 cal/mole. What is the entropy of the gaseous state at
the boiling point?

> [J. Grissard, H. Bernhardt, and G. Oliver, J. Am. Chem.
> Soc., 73, 5725 (1951)]

16-25 The absolute entropy of rhombic sulfur and monoclinic
sulfur was measured at 368.6 K, the transition temperature. A value
of 8.827 eu was obtained for S(r) and 9.042 eu for S(m). What is
the entropy of transition from rhombic to monoclinic sulfur at this
temperature? What is the enthalpy of transition for this same change?

> [E. D. Eastman and W. C. McGavock, J. Am. Chem. Soc., 59,
> 150 (1937)]

16-26 The heat of fusion of molybdenum trioxide is 12,540 cal/mole
at 1068.36 K, the melting point of MoO_3. The absolute entropy of
$MoO_3(s)$ at 1068.36 K was found to be 46.823 cal/deg-mole. What is

the absolute entropy of the liquid state of MoO_3 at this melting
point?

> [L. Cosgrove and P. Snyder, J. Am. Chem. Soc., 75, 1227
> (1953)]

16-27 What is the entropy change of an ideal gas for the expansion
of 2.2 moles of the gas from 100 atm to 1 atm at constant temperature?

16-28 If the process described in Prob. 16-27 is carried out in
such a way that no work is done on or by the gas, what is the entropy
change of the system, the surroundings, and the universe?

16-29 Westrum and Pitzer have determined from heat capacity measurements that the F-H-F⁻ anion in KHF_2 is completely symmetrical. What
would be the expected value of the absolute entropy of KHF_2 at
absolute zero?

> [E. Westrum, Jr., and K. Pitzer, J. Am. Chem. Soc., 71,
> 1940 (1949)]

16-30 The absolute entropy of nitrous oxide was determined at
184.59 K, the boiling point, in two different ways. Using spectroscopic data, a value of 48.50 eu/mole was obtained. From specific
heat data, a value of 47.36 eu/mole was obtained. Specific heat
measurements assume the entropy of nitrous oxide to be zero at
absolute zero, while spectroscopic measurements do not. The difference in these results is due to the different orientations of the
molecule at absolute zero. From the Boltzmann equation, how many
different orientations are necessary to account for this data?

> [R. W. Blue and W. F. Giauque, J. Am. Chem. Soc., 57,
> 991 (1935)]

16-31 The absolute entropy of $ReCl_3(s)$ was measured and found to be
29.61 eu/mole, and the absolute entropy for Re(s) was 8.74 eu/mole.
The heat of formation of $ReCl_3(s)$ was -63.0 kcal/mole. What is the

free energy of formation of $ReCl_3(s)$? Use the entropy value for
Cl_2 in Table 16-1.

> [R. B. Bevan, Jr., R. A. Gilbert, and R. H. Busey, J. Phys.
> Chem., 70, 147 (1966)]

16-32 Using the value of the enthalpy of formation for $PH_3(g)$ from
Table 9-3 and the absolute entropies of $P(s)$ and $H_2(g)$ from
Table 16-1, verify the value for the free energy of formation of
$PH_3(g)$ listed in Table 16-2.

16-33 The free energy change for the reaction

$$Hg(\ell) + H_2S(g) \rightleftarrows HgS(s) + H_2(g)$$

at 25°C, was determined to be -2.33 kcal. Using the values from
Table 16-2, calculate the free energy of formation of HgS.

> [J. Goates, A. Cole, and E. Gray, J. Am. Chem. Soc., 73,
> 3596 (1951)]

16-34 The free energy change for the reaction

$$SmCl_3(s) + H_2O(g) = SmOCl(s) + 2HCl(g)$$

was determined by measuring the equilibrium pressures of H_2O and
HCl. In one experiment at 893 K, the pressures obtained for duplicate
determinations were HCl; 418.8, 406.3 mm; and H_2O: 1.60, 1.60 mm.
What is $\Delta G°$ for this reaction?

> [C. Koch and B. Cunningham, J. Am. Chem. Soc., 75, 797
> (1953)]

16-35 The free energy of formation of Ag_2O was determined by
measuring the pressure of O_2 in equilibrium with Ag and Ag_2O at
various temperatures. At 173°C a pressure of 422 mm was found.
What is $\Delta G°_f$ of Ag_2O at this temperature?

> [A. Benton and L. Drake, J. Am. Chem. Soc., 54, 2188 (1932)]

16-36   The emf of the cell

   Hg, HgS|H$_2$S(1 atm)|HCl (0.1 m)|H$_2$ (1 atm) Pt

was determined at 25°C.   E° was 0.0504 V.   The cell reaction is

   Hg(ℓ) + H$_2$S(g) $\rightleftarrows$ HgS(s) + H$_2$(g)

Calculate $\Delta G°$ for this reaction.

          [J. Goates, A. Cole, and E. Gray, J. Am. Chem. Soc., 73,
          3596 (1951)]

16-37   The free energy of formation of PbS(s) was determined from
500 to 800°C using a proper electrochemical cell.   The E°, in
millivolts, of the cell was determined to be E° = 867.4-0.4884 T.
Here T is degrees Kelvin.   What is $\Delta G°$ for this reaction at 500°C?

          [W. T. Thompson and S. N. Flengas, J. Electrochem. Soc.,
          118, 419 (1971)]

16-38   The E° of the cell

   Ag(s)|AgBr(ℓ)|Br$_2$(g, 1 atm), C (rod)

was found to be 0.7865 V at 500°C.   Write the cell reaction.   What
is $\Delta G°_f$ for this reaction?

          [E. J. Salstrom and J. H. Hildebrand, J. Am. Chem. Soc.,
          52, 4653 (1930)]

16-39   Wachter and Hildebrand have measured E° for the cell

   Pb (ℓ)|PbCl$_2$(ℓ)|Cl$_2$(g, 1 atm).

A tungsten wire was used to contact the molten lead and a graphite
rod for the chlorine electrode.   E° at 500°C was 1.2730 V.   What
is the cell reaction?   What is $\Delta G°$ for this reaction?

          [J. Am. Chem. Soc., 52, 4655 (1930)]

Chapter 17

CHEMICAL KINETICS

17-1 INTRODUCTION

Chemical kinetics includes the study of the rates at which reactions occur, the factors which influence the rates, and detailed mechanisms by which the reactions occur. The factors which influence the rate of a chemical reaction are concentration, temperature, and catalysts.

17-2 RATE OF REACTION

The rate of a chemical reaction is usually defined as the change in concentration of a <u>reactant</u> per unit of time. The units are the units of concentration divided by units of time, for example, mole per liter per second.

17-3 RATE EQUATION

At constant temperature, the rate of a chemical reaction is proportional to the concentration of the reacting substances (law of mass action). An equation describing this relationship for a particular reaction is called the <u>rate equation</u> of the reaction.

In a generalized reaction represented by A → C + D, the rate of the reaction can be represented by the rate equation:

Rate $= k[A]^n$

where [ ] represents the concentration of the species in moles per liter. The value of the exponent n in the rate equation is called the order of the reaction. The constant k is the specific rate constant of the reaction.

The nature of the proportionality between the rate and the concentration of a reactant in a particular reaction is given by the value of n, the order of the reaction, and the value of k. The numerical values of n and k are calculated from experimental data. The usual data are the concentration of a reactant observed at time intervals during the course of the reaction.

Example 17-1

The decomposition of N_2O_5 has been investigated several times, both in the gas phase and in solution. The following data have been calculated from one such investigation by H. C. Ramsperger, M. E. Nordberg, and R. C. Tolman.

Time, sec	Concentration of N_2O_5, moles/liter
110	1.92×10^{-3}
980	1.65×10^{-3}
2,310	1.31×10^{-3}
3,670	1.04×10^{-3}
4,770	0.865×10^{-3}
6,580	0.644×10^{-3}
8,988	0.428×10^{-3}

Source: Proc. Natl. Acad. Sci., U. S., 15, 450 (1929)

These data are graphed in Fig. 17-1. The upper curve is the log of the concentration, and the lower curve is the concentration. The average rate of the decompositon during any time interval may be calculated from the data.

If the concentration of N_2O_5 is represented by $[N_2O_5]$, the average rate of decomposition will be

$$r_{av} = \frac{[N_2O_5]_2 - [N_2O_5]_1}{t_2 - t_1} = \frac{\Delta [N_2O_5]}{\Delta t}$$

Since $[N_2O_5]_2 < [N_2O_5]_1$, $\Delta[N_2O_5]$ will be negative and r_{av} will be negative, signifying that it is the rate of disappearance of N_2O_5.

Since the concentration of N_2O_5 is continually decreasing, the rate of the reaction will continually diminish. The average rate of reaction is not a useful concept, but the instantaneous rate is useful. If the time interval Δt over which the rate is observed is made smaller and smaller, it approaches zero as a limit. By using calculus notation

$$r = \frac{-d[N_2O_5]}{dt} = \lim_{\Delta t \to 0} \frac{-\Delta [N_2O_5]}{\Delta t}$$

From experimental observation, the rate of this reaction is proportional to the concentration of N_2O_5

$$\frac{-d[N_2O_5]}{dt} = k[N_2O_5]$$

This is the rate equation for the decomposition of dinitrogen pentoxide. In the general case of a reaction $A \to C + D$, the rate equation takes the form:

$$\frac{-d[A]}{dt} = k[A]^n$$

As noted at the beginning of this section, the value of n, the exponent in the rate equation, is the order. The rate of decomposition of N_2O_5 is first-order with respect to the N_2O_5 concentration.

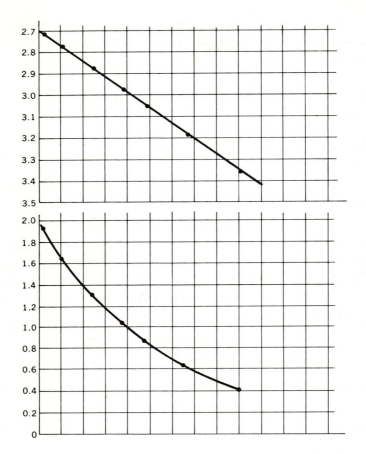

FIGURE 17-1   Log of concentration of N_2O_5 vs time (upper curve);
              concentration of N_2O_5 x 10^3 vs time (lower curve)

17-4   DETERMINATION OF n AND k

The order of a reaction and its specific reaction rate constant
may be determined by means of the integrated form of the rate
equation:

$$\frac{-d[A]}{[A]} = k \ dt$$

by integration

$$- \int_{[A_o]}^{[A]} \frac{d[A]}{[A]} = k \int_o^t dt$$

$$-(\ln[A] - \ln[A_o]) = k(t - 0)$$

$$\ln[A_o] - \ln[A] = kt$$

$$2.303 \, (\log[A_o] - \log[A]) = kt$$

$$\log[A_o] - \log[A] = \frac{k}{2.303} t$$

$[A_o]$ is the initial concentration of the reactant A. The equation may be written in the form:

$$\log[A] = - \frac{k}{2.303} t + \log[A_o]$$

where $\frac{-k}{2.303}$ is the slope and $\log[A_o]$ is the intercept.

Example 17-2

The rate of decomposition of azomethane was determined by H. C. Ramsperger. The reaction which occurred was

$$CH_3NNCH_3 \rightarrow C_2H_6 + N_2$$

Time, min	$[CH_3NNCH_3]$, moles/liter
0	1.21×10^{-2}
10	1.04×10^{-2}
20	0.880×10^{-2}
33	0.706×10^{-2}
46	0.575×10^{-2}
65	0.434×10^{-2}

Source: J. Am. Chem. Soc., 49, 912 (1927)]

Calculate the specific reaction rate constant.

A plot of $\log [CH_3NNCH_3]$ against time gives a straight line, indicating that the reaction is first-order (see Fig. 17-2). (The use of semilogarithmic graph paper minimizes the work involved.)

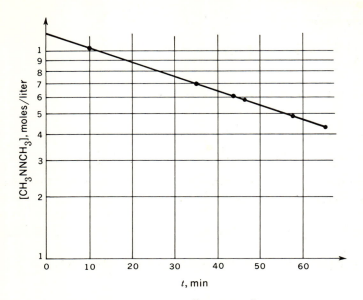

FIGURE 17-2   Log of $[CH_3NNCH_3]$ vs time

From the slope of the curve, using the logarithms of the concentrations

$$k = 1.57 \times 10^{-2}$$

The specific reaction rate constant k may be calculated from the integrated form of the first-order rate equation:

$$\log [A] = -\frac{k}{2.303} t + \log [A_o]$$

$$k = \frac{2.303}{t} \times (\log[A_o] - \log [A])$$

$$= \frac{2.303}{10} \times (\log 1.21 \times 10^{-2} - \log 1.04 \times 10^{-2})$$

$$= 0.230 \times 0.0658 = 1.51 \times 10^{-2} \ min^{-1}$$

In a similar way, other values of k may be calculated from the data.

t, min	k, min^{-1}
20	1.55×10^{-2}
33	1.60×10^{-2}
46	1.60×10^{-2}
65	1.51×10^{-2}

The average of these calculated values is 1.54×10^{-2} min^{-1}.

Example 17-3

F. O. Rice and D. Getz found that at 64.0°C, 0.00091 mole/liter of N_2O_5 had decomposed after 2.474 min from an initial concentration of 0.001943 mole/liter. Calculate k for this first-order decomposition.

$$\log 1.943 \times 10^{-3} - \log 1.032 \times 10^{-3} = \frac{k}{2.303} 2.474$$

$$\frac{2.303 \times 0.2749}{2.474} = k$$

$$k = 0.256 \text{ min}^{-1}$$

[J. Phys. Chem., 31, 1572 (1927)]

17-5 HALF-LIFE

The time required for one-half of the initial quantity of a reactant to be consumed in a reaction is the half-life of that substance. For a first-order reaction, the half-life $t_{\frac{1}{2}}$ is given by:

$$t_{\frac{1}{2}} = \frac{0.693}{k}$$

Example 17-4

In the reaction described in Example 17-2, calculate the half-life by using the average value of k, 1.54×10^{-2}.

$$t_{\frac{1}{2}} = \frac{0.693}{1.54 \times 10^{-2}} = 45.0 \text{ min}$$

($t_{\frac{1}{2}}$ estimated from the curve is 43.5 min)

17-6   OTHER INTEGRATED EQUATIONS FOR FIRST-ORDER REACTIONS

Another, practical form of the first-order rate equation is:

$$- \frac{d(a - x)}{dt} = k(a - x)$$

where a is the initial concentration of the reacting species and x
is the quantity which has reacted in time t.

After integration and rearrangement, the equation may be
expressed as

$$k = \frac{2.303}{t} \log \frac{a}{a - x}$$

Example 17-5
Calculate the value of k for the decomposition of N_2O_5 at 64.4° from
the data of F. O. Rice and D. Getz. After 2.123 min, 0.001018 mole
of N_2O_5 had decomposed when the initial quantity of N_2O_5 was
0.002347 mole. (Assume that the volume was constant.)
          [J. Phys. Chem., 31, 1572 (1927)]

$$k = \frac{2.303}{t} \log \frac{a}{a - x}$$

a   =   0.002347

x   =   0.001018

t   =   2.123 min

a - x   =   0.001329

$$k = \frac{2.303}{2.123} \log \frac{0.002347}{0.001329}$$

log 0.002347   =   -3 + 0.3705   =   -2.6295
log 0.001329   =   -3 + 0.1235   =   -2.8765

$$(-2.6295) - (-2.8765) = 0.2470$$

$$k = \frac{2.303 \times 0.2470}{2.123} = 0.268 \ min^{-1}$$

If the rate equation is integrated between limits, another useful integrated form results:

$$2.303 \ log\frac{a - x_1}{a - x_2} = k(t_2 - t_1)$$

where x_1 and x_2 are the concentrations of the substance which have reacted at times t_1 and t_2.

Example 17-6

Calculate k for the decomposition of N_2O_5 at 45°C from the data of F. Daniels. The pressure of N_2O_5 after 1800 sec was 140 mm, and after 3600 sec it was 58 mm.

> [Chemical Kinetics, Cornell University Press, Ithaca, New York, 1938, p.9]

$$2.303 \ log\frac{a - x_1}{a - x_2} = k(t_2 - t_1)$$

The pressures 140 and 58 mm are measures of the concentrations of N_2O_5 at times t_1 and t_2, $a - x_1$, and $a - x_2$.

$$2.303 \ log\frac{140}{58} = k \times (3600 - 1800)$$

$$2.30 \times 0.382 = k \times 1800$$

$$k = \frac{2.30 \times 0.382}{1.80 \times 10^{+3}} = 4.87 \times 10^{-4} \ sec^{-1}$$

17-7 HIGHER-ORDER REACTIONS

In the general reaction

$$aA + bB \rightarrow products$$

the rate of the reaction may be described by the rate equation:

$$\frac{-d[A]}{dt} = k[A]^m[B]^n$$

The exponent m is the "order of the reaction with respect to reactant A," and n is the "order with respect to the reactant B." The overall order of the reaction is the sum of n and m. As in the case of first-order reactions, the order must be determined by experiment.

Example 17-7

An investigation of the rate of reaction of $2NO_2 + F_2 \rightarrow 2NO_2F$ has shown that it is first-order with respect to NO_2 and first-order with respect to F_2.

$$\frac{-d[F_2]}{dt} = k'[NO_2]$$

$$\frac{-d[F_2]}{dt} = k''[F_2]$$

The overall order of the reaction is second-order, the sum of the two exponents.

$$\frac{-d[F_2]}{dt} = k[NO_2][F_2]$$

[R. L. Perrine and H. S. Johnston, J. Chem. Phys., 21, 2202 (1953)]

Rate equations for higher reactions are complicated by the fact that the number of moles of each reactant involved may be different. In the generalized equation

$$aA + bB \rightarrow cC + dD$$

the rate of the reaction is defined as:

$$-\frac{1}{a}\frac{d[A]}{dt} = -\frac{1}{b}\frac{d[B]}{dt} = \frac{1}{c}\frac{d[C]}{dt} = \frac{1}{d}\frac{d[D]}{dt}$$

Example 17-8

In the reaction described in Example 17-7, the rate may be expressed as:

$$-\frac{d[F_2]}{dt} = -\frac{1}{2}\frac{d[NO_2]}{dt} = k[NO_2][F_2]$$

17-8 SECOND-ORDER REACTIONS

Just as the use of calculus led to a differential equation for a first-order reaction, so a rate equation for a second-order reaction may be found:

$$\frac{-d[A]}{dt} = k[A]^2$$

Integration leads to

$$\frac{1}{[A]} - \frac{1}{[A_o]} = kt$$

This equation holds for second-order reactions in which the reactants are present in equivalent concentrations. More involved equations result from more complex situations. The problems here are limited to this simple type.

A plot of 1/[A] versus t will show that a reaction is second-order if a straight line results.

Example 17-9

What is the order of the reaction

$$CH_3COOCH_3 + OH^- \rightarrow CH_3CO_2^- + CH_3OH?$$

$[CH_3COOCH_3]$ = $[OH^-]$ = 0.0100 M at the start of the reaction.

t, min	[CH$_3$COOCH$_3$]	1/[CH$_3$COOCH$_3$]
0	0.0100	100
3	0.00740	135
4	0.00683	147
5	0.00634	158
6	0.00589	170
7	0.00550	182
8	0.00519	193
10	0.00464	216
12	0.00416	241
15	0.00363	276

Source: J. Walker, Proc. Roy. Soc. (London), Ser. A, 78,
157 (1906)

The plot of 1/[CH$_3$COOCH$_3$] versus t indicates that the reaction
is second-order with respect to [CH$_3$COOCH$_3$] (see Fig. 17-3).
k may be calculated from the equation:

$$\frac{1}{[A]} - \frac{1}{[A_o]} = kt$$

$$135 - 100 = 3k$$

$$\frac{35}{3} = k = 11.7 \ (mole/liter)^{-1} \ min^{-1}$$

$$147 - 100 = 4k$$

$$\frac{47}{4} = k = 11.7 \ (mole/liter)^{-1} \ min^{-1}, \ etc.$$

Unlike k for first-order reactions, the units of concentration
must be included in k for second-order reactions.

17-9   HALF-LIFE OF SECOND-ORDER REACTIONS

The time required for the initial concentration of the reactants
to decrease to one-half of their initial values for second-order
reactions is given by

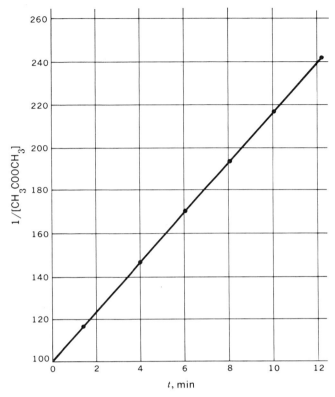

FIGURE 17-3 $1/[CH_3COOCH_3]$ vs time

$$t_{\frac{1}{2}} = \frac{1}{ka}$$

where a is the initial concentration of one of the reactants.

Example 17-10

What is the half-life of the reaction in Example 17-9?

$$t_{\frac{1}{2}} = \frac{1}{ka}$$

$$t_{\frac{1}{2}} = \frac{1}{11.7 \times 0.0100} = 8.55 \text{ min}$$

$$\frac{1}{\dfrac{mole^{-1}}{liter} \times min^{-1} \times \dfrac{mole}{liter}} = min$$

PROBLEMS

17-1 In an experiment in which the thermal decomposition of N_2O_5 was investigated, an initial concentration of 0.002347 mole/liter had decreased to 0.001329 mole/liter in 2.123 min at 64.4°C. Calculate k for the first-order reaction.

[F. O. Rice and D. Getz, J. Phys. Chem., 31, 1572 (1927)]

17-2 In a second Rice and Getz experiment, 0.000902 mole/liter of N_2O_5 decomposed from an initial concentration of 0.001993 mole/liter in 2.226 min. Determine k for the reaction.

[F. O. Rice and D. Getz, J. Phys. Chem., 31, 1572 (1927)]

17-3 In a kinetics study the initial concentration of one of the reactants was 0.188 mM in 4.00 ml of solution. After 25 min, 68.4% of the substance remained. Assuming that the reaction was first-order with respect to the substance, calculate k.

[W. A. Pryor, J. Am. Chem. Soc., 80, 6483 (1958)]

17-4 Calculate the specific reaction-rate constant for the first-order decomposition of azomethane, $(CH_3)_2N_2$, at 320.4 K. The initial pressure was 21.23 cm in a 210-cc container; the pressure after 9 min was 10.20 cm.

[H. C. Ramsperger, J. Am. Chem. Soc., 49, 912 (1927)]

17-5 The vapor phase thermal decomposition of nitroglycerin was
studied at 160°C. It was found to follow first-order kinetics. The
following data were obtained from a series of experiments:

Initial concentration, g cm^{-3} (M/V) nitroglycerin	Time (sec)	% Decomposition
1.11	300	52.0
0.80	300	52.9
0.52	300	53.2
0.41	300	53.9
1.21	180	34.6
0.92	180	35.9
0.67	180	36.0
0.39	180	35.4

What is the average value of the rate constant determined from the
above data?

[C. E. Waring and G. Krastins, J. Phys. Chem., 74,
999 (1970)]

17-6 Determine the order of the reaction and the value of k from
the following data for the decomposition of N_2O_5:

t, sec	$P_{N_2O_5}$, mm
110	0.826
980	0.708
2310	0.562
3670	0.449
4770	0.372
6580	0.277
8988	0.184

Source: H. C. Ramsperger, M. E. Nordberg, and R. C. Tolman, Proc.
Natl. Acad. Sci. U. S., 15, 456 (1929)

17-7 In an investigation of the decomposition of N_2O_5 in carbon
tetrachloride solution, R. H. Lueck found that the specific reaction-
rate constant was 3.21×10^{-4} sec^{-1}. What was the half-life of the
reaction?

[J. Am. Chem. Soc., 44, 757 (1922)]

17-8 The rate of decomposition of azomethane was studied by H. C.
Ramsperger. In one experiment, conducted in a 210-cc chamber at
296.8°C, the following data were obtained:

t, min	Pressure of azomethane, cm
0	3.62
15	3.00
30	2.49
48	1.93
75	1.31

Source: J. Am. Chem. Soc., 49, 912 (1927)

Calculate the specific reaction rate constant and the half-life
of the reaction.

17-9 The rate of hydrolysis of a complex boron cation was followed
by measuring the volume of hydrogen given off by the reaction. The
concentration of the unreacted boron complex is proportional to
$V_\infty - V_t$ (volume of hydrogen after the reaction is completed - volume
of hydrogen at time t.) Determine the order and the rate constant
for this reaction.

	100 x
Time (sec)	$(V_\infty - V_t)$ ml
152	90
248	87
344	83
401	81
540	75
589	73
660	70
775	65
955	58
1024	56
1075	54
1164	51
1254	48
1392	44
1561	39
1799	34
2992	16
3160	14
3715	11.5
4010	8.0
4676	6.0
5330	3.0

[N. E. Miller, J. Am. Chem. Soc., 92, 4564 (1970)]

17-10 The thermal isomerization of cyclopropane occurs according
to the equation:

$$CH_2\text{---}CH_2 \rightarrow CH_3 - CH = CH_2$$
$$\diagdown CH_2 \diagup$$

cyclopropane   propene

T. S. Chambers and G. B. Kistiakowsky found that the specific
reaction-rate constant for this reaction at 499°C was 5.95×10^{-4}
sec^{-1}. What is the half-life of the reaction? What fraction of the
cyclopropane will remain after 1.00 hr at 499.5°C?

[J. Am. Chem. Soc., 56, 399 (1934)]

17-11 The thermal isomerization of cyclobutane, described in
Prob. 17-30, was carried out in a 0.53-liter vessel at 150°C. The
initial pressure was 50 mm. Calculate the concentration of cyclobutane, in moles per liter, after 30 min. k = 2.0×10^{-4} sec^{-1}.

[W. Cooper and W. D. Walters, J. Am. Chem. Soc., 80,
4220 (1958)]

17-12 The initial pressure of cyclopropane, C_3H_6, in a 500-cc
vessel was 702.6 mm at 499.5 K. Calculate the concentration of
cyclopropane which remained after 720 sec if the specific reaction-
rate constant for its isomerization was 5.95×10^{-4} sec^{-1}.

[T. S. Chambers and G. B. Kistiakowsky, J. Am. Chem. Soc.,
56, 399 (1934)]

17-13 Chlorine dioxide reacts with iodide ions in aqueous solutions
at 25°C according to the following equation:

$$2ClO_2 + 2I^- \rightarrow 2ClO_2^- + I_2$$

When a large excess of $[I^-]$ is present, the reaction is pseudo
first-order. A $ClO_2 \cdot I^-$ complex is formed which decomposes following
first-order kinetics. The concentration of this complex can be
measured spectroscopically.

Time(sec)	$[ClO_2 \; I^-]M \times 10^4$
0.00	4.77
1	4.31
2	3.91
3	3.53
4	3.19
5	2.89
10	1.76
15	1.06
20	0.64
30	0.24
40	0.087
50	0.032

What is k for this pseudo first-order reaction?
  [H. Fukutomi, and G. Gordon, J. Am. Chem. Soc., 89,
  1362 (1967)]

17-14 What is the half-life of a second-order reaction where
k = 1.8×10^{-4} (mM/liter)$^{-1}$ min^{-1}?

17-15 Calculate k for the second-order reaction in which the initial
concentration of the reacting substance was 123 mM/liter and the
concentration after 10 min was 90.0 mM/liter.

17-16 The decomposition of NO_2 was studied by M. Bodenstein and I.
Ramstetter, from whose paper the data below have been taken. The
decomposition proceeded according to the equation:

  $2NO_2 \rightleftarrows 2NO + O_2$

t, min	[NO_2], mM/liter	$1/[NO_2]$
0	123	8.13×10^{-3}
1	107.6	9.26×10^{-3}
2	105.1	9.54×10^{-3}
3	102.8	9.74×10^{-3}
4	100.8	9.90×10^{-3}
5	99.1	10.1×10^{-3}
6	97.3	10.3×10^{-3}

Source: I. Ramstetter, Z. Physik. Chem. (Leipzig), 100, 106 (1922)
Determine the order of the reaction.

17-17 The rate of reduction of a series of oxonides with tri-
phenyl phosphine at 25°C was found to follow second-order kinetics.
In a series of experiments, the half-life of the reaction was
measured at various initial concentrations of both reactants. At
all times the initial concentration of each reactant was the same.

Initial concentration of reactants (mole liter^{-1})	$t_{\frac{1}{2}}$(sec)
0.05	302
0.10	152
0.15	101
0.20	75

[J. Charles and S. Flisgar, Can. J. Chem., 48, 1309 (1970)]
Determine k for this reaction.

17-18 The reduction of an ozonide of styrene with triphenylphosphine
was followed spectroscopically. The initial concentration of each
reactant was 0.1 M. The fraction of each substance reacted is
given by D/D_∞ from the following data:

t (sec)	D x 10^2
0	0
355	22
638	29
949	32
1265	34
2130	37.5
2480	38.0
2870	38.5
∞	43.0

[J. Charles and S. Flisgar, Can. J. Chem., 48, 1309 (1970)]
What is the second-order rate constant k for this reaction?

17-19 A study of the acid hydrolysis of organic acetals and ketals
showed they followed pseudo first-order kinetics. The reactions were
first-order with respect to the concentration of acid and first-order
with respect to the acetal concentration. Since the acid acted only
as a catalyst, its concentration did not change with time, making
the observed reaction pseudo first-order. In a dioxane-water
mixture at 34.3°C the following data were obtained:

time (min)	10^3 x [acetal] mmol/ml
0	23.6
9	22.9
21	22.0
48	20.8
81	18.7
128	16.7
168	14.0
226	12.1
296	10.3
348	8.8
416	7.6

[M. S. Newman and R. E. Dickson, J. Am. Chem. Soc., 92, 6880 (1970)]

a. What is k for the pseudo first-order reaction?

b. If the $[H^+]$ was 0.02 M, what is the second-order rate constant?

17-20 The rate of the reaction $NH_4NO_2 \rightarrow N_2 + H_2O$ was investigated by S. K. Ghosh. The following data were obtained:

t, hr	$[NH_4^+]$	$[NO_2^-]$
0	0.500	0.500
3	0.432	0.459
6	0.401	0.430
9	0.373	0.410
24	0.340	0.361
30	0.330	0.351

Source: Z. Physik. Chem. (Leipzig), 206, 321 (1957)

a. What is the order of the reaction with respect to NH_4^+?

b. What is the order of the reaction with respect to NO_2^-?

c. What is the overall order of the reaction?

SUPPLEMENTARY PROBLEMS

17-21 What is the first-order reaction-rate constant for the isomerization of cyclopropane from the observations of E. S. Comer and R. N. Pease? They observed an initial pressure of 910 mm, and the pressure due to the cyclopropane which had reacted in 1200 sec was 450 mm.

[J. Am. Chem. Soc., 67, 2067 (1945)]

17-22 Find k for the thermal decomposition of N_2O_5 at 64.6°C
from the observation that, of an initial quantity of 0.002193 mole,
0.000977 mole had decomposed after 2.194 min.

[F. O. Rice and D. Getz, J. Phys. Chem., 31, 1572 (1927)]

17-23 Calculate k for the decomposition of N_2O_5 at 35°C from the
data of F. Daniels and E. H. Johnston. The initial pressure of
N_2O_5 was 308.2 mm, and the partial pressure due to the N_2O_5 which
had decomposed after 2400 sec was 90.0 mm.

[J. Am. Chem. Soc., 43, 53 (1921)]

17-24 The rate of decomposition of N_2O_5 in CCl_4 solution may be
followed by measuring the amount of decomposition X in terms of the
volume of oxygen liberated after various times. Determine the value
of k for the decomposition from the data of R. H. Lueck, measured
at 40°C.

t, sec	V_{O_2}, cc
0	0.00
600	6.30
900	8.95
1200	11.40
1800	15.53
2400	18.90
3000	21.70
∞	34.75

Source: J. Am. Chem. Soc., 44, 757 (1922)

17-25 Calculate the value of k for the isomerization of cyclopropane
from the observation that, of an initial pressure of 649 mm, 143 mm
had reacted after 420 sec.

[E. S. Comer and R. N. Pease, J. Am. Chem. Soc., 67,
2067 (1945)]

<u>17-26</u> Find k for the decomposition of N_2O_5 at 37.30°C from the pressures of N_2O_5 after 2310 and 6580 sec, which were 0.562 and 0.277 mm, respectively.

[H. C. Ramsperger, M. E. Nordberg, and R. C. Tolman,
Proc. Natl. Acad. Sci. U. S., <u>15</u>, 456 (1929)]

<u>17-27</u> In an experiment of F. O. Rice and D. Getz, a CCl_4 solution containing 8.86 x 10^{-5} mole of N_2O_5 per gram of solution was observed while decomposition of N_2O_5 occurred at 34.2°C. After 77.2 min, 3.67 x 10^{-5} mole of N_2O_5 per gram of solution had decomposed; after 105 min, 4.44 x 10^{-5} mole per gram of solution had decomposed. Find k.

[J. Phys. Chem., <u>31</u>, 1572 (1927)]

<u>17-28</u> The reaction of dimethylmagnesium with 4-methylmercapto-acetophenone follows pseudo first-order kinetics if the $(CH_3)_2Mg$ is present in a large excess. In one experiment with the ketone concentration at 10^{-3} M and $(CH_3)_2Mg$ at 0.185 M, the following data were obtained:

time (sec)	Rel. A
0.050	9.26
0.075	7.08
0.100	5.48
0.125	4.30
0.150	3.24
0.175	2.49
0.200	1.92
0.225	1.46
0.250	1.12
0.275	0.89
0.300	0.70

Rel. A. is proportional to the concentration of the 4-methylmercapto-acetophenone. What is k for this reaction?

[S. Smith and J. Billet, J. Am. Chem. Soc., 89, 6948 (1967)]

17-29 The decomposition of N_2O_5 in carbon tetrachloride at 35° was found to be first-order, and the specific rate constant was 8.67 x 10^{-2} min^{-1}. In one experiment the initial concentration of N_2O_5 was 8.54 x 10^{-5} mole/g of solution. Calculate the concentration of N_2O_5 after 15.0 min.

17-30 The thermal isomerization of cyclobutene:

$$HC = CH$$
$$| \quad | \quad \xrightarrow{150°} \quad CH_2 = CH - CH = CH_2$$
$$H_2C = CH_2$$

is first-order, and its specific reaction-rate constant was observed to be 2.0 x 10^{-4} sec^{-1}. Calculate the half-life of the reaction. What fraction of the cyclobutane will remain after 20.0 min?

[W. Cooper and W. D. Walters, J. Am. Chem. Soc., 80, 4220 (1958)]

17-31 The initial concentration of a carbon tetrachloride solution of N_2O_5 was 5.33 moles/liter. H. Eyring and F. Daniels found that the specific reaction-rate constant was 6.74 x 10^{-4} min^{-1}. Calculate the concentration of N_2O_5 after 126 min.

[J. Am. Chem. Soc., 52, 1472 (1930)]

17-32 Determine the concentration of azomethane remaining after 33 min if the initial concentration was 1.21 x 10^{-2} mole/liter and k = 1.60 x 10^{-2} min^{-1}.

[H. C. Ramsperger, J. Am. Chem. Soc., 49, 912 (1927)]

17-33 The reaction of chlorite ion with peroxymonosulfate ion was shown to be first-order in each of the reactant concentrations:

$$ClO_2^- + HSO_5^- = ClO_3^- + SO_4^{2-} + H^+$$

What is the overall rate equation for the reaction?

> [R. W. Johnson and J. O. Edwards, Inorg. Chem., 5,
> 2073 (1966)]

17-34 The hydrolysis of C_6H_5COCl was studied by B. L. Archer and R. F. Hudson, who found that after 186 sec, 0.00084 mole/liter had reacted, and that after 236 sec, 0.00102 mole/liter had reacted. The initial concentration was 0.0100 mole/liter. Calculate k for the reaction.

> [J. Chem. Soc., 1950, 3259]

17-35 Determine the order of the reaction with respect to the methyl acetate concentration in the reaction

$$CH_3COOCH_3 + OH^- = CH_3CO_2^- + CH_3OH$$

The data are from the work of J. Walker. The initial concentration was 0.01000 M. Calculate k for this reaction.

t, min	X, moles/liter
0	0
3	0.00260
4	0.00317
5	0.00366
6	0.00411
7	0.00450
8	0.00481
10	0.00536
12	0.00584
15	0.00637

Source: J. Walker, Proc. Roy. Soc. (London), Ser. A, 78, 157 (1906)

17-36 The rate of oxidation of I⁻ by ClO⁻ was studied by Y. T.
Chia and R. E. Connick, at 25°C. Starting with 2.00×10^{-3} M I⁻
and ClO⁻, the second-order rate constant was found to be 60.6
$(M/liter)^{-1}$ sec^{-1}. Calculate the concentration of each of the
ionic apecies after 5.00 sec.

   I⁻ + ClO⁻ = IO⁻ + Cl⁻

   [J. Phys. Chem., 63, 1518 (1959)]

17-37 What is the half-life of the reaction in Prob. 17-36?

17-38 Calculate the k for a reaction in which the initial
concentration of the reactant was 0.500 M and the concentration after
360 min was 0.401 M.

   [S. K. Ghosh, Z. Physik. Chem. (Leipzig), 206, 321 (1957)]

Chapter 18

COORDINATION COMPOUNDS

18-1 INTRODUCTION

A coordination compound is a compound containing one or more
coordinate covalent bonds. These compounds consist of a central
ion or atom bonded by coordinate covalent bonds to a definite number
of atoms, molecules, or ions called ligands. The ligands are
distributed about the central atom (or ion) in a definite spacial
configuration.

The coordination number is the total number of coordinate
covalent bonds formed between ligands and the central atom (or ion).
A coordination compound, also called a complex may be a neutral
complex or a charged complex.

18-2 NAMING OF COORDINATION COMPOUNDS

A systematic method of naming complex compounds has been developed
by the International Union of Pure and Applied Chemistry (IUPAC). A
summary of the important rules is as follows:

1. If the complex is an ionic compound, the cation is named
first, then the anion, with a space between the names.

2. Within the complex, first give the names of the ligands using
the Greek prefixes di, tri, tetra, etc., followed by the metal ion
with its oxidation number given in Roman numerals in parentheses.

375

Example 18-1

    a. $[Ag(NH_3)_2]^+$ diamminesilver(I) ion

    b. $[Cu(H_2O)_4]^{2+}$ tetraaquocopper(II) ion

    c. $[Ni(CO)_4]$ tetracarbonylnickel(0)

    d. $[Cr(NH_3)_6]Br_3$ hexaamminechromium(III) bromide

3. When more than one type of ligand is present in the complex, the ligands are named alphabetically according to the _names_ of the ligand. The prefixes di, tri, etc., used to tell how many ligands of a given kind are present in a complex, are ignored in determining the alphabetic order. Negative ligands end in -o, neutral ligands are named as molecules except for H_2O, NH_3, CO, and NO (see Table 18-1), while positive ligands end in -ium. (Prior to 1971 the order of naming ligands was different from the rule listed above. Previously, negative ligands were named first, then neutral ligands, and then positive ligands.)

Example 18-2

    a. $[Cu(NH_3)_2(H_2O)_2]^{2+}$ diamminediaquocopper(II) ion

    b. $[Co(NH_3)_4Cl_2]^+$ tetraamminedichlorocobalt(III) ion

    c. $[Pt(NH_3)_3CN]^+$ triamminecyanoplatinum(II) ion

4. If the ligands are complicated, enclose them in parentheses and use the prefixes _bis_, _tris_, _tetrakis_, etc.

Example 18-3

    a. $[Cu(en)_2]^{2+}$ bis(ethylenediamine)copper(III) ion

    b. $[Co(en)_3]^{3+}$ tris(ethylenediamine)cobalt(III) ion

5. If the complex is an anion, the metal name ends with -ate. For neutral complexes or cation complexes, the metal name is unchanged.

Example 18-4

a. $[Fe(CN)_6]^{4-}$ hexacyanoferrate(II) ion
b. $[Fe(CN)_6]^{3-}$ hexacyanoferrate(III) ion
c. $K_3[FeF_6]$ potassium hexafluoroferrate(III)

TABLE 18-1   Some Common Ligands

	Monodentate		
F^-	fluoro	CN^-	cyano
Cl^-	chloro	SCN^-	S-thiocyanato or thiocyanato (S-bonded)
Br^-	bromo		
I^-	iodo	SCN^-	N-thiocyanato or isothiocyanato (N-bonded)
OH^-	hydroxo		
H_2O	aquo	NO_2^-	nitro (N-bonded)
CO	carbonyl	NO_2^-	nitrito (O-bonded)
NH_3	ammine	NO	Nitrosyl

Bidentate

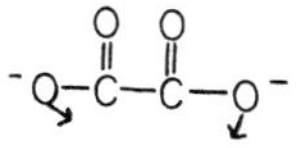

oxalato

$H_2NCH_2CH_2NH_2$

(en)

ethylenediamine

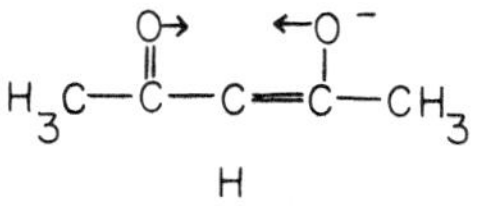

(acac)

acetylacetonato ion

Multidentate

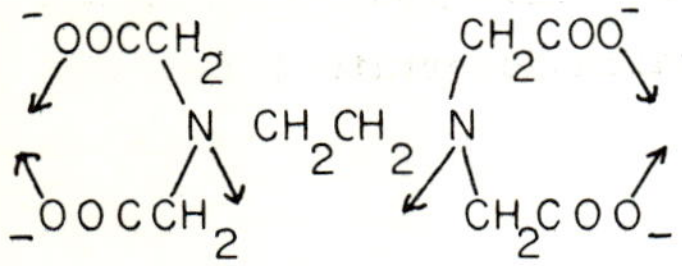

(EDTA)
anion

ethylenediaminetetraacetic acid
(hexadentate)

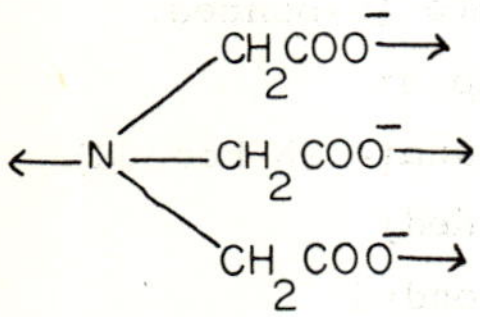

(NTA)
anion

nitrilotriacetic   acid
(tetradentate)

18-3 CHELATES

The number of donor atoms that a ligand contains is called the
denticity of the ligand. Ammonia (NH_3), which has one nitrogen to
form a bond, is a monodentate ligand. Ethylenediamine has two
nitrogen atoms to form coordinate bonds and is called a bidentate
ligand, while ethylenediaminetetraacetic acid, which has four oxygen
and two nitrogen atoms to form coordinate bonds, is called a
hexadentate ligand. Ligands which form two or more coordinate bonds

with a given metal ion are also called polydentate ligands. A
polydentate ligand which is capable of forming two or more coordinate
bonds with the same central metal ion is called a <u>chelating agent</u>
and the resulting complex is known as a <u>chelate</u>. The complexes in
Example 18-3 are chelate complexes. Table 18-1 is a list of some
common ligands. The arrows in the bi- and multidentate ligands show
the donor atoms for complex formation.

<u>Example 18-5</u>
Draw a schematic structure of the complex formed when EDTA chelates
with Ca^{2+}.

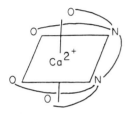

18-4 GEOMETRY AND COMPLEXES

In Chap. 6 the geometry of molecules was discussed. The shape
of the complex for coordination compounds in which the neutral metal
ion has no unpaired electrons is related to the number of ligands
around the metal ion: eg., diamminesilver(I) ion is linear
corresponding to two bonding electron pairs and tetraaquoberyllium(II)
is tetrahedral due to four electron pairs. For those complexes in
which the central metal ion has unfilled d or f orbitals, the
geometry is not directly related to the coordination number (Table
18-2).

TABLE 18-2 Coordination Number, Geometry, and Hybridization
of Some Representative Complex Ions

Coordination Number	Geometry	Hybridization	Example
2	Linear	sp	$Ag(NH_3)_2^+$
4	Tetrahedral	sp^3	$Ni(CO)_4$
4	Square planar	dsp^2	$Ni(DMG)_2$
5	Trigonal bypyramid	d^2sp^2	$Fe(CO)_5$
6	Octahedron	d^2sp^3	$Cr(NH_3)_6^{3+}$

18-5 GEOMETRICAL ISOMERISM

In square planar complexes, isomers are possible when two
different types of ligands are present in a complex, due to different
spacial distributions of the ligands. These are called <u>cis</u> and
<u>trans</u> isomers. In the cis-form, like ligands are adjacent to one
another, while in the trans-form they are opposite one another.

<u>Example 18-6</u>
Draw the cis and trans isomers of square planar diamminedichloro-
platinum(II).

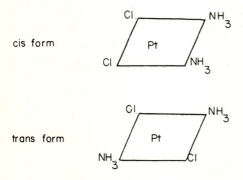

For octahedral complexes, cis and trans isomers can exist if there are two or more different ligands present in the complex.

Example 18-7

B. M. Fung has recently determined the NMR spectra of cis- and trans-dichlorobis(ethylenediamine)cobalt(III) chloride. Draw the structure of each of these molecules.

[J. Am. Chem. Soc., 89, 5788 (1967)]

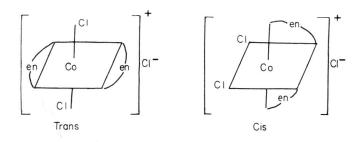

Trans   Cis

18-6 VALENCE BOND THEORY

Valence bond theory describes the behavior and properties of coordination complexes using hybridized orbitals from the central metal ion as given in Table 18-2. Hund's rule is applied to the electrons in nonbonding orbitals. Those structures with unpaired electrons are paramagnetic, while those with no unpaired electrons are diamagnetic.

Example 18-8

Draw the valence-bond diagram for the tetrachloronickelate(II) ion. This complex ion has a magnetic moment corresponding to two unpaired electrons. What is the structure of the ion? Is the ion paramagnetic or diamagnetic?

	3d	4s	4p	4d
Ni^{+2} ion	↑↓ ↑↓ ↑↓ ↑ ↑	—	— — —	— — — — —
$[Ni(Cl)_4]^{2-}$	↑↓ ↑↓ ↑↓ ↑ ↑	↑↓	↑↓ ↑↓ ↑↓	— — — — —

The $[Ni(Cl)_4]^{2-}$ ion is an sp^3 complex hence the structure is tetrahedral. There are two unpaired electrons, so the ion is paramagnetic.

Example 18-9

The tetracyanonickelate(II) ion is diamagnetic. What would be its structure?

	3d	4s	4p	4d
Ni^{+2}	↑↓ ↑↓ ↑↓ ↑ ↑	—	— — —	— — — — —
$[Ni(CN)_4]^{2-}$	↑↓ ↑↓ ↑↓ ↑↓ ↑↓	↑↓	↑↓ ↑↓ —	— — — — —

This ion is a dsp^2 complex and is square planar. There are no unpaired electrons in the structure.

The tetrachloronickelate(II) ion which has the same number of unpaired electrons as does the nickel(II) ion is called a <u>high-spin</u> complex. The tetracyanonickelate(II) ion which has fewer unpaired electrons than the nickel(II) ion is called a <u>low-spin</u> complex.

Example 18-10

The hexacyanoferrate(III) ion has a magnetic moment corresponding to one unpaired electron, while the hexafluoroferrate(III) ion has a magnetic moment corresponding to five unpaired electrons. Draw a valence bond diagram to account for the difference in these two structures.

	3d	4s	4p	4d
Fe^{3+}	↑ ↑ ↑ ↑ ↑	—	— — —	— — — — —
$Fe(CN)_6^{3-}$	↑↓ ↑↓ ↑ ↑↓ ↑↓	↑↓	↑↓ ↑↓ ↑↓	— — — — —
$Fe(F)_6^{3-}$	↑ ↑ ↑ ↑ ↑	↑↓	↑↓ ↑↓ ↑↓	↑↓ ↑↓ — —

The hexacyanoferrate(III) ion is a low spin complex while the hexafluoroferrate(III) ion is a high spin complex. Both are octahedral.

18-7 CRYSTAL FIELD AND LIGAND FIELD THEORY

Crystal field theory, and an extension of this theory known as ligand field theory, describes the properties of coordination complexes in terms of the d-valence orbitals of the central atom.

In crystal field theory the ligands are treated as electrostatic charges. In ligand field theory refinements have been made to account for neutral molecules as ligands. Both give the same type of splitting of the d orbitals.

The d orbitals in the uncoordinated metal ion all have the same energy. Under the influence of ligands the d orbitals are no longer equivalent. The way the d orbitals split depends upon the geometry of the ligands. In an octahedral field the d_{z^2} and $d_{x^2-y^2}$ orbitals (known as e_g orbitals), are equivalent to each other while the d_{xy}, d_{xz}, and d_{yz} orbitals (known as the t_{2g} orbitals) are equivalent to each other. Under the influence of ligands, the e_g orbitals are higher than the t_{2g} orbitals. The difference in energy between these two sets of orbitals is called Δ_o. The stronger the interaction of the ligand with the metal ion, the larger the value of Δ_o.

In a tetrahedral field the relative positions of the d orbitals are reversed. The e_g orbitals now have a lower energy than the t_{2g} orbitals. The difference in energy is called Δ_t.

Example 18-11
Explain the difference in the hexocyanoferrate(III) ion and the hexafluoroferrate(III) ion using ligand field theory.

$$e_g \quad \underline{\quad} \; \underline{\quad}$$

$e_g \; \underline{\uparrow} \; \underline{\uparrow}$

3d $\quad \underline{\uparrow} \; \underline{\uparrow} \; \underline{\uparrow} \; \underline{\uparrow} \; \underline{\uparrow} \qquad t_{2g} \; \underline{\uparrow} \; \underline{\uparrow} \; \underline{\uparrow} \quad \Delta_o \qquad t_{2g} \; \underline{\uparrow\downarrow} \; \underline{\uparrow\downarrow} \; \underline{\uparrow} \qquad \Delta_o$

$\qquad\qquad Fe^{3+} \qquad\qquad Fe(F)_6^{3-} \qquad\qquad Fe(CN)_6^{3-}$

Since energy is needed to pair electrons, in the $Fe(F)_6^{3-}$ complex the splitting (Δ_o) is not large enough to have the electrons pair off, while in $Fe(CN)_6^{3-}$, Δ_o is large enough so that the electrons do force pair.

For equilibrium constants involving complex ions see Chap. 13, Sections 5 and 6.

PROBLEMS

18-1 Name the following complexes:
   a. $[Cu(NH_3)_4]^{2+}$
   b. $K_4[NiF_6]$
   c. $[Hg(SCN)_4]^{2-}$ (S-bonded)
   d. $[Co(SCN)_4]^{2-}$ (N-bonded)
   e. $Ni(CO)_4$
   f. $[Co(en)_3]_2(SO_4)_3$
   g. $K_2[PtCl_3F_3]$
   h. $[Co(NH_3)_3Cl_3]^o$

18-2 Write the formula for each of the following complex compounds.
   a. tetraamminecadmium(II) nitrate
   b. hexamminenickel(II) chloride
   c. hexacarbonylchromium(0)
   d. tetraamminedichlorocobalt(III) nitrate
   e. sodium dichlorodifluoroplatinate(II)
   f. tris(ethylenediamine)rhodium(III) chloride
   g. hexamminechromium(III) pentachlorocuprate(II)

18-3 Name the following:

(a)

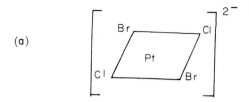

(b)

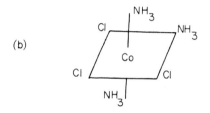

(c)

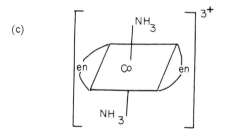

18-4 Draw the structure of each isomer of $[PtClFBrI]^{2-}$.

18-5 Basolo et al. recently gave a method of preparing each of the following compounds. Name them.

a. $[Cr(NH_3)_6][CuCl_5]$

b. $[Cr(NH_3)_6][Ni(CN)_5]$

c. $[Co(NH_3)_6][FeCl_6]$

d. $[Cr(en)_3][Ni(CN)_5]$

   [Inorg. Syn., 11, 47 (1968)]

18-6   When $NbCl_5$ is dissolved in acetonitrile and KCNS is added,
a deep-blue niobium salt is formed. The salt contained K, 8.4%;
Nb, 19.1%; S, 39.8%; C, 15.1% and N, 17.3%. An IR spectrum
showed the ligands are N-bonded. What is the formula and name of
this compound?

        [T. M. Brown and G. F. Knox, J. Am. Chem. Soc., 89,
        5296 (1967)]

18-7   When hexammine cobalt(II) dinitrate is heated under vacuum
at 100°C for 24 hours, a compound having the following composition
is obtained;   Co, 26.93%; N(total), 25.76%; N(ammine), 12.86%;
H, 2.79% and the rest oxygen. What is the empirical formula of
this complex compound? Name the compound.

        [G. L. McPherson, J. A. Weil, and J. K. Kinnaird,
        Inorg. Chem., 10, 1574 (1971)]

18-8   When 1.5 g of dichloroethylenediamineplatinum(II) was treated
with 30% H_2O_2 in water, 1.5 g of dichloroethylenediaminedihydroxy
platinum(IV) was formed. What was the per cent yield?

        [F. Basolo, J. Bailar, and B. Tarr, J. Am. Chem. Soc.,
        72, 2437 (1950)]

18-9   A solution of hexachloroplatinic acid was made by dissolving
4.5 g of pure platinum in aqua regia. This solution was evaporated
almost to dryness and concentrated HCl was added. This was
evaporated to dryness, then 10 ml of distilled water added. 8.0 g
of NaCl was added to the solution and the solution was cooled.
4.5 g of solid yellow-orange disodium hexachloroplatinate(IV) was
recovered. What was the percent yield?

        [L. Cox and D. Peters, Inorg. Syn., 13, 174 (1972)]

<u>18-10</u> Potassium thiocyanate (5.83 g) was added to niobium penta-
chloride (2.70 g) in acetonitrile. The solution was warmed (60-70°)
and stirred. After purification 4.0 g of a deep blue crystalline
solid was obtained. The ligands were N-bonded. Analysis of the
solid: K,8.44%; Nb, 19.06%; N, 17.25%; C, 15.10%; and S, 39.79%.

    a.   What is the formula of the compound?

    b.   What is the name of the compound?

    c.   What was the percent yield?

       [G. Knox and T. M. Brown, <u>Inorg. Syn.</u>, <u>13</u>, 226 (1972)]

<u>18-11</u> $[V(CO)_6]°$ has a magnetic moment corresponding to one unpaired
electron. Draw a valence bond structure for this molecule. What is
the geometrical structure for the complex?

<u>18-12</u> Classify each of the following compounds as low spin or high
spin complexes.

Compound	Number of unpaired electrons
$[Co(NH_3)_5Cl]Cl_2$	0
$K_4[Cr(CN)_6]$	2
$K_3[Fe(CN)_6]$	1
$[Pt(NH_3)_6]Cl_4$	0
$(NH_4)_2[FeF_5H_2O]$	5

       [L. Pauling, <u>J. Am. Chem. Soc.</u>, <u>53</u>, 1367 (1931)] (This is
       the original paper on valence bond theory and is worth
       reading.)

<u>18-13</u> $[Ti(H_2O)_6]^{3+}$ exists in solution as a reddish-violet complex.
Δ_o is 20,300 cm^{-1} for the complex. Magnetic measurements show one
unpaired electron.

    a.   Draw a valence-bond diagram of this complex to explain the
        above information.

b. Draw a ligand-field diagram.

c. Show that the reddish-violet color is reasonable from the Δ_o value.

18-14 The following Δ_t values were obtained for tetrahalocobaltate(II) ions in nitromethane.

Ion	$\Delta_t (cm^{-1})$
$CoCl_4^{2-}$	3214
$CoBr_4^{2-}$	2987
CoI_4^{2-}	2830

Arrange the ligands in order of decreasing ligand-field strength.

[F. A. Cotton, D. M. L. Goodgame, and M. Goodgame, J. Am. Chem. Soc., 83, 4690 (1961)]

SUPPLEMENTARY PROBLEMS

18-15 Name the following complexes:

a. $[Cd(NH_3)_4]^{2+}$

b. $[Ag(CN)_2]^-$

c. $[Co(NH_3)_2Cl_4]^-$

d. $K_4[Mo(CN)_8]$

e. $Fe(CO)_5$

f. $[Co(NH_3)_5(H_2O)](ReO_4)_3$

g. $[Ir(NH_3)_5Cl]Cl_2$

h. $[Ru(en)_2Cl_2]Cl$

i. $[Cr(en)_3][Ni(CN)_5]$

18-16 Write the formula for each of the following complex compounds.

    a.   tetraamminechloronitrocobalt(III) chloride

    b.   tetraamminedichlorocobalt(III) nitrite

    c.   Tricarbonylnitrosylcobalt(0)

    d.   Potassium tetracyanozincate(II)

    e.   Pentaamminechlororuthenium(III) chloride

    f.   Tris(ethylenediamine)chromium(III) sulfate

    g.   Potassium octacyanomolybdate(IV)

18-17 Name the following:

(a)

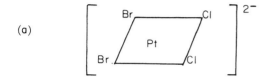

(b)

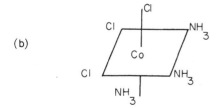

(c)

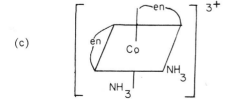

18-18   Name the following compounds prepared by McCutcheon and Schuel

    a. $[Co(H_2O)(NH_3)_5][Co(C_2O_4)_3]$

    b. $[Co(H_2O)(NH_3)_5][Cr(SCN)_4(NH_3)_2]_3$ (S-bonded)

    c. $[Co(H_2O)(NH_3)_5][Co(NO_2)_4(NH_3)_2]_3$

    [J. Am. Chem. Soc., 75, 1845 (1953)]

18-19   A complex cobalt compound was recently isolated having the
following per cent composition: Co, 16.2; C, 20.1; H, 5.60;
N, 26.9, and the rest O. What is the empirical formula of the
compound?

    [D. Lambert and J. Mason, J. Am. Chem. Soc., 88, 1637 (1966)

18-20   When $TaCl_5$ is dissolved in acetonitrile and KCNS is added,
an orange compound is formed having the following per cent
composition; K, 7.2; Ta, 32.1; S, 33.6; C, 12.5; N, 14.6. Infrared
analysis showed the ligands are N-bonded. What is the formula and
name of this compound?

    [T. M. Brown and G. F. Knox, J. Am. Chem. Soc., 89,
    5296 (1967)]

18-21   A blue solid precipitated upon the addition of a solution of
cobaltous thiocyanate (1.54 g) in hot absolute ethanol to a solution
of tetramethylammonium thiocyanate (2.46 g) in hot absolute ethanol.
The yield of the complex was 3.50 g. Analysis found: C, 32.62%;
H, 5.52%; N, 19.05%; Co, 13.4%, and the rest S. X-ray analysis
showed the ligands were N-bonded.

    a.   What is the formula of the compound?

    b.   What is the name of the compound?

    c.   What was the per cent yield?

      [F. A. Colton, D. M. L. Goodgame, M. Goodgame, and A. Sacco
      J. Am. Chem. Soc., 83, 4157 (1961)]

18-22   When 2.5 g of ammonium aquopentachloromolybdate(III) and
0.5 g of ammonium chloride are added to 12 M HCl, and HCl is bubbled

through the solution, 1.95 g of ammonium hexachloromolybdate(III)
are recovered. What is the percent yield?

> [S. V. Brencic and F. A. Cotton, Inorg. Syn., 13, 172 (1972)]

18-23   When chlorine is bubbled through 3.9 g of dichloroethylene-
diamineplatinum(II) for 45 min, 4.4 g of yellow tetrachloroethylene-
diamineplatinum(IV) is produced. What is the percent yield?

> [F. Basolo, J. Bailor, and B. Tarr, J. Am. Chem. Soc., 72,
> 2437 (1950)]

18-24   According to magnetic susceptibility measurements, $[CoF_6]^{3-}$
is a high spin complex, while $[Co(H_2O)_6]^{3+}$ is a low spin complex.
How many unpaired electrons are present in each complex?

> [H. L. Friedman, J. P. Hunt, R. H. Plane, and H. Taube,
> J. Am. Chem. Soc., 73, 4028 (1951)]

18-25   Recently the visible and uv spectra of a series of ruthenium
compounds were measured. Δ_o for the octahedral complexes were as
follows:

Complex	$\Delta_o (cm^{-1})$
$[Ru(H_2O)_6]^{2+}$	19,800
$[Ru(en)_3]^{2+}$	28,100
$[Ru(pn)_3]^{2+}$	28,400
$[Ru(dien)_2]^{2+}$	28,800
$[Ru(NH_3)_6]^{2+}$	27,100

where en = ethylenediamine, pn = propylenediamine, and dien =
diethylenediamine. Name each of these complexes and arrange the
ligands in order of decreasing ligand field strength. Such an order
is known as a spectrochemical series.

> [H. Schmidtke and F. Garthoff, Helv. Chem. Acta, 49,
> 2039 (1966)]

Chapter 19

CHEMISTRY OF THE ELEMENTS

HYDROGEN

19-1 The average abundance of deuterium in natural hydrogen is 0.0145%. The masses of ^{1}H and ^{2}H are 1.007825 and 2.014102 amu, respectively. Calculate the atomic weight of hydrogen.

   [A. E. Cameron and E. Wickers, J. Am. Chem. Soc., 84, 4181 (1962)]

19-2 Calculate the quantity of electricity required to produce 1.00 m^3 of hydrogen gas (at STP) by the electrolysis of water.

19-3 The equilibrium constant K_p for the reaction $CO_2(g) + H_2(g) \rightleftarrows CO(g) + H_2O(g)$ was found to be 1.57 by W. Nernst and F. Haber. Calculate the equilibrium partial pressure of each gas if the initial pressure of CO_2 was 1.00 atm and that of H_2 was 1.00 atm.

   [Z. Anorg. Chem., 38, 5 (1904)]

19-4 Calculate the pH of a 0.0316 M solution of formic acid, HCOOH.

19-5 What volume of hydrogen (at 27°C and 190 torr) would result from the hydrolysis of 2.10 g of calcium hydride?

19-6 Calculate the average bond-dissociation energy of the Si—H bonds in SiH_4.

19-7 E. W. Morley obtained 34.1559 g of water from the reaction of
3.8211 g of hydrogen with oxygen. What mass of oxygen was required
for the reaction?

      [Am. Chem.J., 17, 267 (1895)]

19-8 Determine the average O—H bond-dissociation energy in water.
Compare the average bond-dissociation energy with the bond-dissociation energy $H_2O(g) \rightarrow H(g) + OH(g)$ and $OH(g) \rightarrow O(g) + H(g)$

19-9 A bulb containing 0.0470 g of acetic acid vapor at 60.6°C had
a pressure of 28.22 mm. The volume of the bulb was 359.8 ml.
Calculate the molecular weight of the vapor and account for the
result.

      [M. D. Taylor, J. Am. Chem. Soc., 73, 315 (1951)]

19-10 What is the hydronium ion concentration and pH of a 0.125%
solution of hydrochloric acid the density of which is 1.0623 g/ml?

19-11 The density of trifluoroacetic acid vapor at 81.2°C and
207.5 mm was 1.555 g/liter. Calculate the molecular weight of the
vapor and account for the result.

      [R. E. Lundin, F. E. Harris, and L. K. Nash, J. Am. Chem. Soc.,
      Soc., 74, 4654 (1952)]

19-12 Calculate the emf of each of the following cells:
   a. H_2(1 atm), Pt$|$H$_3$O$^+$(1 m)$||$Cl$^-$(1 m), AgCl(s)$|$Ag
   b. H_2(0.100 atm), Pt$|$H$_3$O$^+$(0.0200 m)$||$Cl$^-$(0.0200 m), AgCl(s)$|$Ag

19-13 Calculate the pH of a 0.212 M solution of nitrous acid.

19-14 Calculate the emf of a cell consisting of a normal hydrogen
electrode and a silver-saturated solution of silver chloride electrode.

19-15   The equilibrium constant for the formation of ammonia from the elements $1/2N_2(g) + 3/2H_2(g) \rightleftharpoons NH_3(g)$ is 6.59×10^{-3} atm^{-1}. Find the partial pressure of ammonia if the pressure of $H_2 = N_2$ is 2.12 atm.

OXYGEN

19-16   Calculate the atomic weight of oxygen from the isotopic abundances of atmospheric oxygen: ^{17}O = 0.0374%, ^{18}O = 0.2039%. The nuclidic masses are ^{16}O = 15.994915, ^{17}O = 16.999134, and ^{18}O = 17.999160 amu.

[A. O. C. Nier, Phys. Rev., 77, 789 (1950)]

19-17   Oxygen gas was passed over copper metal. The volume of oxygen (at STP) which was consumed was 17.8 ml, and the weight of copper used was 0.1019 g. Was the reaction quantitative?

[M. G. Habashy, Anal Chem., 36, 1782 (1964)]

19-18   When oxygen gas was passed over 0.1013 g of pulverized tin, 13.5 ml of oxygen was consumed. What percent of the tin reacted?

[M. G. Habashy, Anal. Chem., 36, 1782 (1964)]

19-19   For the reaction $2H_2O(g) \rightleftharpoons 2H_2(g) + O_2(g)$, K_p = 4.5×10^{-21} at 1000 K. Calculate the equilibrium partial pressures of hydrogen and oxygen if 1.00 mole of water is heated to 1000 K in a 1.00-liter vessel.

19-20   A hydrate was analyzed and found to contain 18.05% CaO, 64.66% CrO$_3$, and 17.43% H$_2$O. What is the formula of the hydrate?

[W. H. Hartford, K. A. Lane, and W. A. Meyer, Jr.,
J. Am. Chem. Soc., 72, 3353 (1950)]

19-21 Calculate the standard enthalpy of the reaction
$2H_2O_2(\ell) \rightarrow 2H_2O(\ell) + O_2(g)$.

19-22 Calculate the average bond-dissociation energy of the O—F
bonds in OF_2.
        [S. W. Benson, J. Chem. Educ., 42, 517 (1965)]

19-23 Calculate the standard enthalpy for the reaction
$2F_2(g) + H_2O(\ell) \rightarrow 2HF(g) + OF_2(g)$.

19-24 Find E° at 25°C for the cell $Pt|Fe^{2+}, Fe^{3+}||H_3O^+Pt,O_2$.

19-25 Calculate the volume of oxygen expected from the passage of a
current of 1.25 amp for 4.00 hr through a solution of sodium
fluoride.

19-26 An equimolar mixture of $CO(g)$ and $O_2(g)$ is contained in a
bulb at a total pressure of 1.00 atm. A spark is passed through
the gas and a reaction occurs to form $CO_2(g)$. What is the final
pressure in the bulb?

19-27 The standard oxidation potential of water is -1.23 V;
$2H_2O = O_2(g) + 4H^+ + 4e^-$. Calculate the potential for the half
cell $O_2(1 \text{ atm})| H_2O$ (pH = 7.00).

19-28 Calculate the mass of copper which can be obtained from the
reduction of 1.25181 g of copper(II) oxide with hydrogen.
        [R. Ruer and K. Bode, Z. Anorg. Allgem. Chem., 137,
         101 (1924)]

19-29 What is the formula of the oxide of tungsten formed by the
reaction of 2.44588 g with oxygen if the product weighed 3.08318 g?
        [G. E. Thomas, J. Am. Chem. Soc., 21, 373 (1899)]

19-30 A sample of nitric oxide weighing 0.61862 g was catalytically
decomposed with nickel. Calculate the weight of oxygen and
nitrogen anticipated.

> [R. W. Gray, J. Chem. Soc., 87, 1616 (1905)]

ALKALI METALS

19-31 Calculate the atomic weight of lithium from the ratio LiCl/Ag,
0.392992.

> [T. W. Richards and H. H. Willard, J. Am. Chem. Soc., 32,
> 4 (1910)]

19-32 A 0.0779 m solution of C_2H_5Li in benzene had a freezing-point
depression of 0.065°. What conclusion may be drawn from this
experiment?

> [T. L. Brown and M. T. Rogers, J. Am. Chem. Soc., 79,
> 1861 (1957)]

19-33 In a determination of atomic weights, 4.24171 g of lithium
chloride was dissolved in perchloric acid. When the solution was
evaporated to dryness, 10.64596 g of lithium perchlorate was
collected. Compare the result of this experiment with the theoretical
yield of lithium perchlorate expected.

> [T. W. Richards and H. H. Willard, J. Am. Chem. Soc., 32,
> 47 (1910)]

19-34 The measured enthalpy of hydrolysis of sodium hydride was
found to be -30.63 kcal/mole. The measured enthalpy of reaction of
sodium metal with water was -44.23 kcal/g atom. Calculate the
enthalpy of formation of sodium hydride.

> [C. E. Messer, L. G. Fasolino, and C. E. Thalmayer,
> J. Am. Chem. Soc., 77, 4525 (1955)]

19-35 The measured heat of hydrolysis of lithium hydride was -31.76
kcal/mole. The measured heat of the reaction of lithium metal with
water was -53.10 kcal/g atom. Calculate the heat of formation of
lithium hydride.

  [C. E. Messer, L. G. Fasolino, and C. E. Thalmayer,

  J. Am. Chem. Soc., 77, 4525 (1955)]

ALKALINE-EARTH METALS

19-36 Determine the atomic weight of calcium from an observation
that 0.387200 g of calcium chloride yields 1.00000 g of silver
chloride.

  [O. Honigschmid and K. Kempter, Z. Anorg. Allgem. Chem.,

  195, 1 (1931)]

19-37 Calculate the volume of carbon dioxide, measured at 17°C
and 280 torr, which can be obtained by heating 0.480 g of calcium
carbonate.

19-38 Find the barium ion concentration of a solution which is
also 0.100 M in sodium sulfate.

19-39 Calculate the standard enthalpy of the reaction

   $MgO(s) + CO_2(g) = MgCO_3(s)$ ($\Delta H°_f$ = -144, -94.0, -266 kcal/mole)

19-40 What volume of hydrogen gas (at STP) can be formed by the
reaction of 0.682 g of calcium hydride with water?

BORON, ALUMINUM

19-41 The isotopic composition of boron in nature is variable.
For a series of minerals the $^{11}B/^{10}B$ ratio varied from 3.95 to 4.10.

Calculate the atomic weights corresponding to these ratios and compare
them with the accepted value for the atomic weight of boron.

> [A. E. Cameron and E. Wichers, J. Am. Chem. Soc., 84,
> 4182 (1962)]

19-42 A compound containing boron and chlorine was prepared by
A. Stock and O. Priess. A sample weighing 0.1319 g gave 0.4850 g
of silver chloride after reaction with silver nitrate. A sample
weighing 0.3671 g occupied a volume of 149.8 cc at 379 mm and 19°C.
What were the name and structure of the compound?

> [Chem. Ber., 47, 3109 (1914)]

19-43 Boron was heated in a stream of bromine vapor at red heat.
A sample of the product weighing 0.4398 g was dissolved in 15.99 g
of benzene. The solution had a freezing-point depression of 0.542°C.
Calculate the molecular weight of the compound.

> [A. Stock and E. Kuss, Chem. Ber., 47, 3113 (1914)]

19-44 A sample of the compound prepared as in Prob. 5-23 which
weighed 67.0 mg exerted a pressure of 435 mm in a container whose
volume was 93.2 cc at 0°C. What was the molecular weight of the
vapor?

> [A. Stock, E. Kuss, and O. Priess, Chem. Ber., 47,
> 3118 (1914)]

19-45 A boron hydride, mp 177 to 178.5°, which was stable in air,
was prepared by A. R. Pitochelli. The compound contained 89.11% B
and 10.44% H. Its molecular weight from boiling-point-elevation
experiments was 212.7. When it was decomposed to form the elements,
1.98×10^{-4} mole gave 2.21×10^{-3} mole of hydrogen. What is the
formula of the hydride?

> [J. Am. Chem. Soc., 84, 3218 (1962)]

19-46 A mixture of boric acid and urea was heated in an atmosphere
of ammonia over the temperature range 500 to 950°C. The product
formed contained 44.2% B and 56.0% N. What was it?

[J. Thomas, N. E. Weston, and T. E. O'Connor, *J. Am. Chem.*
Soc., 84, 4619 (1962)]

19-47 Calculate the molecular weight of the fluoride of boron whose
vapor density at standard conditions was 3.00 g/liter as determined
by J. Dumas.

[Traite Chim., 1, 382 (1828)]

19-48 What is the shape of the molecules of the compound in 19-47?

19-49 BF_3 and NH_3 gases were mixed in a glass bulb. A white solid
formed and upon analysis was found to contain 19.93% NH_3 and 12.72% B.
A solution containing 8.15 g of the white solid in 1000 g of water
had a freezing-point depression of 0.00179°C. What were the formula
and structure of the white solid?

[A. W. Laubengayer and G. F. Condike, *J. Am. Chem. Soc.*,
70, 2274 (1948)]

19-50 Boron trichloride was passed through an electric discharge,
and a product which had a melting point of -92.6°C was formed. A
sample of the compound was found to contain 0.0878 g of Cl and
0.0131 g of B. A 0.0810 g sample exerted a pressure of 48.9 mm at
26.5° in a volume of 191.6 cc. What was the formula of the compound?

[G. Urry, T. Wartik, R. E. Moore, and H. I. Schlesinger,
J. Am. Chem. Soc., 76, 5296 (1954)]

19-51 A 0.0995-g sample of the compound described in Prob. 19-50
produced 13.26 cc of hydrogen gas (at STP) when it reacted with
sodium hydroxide. Sodium borate, sodium chloride, and water were

the other products. Was the quantity of hydrogen produced consistent with the formula of the compound as found in Prob. 19-50?

> [G. Urry, T. Wartik, R. E. Moore, and H. I. Schlesinger, J. Am. Chem. Soc., 76, 5296 (1954)]

19-52 Gallium metal and iodine were heated at 350° for 24 hr; a lemon-yellow solid containing 15.56% gallium and 84.45% iodine was formed. The lemon-yellow compound was then heated in an atmosphere of oxygen for 2 hr at 300 to 350°; a white solid containing 32.34% gallium and 59.20% iodine was formed. When fluorine was passed over the white solid for 30 min at room temperature, a compound containing 66.42% gallium and 18.90% fluorine was formed. Write equations for the three reactions which occurred.

> [F. M. Brewer, P. L. Goggin, and G. S. Reddy, J. Inorg. Nucl. Chem., 28, 361 (1966)]

19-53 The enthalpy of formation of aluminum fluoride (-355.7 kcal/mole) was determined from the reaction of aluminum with lead(II) fluoride; calculate the enthalpy change of the reaction. $\Delta H°_f$ (PbF_2) = -158.5 kcal/mole.

> [P. Gross, C. Hayman, and D. L. Levi, Trans. Faraday Soc., 50, 477 (1954)]

19-54 Assume ideal behavior and calculate the pressure exerted by 1.34 g of aluminum chloride vapor in a 200-ml bulb at 180°.

19-55 A. Stock and C. Massanex prepared a boron hydride which contained 81.7% B and 18.3% H. When 2.0 cc of the gas reacted with water, 21.6 cc of H_2 was formed. Do these observations support the opinion that the compound was B_4H_{10}, which reacted with water according to the equation

$$B_4H_{10} + 12H_2O = 4B(OH)_3 + 11H_2$$

> [Chem. Ber., 45, 3559 (1912)]

CARBON, SILICON

19-56 A sample of a compound containing 69.7% sulfur and weighing
0.1221 g was allowed to react with water. The white precipitate
of silicon dioxide, which was formed, weighed 0.0794 g after drying.
What was the compound?

  [H. Gabriel and C. Alvarez-Tostado, J. Am. Chem. Soc., 74,
  263 (1952)]

19-57 What are the formulas, Lewis structures, and shapes of the
molecules of the two compounds whose analyses were 5.28% Si and
94.66% I, and 34.05% Ge and 66.0% Cl?

  [W. C. Schumb and D. W. Breck, J. Am. Chem. Soc., 74,
  1758 (1952)]

19-58 A solution of 350 g of phosgene ($COCl_2$) in acetonitrile as a
solvent was added to a mixture of 380 g of sodium fluoride and
acetonitrile, and the entire mixture was heated at 30 to 40°. The
volatile product, COF_2, weighed 190 g, and the other product was
sodium chloride.

  a. Which of the reactants was in excess?
  b. What was the percent yield of carbonyl fluoride?
  c. Write a Lewis structure of carbonyl fluoride.
  d. What is the shape of the molecules of carbonyl fluoride?
      [F. S. Fawcett, C. W. Tullock, and D. D. Coffman, J. Am.
      Chem. Soc., 84, 4278 (1962)]

19-59 Write a Lewis structure for carbon suboxide, C_3O_2, which is
linear. The C—C distance is 1.28 Å.

  [R. L. Livingston and C. N. R. Rao, J. Am. Chem. Soc., 81,
  285 (1959)]

19-60 Predict the shape of the molecules of formylfluoride, HCOF.

  [M. E. Jones, K. Hedberg, and V. Schomaker, J. Am. Chem.
  Soc., 77, 5278 (1955)]

19-61   A hydride of silicon was prepared by A. Stock and C. Somieski.
A sample weighing 0.0751 g exerted a pressure of 509 mm at 21° in a
bulb whose volume was 43.3 cc.   What was the molecular weight of
the hydride?

      [Chem. Ber., 49, 130 (1916)]

19-62   Calculate the volume of silane (at STP) expected from the
reaction of 0.390 g of Mg_2Si with HCl(aq), assuming that no other
hydride is formed.

19-63   Calculate the volume of carbon tetrafluoride, measured at
30°C and 742 torr, which can be obtained from the reaction of fluorine
with 0.940 g of silicon carbide.   What weight of sodium fluorosilicate
would be formed if the mixture of gaseous products was passed over
sodium hydroxide?

19-64   Write a Lewis structure and predict the shape of the molecules
of thiocarbonyl fluoride, F_2CS.

19-65   Calculate the weight of silicon tetrachloride which can be
formed by treatment of a mixture of 10.0 g of silica and excess
carbon with chlorine gas at an elevated temperature assuming 100%
yield.   The other product of the reaction is carbon monoxide.

19-66   For the reaction C(graphite) + O_2(g) → CO_2(g) at 25°C,
$\Delta H°$ = -94.05 kcal/mole; for the reaction C(diamond) + O_2(g) → CO_2(g),
$\Delta H°$ = -94.50 kcal/mole.   Calculate $\Delta H°_{trans}$ for the change
C(graphite) → C(diamond).

19-67   Calculate E° for the cell $Sn|Sn^{2+}||Cu^{2+}|Cu$.

19-68   Write a Lewis structure and predict the shape of the molecules
of tin tetrachloride.   In hydrochloric acid solution, tin
tetrachloride forms $SnCl_6^{2-}$ ions.   What is the expected shape of
these ions?

19-69 The electrolytic reduction of 2.194802 g of $SnCl_4$ produced 1.000000 g of tin. Calculate a value for the atomic weight of tin from this observation.

      [G. P. Baxter and H. W. Starkweather, J. Am. Chem. Soc., 42, 905 (1920)]

19-70 Determine the lead ion concentration in a solution which is 0.300 M in hydrochloric acid.

NITROGEN, PHOSPHORUS, ARSENIC, ANTIMONY

19-71 Calculate the atomic weight of nitrogen from the $^{14}N/^{15}N$ ratio in atmospheric nitrogen 272. The nuclidic masses are $^{14}N = 14.003074$ and $^{15}N = 15.000108$.

      [A. E. Cameron and E. Wichers, J. Am. Chem. Soc., 84, 4182 (1962)]

19-72 A sample of phosphorus vapor weighing 10.46 mg in a bulb whose volume was 18.08 cc exerted a pressure of 246.5 torr at 575°C. What was the molecular weight of the vapor?

      [A. Stock, G. E. Gibson, and E. Stamm, Chem. Ber., 45, 3529 (1912)]

19-73 One liter of a compound of arsenic and hydrogen weighed 3.50 g at 0°C and 1.00 atm. Calculate the molecular weight of the gas.

      [J. Dumas, Ann. chim. phys., 33, 357 (1826)]

19-74 What is the shape of the molecules of the compound in Prob. 19-73?

19-75 In an investigation of the decomposition of hydroxylamine in alkaline solution, 1.4848 g of hydroxylamine hydrochloride produces

7.189 mM of ammonia. Write an equation for the decomposition which
is consistent with the results of this experiment.

[K. B. Morris, D. F. Johnson, and J. B. Morris,

J. Am. Chem. Soc., 72, 2776 (1950)]

19-76 The reaction of hydrazine with hydrochloric acid led to the
formation of a compound containing 26.9% N, 6.2% H, and 67.2% Cl.
What is the formula of the compound?

[T. Curtius, Chem. Ber., 20, 1633 (1887)]

19-77 Nitrogen trifluoride gas was passed over copper metal at
375°, and copper(II) fluoride and a gas containing 71.92% fluorine
were obtained. The molecular weight of the gas was found to be 107
by gas-density measurements.

a. Write an equation for the reaction.

b. What weight of NF_3 would be required to produce 1.00 liter
of the gaseous product if the yield was 62.0%?

[C. B. Colburn and A. Kennedy, J. Am. Chem. Soc., 80,
5004 (1958)]

19-78 A sample of gas, believed to be NF_3, was allowed to react
with water. If 23.0 cc of the gas produced 11.4 cc of NO, is the
result consistent with the equation $2NF_3 + 3H_2O = 6HF + NO_2 + NO$?

[O. Ruff, J. Fischer, and F. Luft, Z. Anorg. Allgem. Chem.,
172, 423 (1928)]

19-79 The electrolysis of molten ammonium hydrogen fluoride produced
a gas at the anode. A sample of the gas (20.5 cc) was mixed with
hydrogen gas until the total volume was 57.5 cc and a spark was
passed through the gas mixture. One of the products was HF, which
was removed by absorption in sodium hydroxide solution. The volume
of gas remaining was 18.2 cc. What was the formula of the
nitrogen-fluorine compound obtained in the electrolysis?

[O. Ruff, J. Fischer, and F. Luft, Z. Anorg. Allgem. Chem.,
172, 421 (1928)]

19-80 Another sample of the gas produced in the electrolysis in
Prob. 19-79, which weighed 0.5532 g, exerted a pressure of 730.6 torr
at 0°C in a flask whose volume was 180.85 cc.

    a. What is the molecular weight of the gas? Analysis indicated
        the presence of 80.2% fluorine in the compound.

    b. What is the formula of the compound?

        [O. Ruff, J. Fischer, and F. Luft, Z. Anorg. Allgem. Chem.,
        172, 421 (1928)]

19-81 Write a Lewis structure for the P_4 molecule. Electron and
X-ray diffraction studies show that it is tetrahedral.

19-82 Dimethylaminodifluorophosphine, $(CH_3)_2NPF_2$, reacted with
hydrogen iodide at room temperature to yield a compound containing
15.37% P, 18.09% F, and 64.34% I. Its observed molecular weight
was 192.9.

    a. What is the formula of the compound?

    b. What is the shape of the molecules of the compound?

        [R. W. Rudolph, J. G. Morse, and R. W. Parry, Inorg. Chem.,
        5, 1464 (1966)]

19-83 The compound prepared in Prob. 19-82 reacted with mercury at
room temperature to produce a compound containing 44.1% phosphorus
and 53.8% fluorine. Its molecular weight, calculated from
vapor-density measurements, was 138.7.

    a. What is the compound?

    b. Write a Lewis structure for the molecules.

    c. Describe the geometrical structure of the molecule.

    d. Write a balanced equation for the preparation.

        [M. Lustig, J. K. Ruff, and C. B. Colburn, J. Am. Chem. Soc
        88, 3875 (1966)]

19-84 Phosphorus pentachloride (98 g) reacted with calcium fluoride at 400° to give 41 g of phosphorus pentafluoride. Calculate the percent yield.

[E. L. Muetterties et al., J. Inorg. Nucl. Chem., 16, 54 (1961)]

19-85 In the same paper as Prob. 19-84 it was reported that 87 g of phosphorus trifluoride was prepared by the reaction of 180 g of phosphorus trichloride with 234 g of calcium fluoride at 350° for 8 hr.

a. Which reactant was in excess?

b. What was the percent yield?

[E. L. Muetterties et al., J. Inorg. Nucl. Chem., 16, 54 (1961)]

19-86 The action of anhydrous hydrogen fluoride on KH_2AsO_4 gave a solid which contained 32.2% As, 48.8% F, and the remainder potassium. Predict the shape of the anion in the compound.

[H. M. Bess and R. W. Parry, J. Am. Chem. Soc., 79, 1591 (1957)]

19-87 Account for the fact that the three Sb-Cl basal distances in antimony pentachloride are 2.29 Å and the two Sb-Cl apical distances are 2.34 Å.

[S. M. Ohlberg, J. Am. Chem. Soc., 81, 811 (1959)]

19-88 Fluorine, diluted with nitrogen, was passed over a mixture of arsenic trichloride and arsenic(III) oxide. Among the products was a compound containing 49.91% As and 38.21% F; the remainder was oxygen. The vapor density of the compound was about 148 g per molar volume. What is the structure of the compound?

[G. Mitra, J. Am. Chem. Soc., 80, 5639 (1958)]

19-89 A. Stock and O. Guttermann prepared a hydride of antimony.
A sample weighing 0.3333 g exerted a pressure of 754 mm in a
62.86-cc bulb at 15°C. What was the molecular weight of the hydride?
        [Chem. Ber., 37, 887 (1904)]

19-90 In a determination of the atomic weight of nitrogen, R. W.
Gray found that 0.35851 g of nitric oxide occupied 267.43 cc at
759.91 torr and 0°C. Calculate a value for the atomic weight of
nitrogen by using the value 15.999 for oxygen.
        [J. Chem. Soc., 87, 1608 (1905)]

SULFUR, SELENIUM, AND TELLURIUM

19-91 At 27°C and 420 mm, 0.170 g of H_2S was collected in a 223-ml
vessel which also contained 0.622 g of tin. The tin was heated until
a reaction occurred. What was the pressure after the reaction
occurred (measured at 27°C), if tin(II) sulfide was formed?

19-92 The solubility of rhombic sulfur in carbon disulfide is
46.1 g/100 g of solvent at 22°C. What are the molar concentration
and the mole fraction of sulfur?

19-93 Sulfur and silver fluoride were heated to 140°C, and a blue
product from the reaction was collected at -70°C. A sample of the
product contained 0.4327 g of sulfur and 0.5055 g of fluorine.
What is the empirical formula of the compound?
        [D. K. Padma and S. R. Satyanarayana, J. Inorg. Nucl. Chem.,
        28, 2432 (1966)]

19-94 A 1.290-g sample of the compound prepared in Prob. 19-93
was treated with HI. Hydrogen fluoride, 9.378 g of iodine, and
0.6274 g of H_2S were obtained as products.
    a. What is the formula of the compound?

b.   Write a Lewis structure of the compound.

c.   What is the shape of the molecules of the compound?

19-95 $SeOCl_2$ reacted with fluorine at 50° to give a compound whose analysis was Se, 37.23, 37.01; F, 54.68, 54.31.   A molecular weight determination gave 208.5 g/mole.

a.   What is the formula of the compound?

b.   Propose a structure for the molecules of the compound.

[G. Mitra and G. H. Cady, J. Am. Chem. Soc., 81, 2646 (1959)]

19-96   A 0.1865-g sample of a compound containing 56.4% P and 43.6% S raised the boiling point of 17.40 g of carbon disulfide 0.110°. What was the compound?

[A. Heiff, Z. Physik. Chem., 12, 209 (1893)]

19-97   The conversion of 1.0000 g of sodium carbonate to sodium sulfate gave 1.3402 g of the latter.   Calculate a value for the atomic weight of sulfur from this observation.

[T. W. Richards and C. R. Hoover, J. Am. Chem. Soc., 37, 108 (1915)]

19-98   Determine the sulfide ion concentration of a saturated solution of H_2S (0.1 M) which is also 0.3 M in HCl.

19-99   Calculate the Cu^{2+} ion concentration and the Zn^{2+} ion concentration in the solution in Prob. 19-98.

19-100   An oxyfluoride of sulfur was prepared by F. B. Dudley, G. H. Cady, and D. F. Eggers, Analysis found: S, 26.1, 27.03, 26.6, 26.4; F, 31.5, 31.3, 32.0, 31.4.   The molecular weight from vapor-density measurements was 117 to 119.5 in eight determinations. What is the formula of the compound?   Propose a structure.

[J. Am. Chem. Soc., 78, 291 (1956)]

19-101   The compound CH_3SH was fluorinated, and the principal
product contained 16.1% sulfur, 77.3% fluorine, and the remainder
carbon. The molecular weight from gas-density measurements was 197.
What is the formula of the compound? The carbon and sulfur atoms
are joined by a single bond. Draw a diagram representing the
structure of the molecules of the compound.

> [G. A. Silvey and G. H. Cady, J. Am. Chem. Soc., 72,
> 3624 (1950)]

19-102   The reaction of a stream of HCl gas with S_4N_4 gave a compound
containing 60.0% sulfur, 20.4% nitrogen, and 17.5% chlorine. What
is the formula of the compound?

> [A. G. MacDiarmid, J. Am. Chem. Soc., 78, 3872 (1956)]

19-103   In 1884 it was claimed that the reaction of P_4S_3 with S led
to the formation of P_2S_4 and P_8S_{11}. To confirm this claim, A. R.
Pitochelli and L. F. Audrieth dissolved 27.5 g of P_4S_3 and 4 g of
S_8 in 100 ml of CS_2 and heated the solution under pressure at 170 to
200° for 3 hr. A mixture of two solids resulted. One contained
43.6% P and 57.9% S; the other contained 35.99% P and 64.4% S.
What were the products? Write an equation for the reaction which
occurred.

> [J. Am. Chem. Soc., 81, 4458 (1959)]

19-104   A solution of 3.795 g of sulfur in 100 g of carbon disulfide
boils at 46.66°C. The boiling point of carbon disulfide is 46.30°C;
K_b is 2.41. What is the formula of the sulfur molecules in the
solution?

19-105   What is the formula of the oxide of tellurium if 743.2 mg
gave 148.8 mg of oxygen when decomposed to the elements?

> [A. Metzner, J. Am. Chem. Soc., 21, 206 (1899)]

HALOGENS

<u>19-106</u> Calculate a value of the atomic weight of bromine from the isotopic abundances and isotopic masses: ^{79}Br = 50.53%, 78.9434 amu; ^{81}Br = 49.463%, 80.9421 amu.

> [A. E. Cameron and E. Wichers, <u>J. Am. Chem. Soc.</u>, <u>84</u>, 4187 (1962)]

<u>19-107</u> Determine the mass of sodium bromide required to react with an acidified solution containing 0.755 g of sodium bromate.

<u>19-108</u> Calculate the equilibrium concentration of HOCl from the equilibrium constant K_c = 4.66 x 10^{-4} for the reaction of chlorine with water. The initial concentration of chlorine was 0.0100 mole/liter.

<u>19-109</u> Find the standard electrode potential of the cell
$$Cl_2, \ Pt|Cl^-||Mn^{2+}, \ MnO_4^-, \ H_3O^+|Pt.$$

<u>19-110</u> Calculate the potential of the cell
$$Cl_2(1 \ atm), \ Pt|Cl^-(1 \ m)||Mn^{2+}(0.01 \ m), \ MnO_4^-(2.0 \ m)H_3O^+(2.0 \ m)|Pt.$$

<u>19-111</u> The equilibrium $BrF_3(g) + Br_2(g) \rightleftarrows 3BrF(g)$ was studied by R. K. Steunenberg, R. C. Vogel, and J. Fischer. At 65.2°C, initial partial pressures of Br_2 and BrF_3 of 30.3 and 30.2 mm, respectively, led to a total pressure of 79.0 mm at equilibrium. What is K_p for the equilibrium?

> [<u>J. Am. Chem. Soc.</u>, <u>79</u>, 1320 (1957)]

<u>19-112</u> Calculate the value of K_c for the dissociation of I_3^-, where the triiodide ion concentration is 1.032 x 10^{-2} M, the iodide ion

concentration is 1.019×10^{-2}M, and the iodine concentration is 1.320×10^{-3} M.

> [L. I. Katzin and E. Gebert, J. Am. Chem. Soc., 77,
> 5814 (1955)]

19-113 Find the iodine concentration of a solution prepared by dissolving 0.100 mole of iodine in 1.00 liter of 0.100 M KI solution. K_c for the formation of I_3^- is 0.752.

19-114 What is the shape of the molecules of each of the compounds in Prob. 19-111?

19-115 The magnetic susceptibility of a sample of iodine dioxide was measured and the compound was found to be diamagnetic. What is the structure of the compound?

> [W. K. Wilmarth and S. S. Dharmatti, J. Am. Chem. Soc., 72,
> 5789 (1950)]

19-116 Predict the shapes of the molecules of IF_5 and IF_7 and compare your predictions with the shapes proposed to account for their infrared and raman spectra.

> [R. C. Lord, M. A. Lynch, W. C. Schumb, and E. J.
> Slowinski, Jr., J. Am. Chem. Soc., 72, 522 (1950)]

19-117 The equilibrium $BrF_3 + Br_2 \rightleftarrows 3BrF$ was studied at 74.5°C. The initial pressure of BrF_3 and Br_2 was 49.9 mm for each. At equilibrium the total pressure was 129.1 mm. Calculate K_p for the equilibrium at 74.5°C.

> [R. K. Steunenberg, R. C. Vogel, and J. Fischer, J. Am.
> Chem. Soc., 79, 1320 (1957)]

19-118 At 84.8°C, K_p for the equilibrium $BrF_3 + Br_2 \rightleftharpoons 3BrF$ was found to be 2430. Calculate the partial pressure of BrF at equilibrium if the partial pressures of BrF_3 and Br_2 were each equal to 18.7 mm.

> [R. K. Steunenberg, R. C. Vogel, and J. Fischer,
> J. Am. Chem. Soc., 79, 1320 (1957)]

19-119
   a. Predict the shape of the molecules of bromine pentafluoride.
   b. Why is the bromine atom below the basal plane of the molecule?
       [R. S. McDowell and L. B. Asprey, J. Chem. Phys., 37, 165 (1962)]

19-120 A gaseous mixture of IF_5 and F_2 in which the initial pressure of fluorine was 562.1 mm and that of IF_5 was 99.4 mm was prepared. After reaction was complete, what was the total pressure?

> [J. Fischer and R. K. Steunenberg, J. Am. Chem. Soc., 79, 1876 (1957)]

19-121 The reaction of iodine with fluorine was studied by E. B. R. Prideaux. He obtained a sample of a compound which contained 0.0770 g of iodine and 0.0564 g of fluorine. What was the empirical formula?

> [J. Chem. Soc., 89, 317 (1906)]

19-122 A saturated solution of chlorine in water at 25°C under a pressure of 1 atm is about 0.06 M in total chlorine. Calculate the concentration of free chlorine if the equilibrium constant for the reaction of chlorine with water is 4.66×10^{-4}.

19-123 What are the shapes of the chlorite ions and the hypochlorous acid molecules?

<u>19-124</u> When O_2F_2 liquid (5.468 g) decomposed to oxygen and fluorine
at 190 K in a calorimeter, 88.0 cal of heat was evolved. Calculate
ΔE for the decomposition $O_2F_2(\ell) \rightarrow O_2(g) + F_2(g)$ at 190 K.
        [A. D. Kirshenbaum, A. V. Grosse, and J. G. Aston,
        J. Am. Chem. Soc., <u>81</u>, 6400 (1959)]

<u>19-125</u> Use ΔE from Prob. 19-124 to calculate ΔH for the reaction
$O_2(g) + F_2(g) \rightarrow O_2F_2(\ell)$ at 190 K.
        [A. D. Kirshenbaum, A. V. Grosse, and J. G. Aston,
        J. Am. Chem. Soc., <u>81</u>, 6400 (1959)]

NOBLE GASES

<u>19-126</u> A mixture of fluorine and xenon was heated to 400° in a
nickel reactor for 1 hr. A colorless solid, thermally stable at room
temperature, was formed. In one experiment 0.2952 g of xenon
reacted with 0.1733 g of fluorine. Calculate the simplest formula
of the compound.
        [H. H. Claasen, H. Selig, and J. G. Malin, J. Am. Chem. Soc.,
        <u>84</u>, 3593 (1962)]

<u>19-127</u> A sample of the compound described in Prob. 19-126 and
weighing 0.4006 g reacted with an excess of hydrogen at 400° to
form hydrogen fluoride and 0.2507 g of xenon. The hydrogen fluoride
was absorbed in a solution of sodium hydroxide of known concentration.
Analysis of the solution showed that 0.1435 g of fluoride ion was
present. Are these observations consistent with the formula
proposed in Prob. 19-126?
        [H. H. Claasen, S. Selig, and J. G. Malin, J. Am. Chem. Soc.,
        <u>84</u>, 3595 (1962)]

<u>19-128</u> Write a Lewis structure and draw a diagram representing
the geometrical structure of the compound formed in Prob. 19-126.

19-129 Calculate the average bond-dissociation energy of XeF_4. The standard enthalpy of formation of $XeF_4(g)$ is -51.5 kcal/mole.

19-130 Calculate the molecular weight of argon from the observation that 2111.0 ml weighed 2.5094 g at 0°C and 506.67 torr.

       [G. P. Baxter and W. W. Starkweather, Proc. Natl. Acad. Sci.

       U. S., 14, 60 (1928)]

19-131 A solution was described as being 3.85×10^{-3} M in a xenon(VI) species. If the species was $XeO_4{}^{2-}$, what was the concentration in grams per milliliter?

19-132 A fluoride of xenon was synthesized at the University of Washington. In one experiment 0.409 g of xenon produced 0.758 g of the fluoride. Calculate the empirical formula of the fluoride synthesized.

       [F. B. Dudley, G. Gard, and G. H. Cady, Inorg. Chem., 2,

       229 (1963)]

19-133 What was the simplest formula of the fluoride from which 0.0327 g of xenon was recovered from a sample weighing 0.0607 g?

       [F. B. Dudley, G. Gard, and G. H. Cady, Inorg. Chem., 2,

       229 (1963)]

19-134 A 0.3934 g sample of the fluoride described in Prob. 19-132 was treated with hydrogen. Xenon (0.2125 g) and hydrogen fluoride (0.1770 g) were formed. Do the results of this experiment support the formula calculated in Prob. 19-132?

       [J. G. Maim, I. Sheft, and C. L. Chernick, J. Am. Chem.,

       Soc., 85, 110-111 (1963)]

19-135

    a. Write a Lewis structure for the compound in Prob. 19-134.

    b. Predict the shape of the molecules of the compound.

19-136   A white solid, formed by the reaction of aqueous barium
hydroxide with aqueous XeO_3, contained 71.75% BaO, 20.60% Xe,
and 7.55% oxygen.   What is the formula of the compound?

[A. D. Hirschenbaum and A. V. Grosse, Science, 142,
580 (1963)]

19-137   A sample of xenon hexafluoride weighing 0.0911 g reacted
with mercury at 100°.   Calculate the mass of mercury(II) fluoride
formed.

[E. E. Weaver, B. Weinstock, and C. P. Knop, J. Am. Chem.
Soc., 85, 111 (1963)]

19-138   A sample of a fluoride of xenon (0.0607 g) reacted with
aqueous KI solution to yield 0.0327 g Xe.   What is the formula of
the xenon fluoride?

[F. B. Dudley, G. Gard, and G. H. Cady, Inorg. Chem., 2,
229 (1963)]

19-139   What is the shape of the molecules of XeF_2?

19-140   What is the structure of the molecules of XeO_3?

19-141   The first compound of xenon reported resulted from the
reaction of the gas with PtF_6.   In one experiment an initial pressure
of PtF_6 in a silica bulb was 93.0 mm.   Xe was added until the total
pressure was 201 mm.   After reaction, the residual pressure of
xenon was 17.0 mm.   Assuming that direct combination occurred,
calculate the formula of the product.

[N. Bartlett, Proc. Chem. Soc., 1962, 218]

19-142   A sample of xenon hexafluoride weighing 0.3934 g was treated
with hydrogen by J. G. Malin, I. Selig, and C. L. Chernick, who
obtained 0.2125 g of Xe and 0.1864 g of HF as products.   Compare the
results with the theoretical yields of the products.

[J. Am. Chem. Soc., 85, 110 (1963)]

19-143   A sample of a xenon fluoride weighing 158.9 mg exerted a pressure of 4.61 mm at 24.8°C in a container the volume of which was 2609.5 cc. Calculate the molecular weight.

[B. Weinstock, E. E. Weaver, and C. P. Knop, Inorg. Chem., 5, 2196 (1966)]

19-144   Calculate the average bond dissociation energy of XeF_6 from the heat of formation at 298.15 K, -70.4 kcal/mole.

[B. Weinstock, E. E. Weaver, and C. P. Knop, Inorg. Chem., 5, 2202 (1966)]

19-145   The $\Delta H°_f$ of $KrF_2(g)$ was determined by S. R. Gunn as +13.7 kcal/mole, and the $\Delta H°_f$ of $KF_2(s)$ as +3.7 kcal/mole. Compute the molar enthalpy of sublimation of $KrF_2(s)$ at 25°C.

[J. Am. Chem. Soc., 88, 5924 (1966)]

COPPER, SILVER, AND GOLD

19-146   Calculate the mass of copper deposited by the passage of 2.02 amp for 2.25 hr through a solution of copper(II) sulfate.

19-147   Determine the potential of the cell $Fe|Fe^{2+}||Cu^{2+}|Cu$.

19-148   What volume of nitric oxide can be produced by the reaction of 0.705 g of copper with concentrated nitric acid?

19-149   Calculate the silver ion concentration in a solution which is 0.100 M in NH_3 and in equilibrium with solid silver chloride.

19-150   Calculate the standard enthalpy of decomposition of copper(II) sulfate to copper(II) oxide and sulfur trioxide.

19-151   A compound was prepared which contained 21.63% K, 35.74% Te, 25.15% H_2O, and the remainder oxygen. A sample of the compound reacted with hot hydrochloric acid to form Cl_2 and $TeCl_4$. Write an equation for the reaction of the compound with hydrochloric acid.

[E. B. Hutchins, Jr., J. Am. Chem. Soc., 27, 1172 (1905)]

19-152 Calculate the potential of the cell H_2, $Pt|HBr$ (0.97 m) $AgBr(s)|Ag$.

[W. J. Biermann and R. S. Yamasaki, J. Am. Chem. Soc., 77, 241 (1955)]

19-153 Find the potential of the cell $Pb(Hg)|PbCl_2$, Cl^- (1 m), $AgCl|Ag$.

W. R. Carmody, J. Am. Chem. Soc., 51, 2909 (1929)]

19-154 Calculate the copper(II) ion concentration of a 0.100 M solution of $[Cu(NH_3)_4](NO_3)_2$.

19-155 How many grams of sulfur can be formed by the reaction of concentrated nitric acid with 0.125 mole of CuS?

ZINC, CADMIUM, AND MERCURY

19-156 A sample of mercury(II) chloride weighing 9.8500 g was dissolved in 32.75 g of ethanol. The boiling point elevation was 1.275°. Calculate the molecular weight of the solute.

[E. Beckmann, Z. Physik. Chem. (Leipzig), 6, 453 (1890)]

19-157 A boiling-point-elevation experiment was run to determine the extent of dissociation of mercury(II) chloride in water. The boiling-point-elevation of a solution containing 3.8113 g of the chloride in 43.69 g of water was 0.165°.

a. What species were present in the solution?
b. What is the type of bonding present in $HgCl_2$?
c. What is the structure of $HgCl_2$?

[E. Beckmann, Z. Physik. Chem. (Leipzig), 6, 453 (1890)]

19-158 The boiling point elevation of an aqueous solution of 5.6977 g
of cadmium iodide in 44.69 g of water was 0.181°. What solute
species were present in the solution?

> [E. Beckmann, Z. Physik, Chem. (Leipzig), 6, 453 (1890)]

19-159 The microwave spectrum of the volatile compound prepared by
the action of chlorine gas on mercury(II) oxide was consistent with
a Cl—O bond distance of 1.700 Å and a Cl—O—Cl angle of 110.96°.
Do these results support the prediction of the shape of the
molecules from the electron-pair repulsion principle?

> [G. E. Herberich, R. H. Jackson, and D. J. Miller,
>
> J. Chem. Soc., 1966, 336]

19-160 Calculate the potential of the cell $Zn|Zn^{2+}||Sn^{2+}|Sn$.

19-161 What is the potential of the cell $Zn(Hg)|ZnSO_4(0.02$ m),
$PbSO_4(s)|Pb(Hg)$?

> [I. A. Cowperthwaite and V. K. Lamer, J. Am. Chem. Soc., 53,
>
> 4333 (1931)]

19-162 The E° of the cell $Cd(Hg)|CdCl_2,AgCl|Ag$ is 0.573 V.
Calculate the standard potential E° of the half cell $Cd(Hg)|Cd^{2+}$.

> [H. S. Harned and M. E. Fitzgerald, J. Am. Chem. Soc., 58,
>
> 2624 (1936)]

19-163 Calculate a value of the atomic weight of mercury from the
mercury(II) chloride-silver ratio, 1.25847.

> [O. Honigschmid, L. Birkenbach, and M. Steinhell, Chem. Ber.,
>
> 56B, 1212 (1923)]

19-164 Determine the mercury(I) ion concentration in a solution
which is 0.0873 M in KCl.

19-165 Calculate the quantity of ammonium nitrate formed by the
action of very dilute nitric acid on 0.261 g of zinc metal.

TRANSITION ELEMENTS

19-166 A 0.10 M Na_2HPO_4 solution was added dropwise to a 20%
excess of 0.10 M $CrCl_3$, and the violet, crystalline precipitate
that was formed contained 42.5% water. Further analysis showed the
presence of 20.15% Cr and 12.26% P. What was the compound?

      [A. T. Ness, R. E. Smith, and R. L. Evans, J. Am. Chem. Soc.,
      74, 4688 (1952)]

19-167 A black material containing 35.3% Cr was obtained by heating
the product in Prob. 19-166 to 800°C. What was the compound?

      [A. T. Ness, E. E. Smith, and R. L. Evans, J. Am. Chem. Soc.,
      74, 4688 (1952)]

19-168 In a novel method of preparing metal chlorides, T. E. Austin
and S. Y. Tyree, Jr. obtained a compound containing 34.52% Fe and
65.51% Cl. What was the compound?

      [J. Inorg. Chem., 14, 141-142 (1960)]

19-169 A nickel ferrite was prepared by heating the compound
$Ni_3Fe_6(CH_3CO_2)_{17}O_3OH$ to 1000°. The ferrite analysis was Ni, 25.03;
Fe, 47.54. What is the formula of the ferrite?

      [D. G. Wickham, E. R. Whipple, and E. G. Larson, J. Inorg.
      Nucl. Chem., 14, 217 (1960)]

19-170 What is the formula of the fluoride of niobium which melts
at 78.9° and contains 49.7% Nb and 50.2% F?

      [J. H. Junkins, R. L. Farrar, Jr., E. J. Barber, and H. A.
      Bernhardt, J. Am. Chem. Soc., 74, 3464 (1952)]

19-171 A 3.66098-g sample of titanium(IV) chloride was dissolved in
nitric acid. A solution made by dissolving silver in nitric acid
was added to determine the atomic weight of titanium. Using the

accepted value of the atomic weight of titanium, calculate the mass of silver required to react with the titanium tetrachloride.

> [G. P. Baxter, and A. Q. Butler, J. Am. Chem. Soc., 48,
> 3117 (1926)]

19-172 Calculate the atomic weight of Ti from the titanium(IV) bromide-silver ratio, 0.851788.

> [G. P. Baxter and A. Q. Butler, J. Am. Chem. Soc., 50,
> 408 (1928)]

19-173 What is the emf of the cell $Pt|Fe^{2+}(0.10 \text{ m}), Fe^{3+}(0.0010 \text{ m})||$ $Ce^{3+}(0.0010 \text{ m}), Ce^{4+}(0.10 \text{ m})|Pt$?

19-174 Pt reacted with fluorine and oxygen to give a product which contained 32.7% F, 10.4% O, and 57.5% Pt. What was the compound?

> [N. Bartlett and D. H. Lohmann, Proc. Chem. Soc., 1962, 115]

19-175 Calculate E° for the cell $Pt,O_2|H_2O_2||Mn^{2+}, MnO_2(s)|Pt$.

19-176 Determine the standard emf of the following cell: $Pt|Fe(CN)_6^{4-}, Fe(CN)_6^{3-}||I^-, I_2(s)|Pt$.

19-177 Calculate the emf of the following cell: $Pt|Fe^{2+}(0.1 \text{ m}), Fe^{3+}(0.01 \text{ m})||Fe^{2+}(0.1 \text{ m})|Fe$.

19-178 Balance the following equations:

a. $Cr_2O_7^{2-} + I^- + H^+ = I_2 + Cr^{3+} + H_2O$

b. $CrO_4^{2-} + SO_2 + H^+ + Cr^{3+} + SO_4^{2-} + H_2O$

c. $MnO_4^- + H_2S = MnO_2 + S + H_2O + OH^-$

d. $MnO_4^- + H_2S + H^+ = Mn^{2+} + S + H_2O$

e. $Fe^{2+} + O_2 + H_2O = Fe^{3+} + OH^-$

19-179 Calculate the solubility of silver chromate in a saturated solution $K_s = 9 \times 10^{-12}$.

19-180 Find the Fe^{2+} ion concentration in a 0.10 M H_2S solution which is also 0.30 M in HCl. Calculate the Fe^{2+} ion concentration in a solution in which the S^{2-} ion concentration is 0.010 M.

19-181 Calculate a value of the atomic weight of vanadium from the experimental observation that 4.8564 g $NaVO_3$ gave 2.3277 g NaCl when treated with hydrochloric acid.

[D. J. McAdam, Jr., J. Am. Chem. Soc., 32, 1603 (1910)]

19-182 A sample of a chloride of iron weighing 0.1404 g gave 0.0695 g of Fe_2O_3 and 0.3732 g of AgCl. Calculate the simplest formula of the chloride.

[W. Grunewald and V. Meyer, Chem. Ber., 21, 690 (1888)]

19-183 A chloride of iron contained 34.65% Fe and 65.76% Cl. A sample weighing 0.0864 g exerted a pressure of 742 mm in a bulb whose volume was 6.7 cc at a temperature of 29°C. What was the formula of the vapor?

[W. Grunewald and V. Meyer, Chem. Ber., 21, 690 (1888)]

19-184 The standard energy of combustion of NbC was found to be -283.93 kcal/mole. Calculate the standard enthalpy of formation of NbC from this observation and the standard enthalpies of formation of Nb_2O_5 and CO_2.

[A. D. Mah and B. J. Boyle, J. Am. Chem. Soc., 77, 6513 (1955)]

19-185 Find the standard enthalpy of formation of $Mo(CO)_6$(s) from the standard enthalpy of combustion, -507.52 kcal/mole.

[F. A. Cotton, A. K. Fischer, and G. Wilkinson, J. Am. Chem. Soc., 78, 5169 (1956)]

19-186 Calculate the number of faradays of electricity required to deposit 100 g of chromium metal from a solution of $H_2Cr_2O_7$.

19-187 To 10.0 ml of a 0.100 M $K_2Cr_2O_7$ solution was added 10.0 ml of a 0.100 M $Pb(NO_3)_2$ solution. What is the mass of the precipitate which was formed?

19-188 An attempt was made to dissolve 0.100 mole of K_2MnO_4 in a dilute H_2SO_4 solution. Calculate the quantity of each of the products formed.

19-189 Find the volume of 0.100 M $KMnO_4$ solution required to oxidize the iron in 25.0 ml of 0.100 M $FeSO_4$ solution which also contains H_2SO_4.

19-190 Calculate the volume of chlorine (at STP) which will result from the action of excess hydrochloric acid on 0.268 g of MnO_2.

19-191 Balance the following equations:

   a. $Co^{2+} + NH_3 + O_2 + H_2O = ? + OH^-$

   b. $[Fe(CN)_6]^{4-} + Cl_2 = ? + Cl^-$

   c. $Cr(OH)_3 + Na_2O_2 = ? + NaOH + H_2O$

19-192 Yttrium oxide was dissolved in concentrated hydrochloric acid, and the solution was evaporated until crystals formed. Analysis found: Y, 29.08; Cl, 34.75; H_2O, 36.17. Calculate the formula.

       [H. J. Nalting, C. R. Simmons, and J. J. Klingenberg,
       J. Inorg. Nucl. Chem., 14, 208, (1960)]

19-193 Chromium(III) oxide and lampblack were heated in a hydrogen furnace at 1525°. The solid product contained 13.2% C and 86.2% Cr. What is the formula of the carbide?

       [R. A. Oriani and W. K. Murphy, J. Am. Chem. Soc., 76,
       3345 (1954)]

19-194 Calculate the time required to deposit 125 g of Ni by the passage of a current of 1.60 amp through a solution of nickel sulfate

19-195 What volume of nickel tetracarbonyl, measured at 87°C and 730 torr, can be formed by action of excess carbon monoxide on 10.0 g of nickel.

19-196 A sample of an oxide or uranium, which contained 88.12% uranium, weighed 0.3363 g. It was heated in an atmosphere of oxygen, and the product weighed 0.3493 g. Write an equation for the reaction which occurred.

> [E. F. Smith and J. M. Matthews, J. Am. Chem. Soc., 17, 686 (1895)]

19-197 Protactinium(V) oxide was mixed with carbon, and an excess of bromine vapor was passed over the mixture at 750° for several days. In addition to carbon monoxide, two products were formed. One contained 36.4% Pa and 63.9% bromine and sublimed at 300 to 350° to leave the other as a yellow-green solid. The second contains 47.8% Pa and 49.07% Br. Write an equation for the reaction of Br_2 and C with Pa_2O_5.

> [D. Brown and P. J. Jones, J. Chem. Soc., 1966, 262]

19-198 Calculate the enthalpy of formation of NdOCl from the equation $NdCl_3(s) + H_2O(g) \rightleftharpoons NdOCl(s) + 2HCl(g)$. The experimental value of the heat of hydrolysis was 242.4 kcal/mole.

> [J. Am. Chem. Soc., 76, 1473 (1954)]

19-199 Calculate the formula of the volatile compound containing 66.5% Pu and 30.9% F.

> [B. Weinstock and J. G. Malm, J. Inorg. Nucl. Chem., 2, 380 (1956)]

19-200 Calculate the volume of hydrogen, measured at standard conditions, which can be formed by the reaction of 1.39 g of lanthanum with water.

Appendix A

LOGARITHMS

Logarithms are the exponents to which a base, commonly 10, is raised to equal a number in question. The logarithm of 100 is 2, since $10^2 = 100$; log 1000 = 3 since $10^3 = 1000$. Although $10^{0.3010} = 2.000$ is not so obvious, log 2.000 = 0.3010.

The values of logarithms of numbers are found in tables such as Table D-2, which supplies four-place logarithms. The use of the table is illustrated in the examples which follow.

The log of a number has two parts, called the mantissa and the characteristic. The purpose of the characteristic is to locate the decimal point of the number. The mantissa establishes the identity of the digits of the number. The value of the mantissa of a number is found in a table of logarithms.

The characteristic is determined by the application of two rules. For a number greater than 1 the characteristic is one less than the number of digits to the left of the decimal point.

For numbers less than 1 the characteristic is negative and is one more than the number of zeros between the decimal point and the first digit to the right of the decimal point. Negative characteristics are written in either of two ways. In the first way, the negative sign is placed above the characteristic, $\bar{1}$, $\bar{2}$, $\bar{3}$, etc. The other way is to represent the negative characteristic by 9, mantissa -10, 8, mantissa -10, 7, mantissa -10, respectively, and so on (see examples below).

Examples

Number	Characteristic	Mantissa	Logarithm
2	0	.3010	0.3010
20	1	.3010	1.3010
200	2	.3010	2.3010
27	1	.4314	1.4314
270	2	.4314	2.4314
2.7	0	.4314	0.4314
2700	3	.4314	3.4314
832	2	.9201	2.9201
8.32	0	.9201	0.9201
83.2	1	.9201	1.9201
8320	3	.9201	3.9201

Examples

Number	Characteristic	Mantissa	Logarithm
0.2	-1	.3010	$\bar{1}$.3010 or 9.3010 - 10
0.02	-2	.3010	$\bar{2}$.3010 or 8.3010 - 10
0.27	-1	.4314	$\bar{1}$.4314 or 9.4314 - 10
0.027	-2	.4314	$\bar{2}$.4314 or 8.4314 - 10
0.00832	-3	.9201	$\bar{3}$.920, or 7.9201 - 10

PROPORTIONAL PARTS

The mantissas in the examples given thus far can be found directly in the table. A number containing four significant figures requires the use of the table of proportional parts. The logarithm of 2764 has a characteristic of 3. The nearest mantissas to be found in the table are for 2.760 and 2.770, or .4409 and .4425. Under 4 in the table of proportional parts is the number 6, which is added to .4409 to give the mantissa or 2764, or .4415. The log of 2674 is then 3.4415.

Examples

Find the logarithms of the following numbers:

log 2.816   =   0.4496

log 471.2   =   2.6732

log 0.8138   = $\overline{1}$.9105 or 9.9105 - 10

log 0.006418 = $\overline{3}$.8074 or 7.8074 - 10

log 76.44   =   1.8833

The procedure for reversing the process and finding the number corresponding to a logarithm is as follows:

1. log x = 1.6018

2. Because the characteristic is 1, there will be two digits to the left of the decimal point.

3. The mantissa .6018 lies between .6010 and .6021, which are mantissas for 39.90 and 40.00. The first three digits of the number are therefore 39.9.

4. The difference between .6010 and .6018 is 8. In the table of proportional parts the 8 is found under 7, which is added to 39.9 to produce the number 39.97.

USE OF LOGARITHMS IN MULTIPLICATION AND DIVISION

The use of logs for multiplying and dividing is based on the relationships:

log ab = log a + log b

and

$\log \frac{a}{b}$ = log a - log b

Examples

Use logs to compute the following:

    a. $236.0 \times 4.61 \times 21.2 = x$

        $\log 236.0 = 2.3729$

        $\log 4.610 = 0.6637$

        $\log 21.20 = 1.3263$

           $\log x = 4.3629$

               $x = 23,070$

    b. $2.42 \times 0.3160 \times 0.01750 = x$

        $\log 2.420 = 0.3838 \qquad 0.3838$

        $\log 0.3160 = \bar{1}.4997$ or $9.4997 - 10$

        $\log 0.01750 = \bar{2}.2430$ or $8.2430 - 10$

           $\log x = \bar{2}.1265 \qquad 18.1265 - 20$

             $x = 1.338 \times 10^{-2}$ or 0.01338

    c. $\dfrac{16.82}{0.8436} = x$

    $\log 16.82 = 1.2258$ or $11.2258 - 10$

    $-\log 0.8436 = \bar{1}.9261$ or $9.9261 - 10$

        $\log x = 1.2997 \qquad 1.2997$

           $x = 19.94$

A very important point to remember is that a logarithm, as usually written, is the <u>sum</u> of two parts, the characteristic and the mantissa.

POWERS AND ROOTS

The use of logs gives a convenient means of determining <u>nth</u> powers and <u>nth</u> roots of numbers. The relationships are

$$n \log a = \log a^n \quad \text{and} \quad \frac{\log a}{n} = \log \sqrt[n]{a}$$

Example

Find the values of x by means of logs.

    a. $x = 2.416^2$

        $\log 2.416 = 0.3831$

        $2 \log 2.416 = 0.7662 = \log x$

                    $x = 5.847$

    b. $x = 21.61^3$

        $\log 21.61 = 1.3347$

        $3 \log 21.61 = 4.0041 = \log x$

                    $x = 10,100$

    c. $x = 0.003170^3$

        $\log 0.003170 = \overline{3}.5011$ or $7.5011 - 10$

        $3 \log 0.003170 = \overline{8}.5033$ or $22.5033 - 30$

                $x = 3.187 \times 10^{-8}$

    d. $x = \sqrt{61.42}$

        $\log 61.42 = 1.7883$

        $\frac{1}{2} \log 61.42 = 0.8942 = \log \sqrt{61.42}$

                $x = 7.838$

    e. $x = 481.7^{1/3}$

        $\log 481.7 = 2.6827$

        $1/3 \log 481.7 = 0.8942 = \log x$

                $x = 7.838$

    f. $x = 0.003605^{1/3}$

        $\log 0.003605 = \overline{3}.5569$ or $6.5569 - 9$

        $1/3 \log 0.003605 = \overline{1}.1856$ or $2.1856 - 3$

                $x = 0.1533$

EXPONENTIAL NUMBERS

Very large and very small numbers are often best expressed as powers of 10. For example, the number 2,610,000 is 2.61×10^6 with one digit to the left of the decimal point. The exponent 6 is

found by counting the number of places the decimal point must be
moved to the left in order to leave one digit to the left of the
decimal point.

The number 0.00375 is written 3.75×10^{-4}; the exponent is the
number of places the decimal point must be moved to the right in
order to leave one digit to the left of the decimal point.

Examples
Express the following numbers in the exponential form:
 a.   128,000 = 1.28×10^5
 b.   7860 = 7.86×10^3
 c.   0.000654 = 6.54×10^{-4}
 d.   0.00101 = 1.01×10^{-3}
 e.   10,500 = 1.05×10^4

When numbers raised to powers are multiplied, the exponents
are added; when divided, the exponents are subtracted.

Examples
Carry out the following multiplications by using the exponential
forms of the numbers.
 a. $2.00 \times 10^{-4} \times 3.01 \times 10^{23}$ = 6.02×10^{19}
 b. $4.00 \times 10^{-2} \times 1.50 \times 10^8$ = 6.00×10^6
 c. 0.0025×0.040 = 1.0×10^{-4}
 d. $0.0103 \times 2.00 \times 10^8$ = 2.06×10^6
 e. $86,000 \times 4000$ = 3.44×10^8

Examples
Use exponential forms to compute

 a. $\dfrac{0.00480}{1.20 \times 10^{-3}}$ = $\dfrac{4.80 \times 10^{-3}}{1.20 \times 10^{-3}}$ = 4.00×10^0 = 4.00

    since 10^0 = 1

b. $\dfrac{0.000648}{324} = \dfrac{6.48 \times 10^{-4}}{3.24 \times 10^{2}} = 2.00 \times 10^{-6}$

c. $\dfrac{197}{6.02 \times 10^{23}} = 32.7 \times 10^{-23} = 3.27 \times 10^{-22}$

When a number written in exponential form is raised to a power, the exponent is multiplied by the power. For example,

$(2.00 \times 10^{4})^{2} = 4.00 \times 10^{8}$

$(3.00 \times 10^{-4})^{3} = (27.0 \times 10^{-12}), (8.00 \times 10^{-6})^{1/3} = 2.00 \times 10^{-2}$

When numbers are written in the exponential form, the characteristics of their logarithms are equal to the exponents. The log of 3.000×10^{6} is 6.4771; the log of 3.000×10^{-6} is $\overline{6}.4771$ or 4.4711 - 10.

Looked at from another point of view, the log of a number written in the exponential form is the sum of two logs. The log 3.000×10^{6} is the log of 3.000 plus the log of 10^{6}, 0.4771 + 6.0000.

$\log 3.000 \times 10^{-6} = 0.4771 - 6.000$

Examples

Write the logarithms of the following:

a. $\log 2.000 \times 10^{4} = 4.3010$
b. $\log 2.000 \times 10^{-3} = \overline{3}.3010$
c. $\log 4.710 \times 10^{6} = 6.6730$
d. $\log 8.565 \times 10^{-4} = \overline{4}.9323$
e. $\log 28.74 \times 10^{-2} = 1.4585 + \overline{2}.000$

PROBLEMS

A-1 Use logs to compute the following:

a. $\dfrac{6.28}{44.0}$ b. $\dfrac{16.0}{6.02}$ c. $\dfrac{0.133}{117}$

d. $\dfrac{49.7}{92.9}$ e. $\dfrac{50.2}{19.0}$ f. $\dfrac{23.1}{39.1}$

g. $\dfrac{14.2}{47.9}$ h. $\dfrac{62.7}{35.5}$ i. $\dfrac{22.36}{14.01}$

A-2 Use logs to calculate:

a. $\dfrac{22.36}{14.01}$ b. $\dfrac{40.20}{95.94}$ c. $\dfrac{58.98}{35.45}$

d. $\dfrac{39.46}{69.72}$ e. $\dfrac{60.46}{35.45}$

A-3 Solve with logs:

a. $\dfrac{20.82}{34.83}$ x 100   b. $\dfrac{14.01}{34.83}$ x 100   c. $\dfrac{28.21}{205.7}$ x 100

d. $\dfrac{42.03}{205.7}$ x 100   e. $\dfrac{35.45}{205.7}$ x 100

A-4 Solve the following with logs:

a. $268 \times \dfrac{17.9}{760} \times \dfrac{273}{342}$

b. $\dfrac{0.0516 \times 82.0 \times 342 \times 760}{17.9 \times 268}$

c. $\dfrac{2.66 \times 0.0820 \times 293 \times 760}{749} =$

d. $\dfrac{736 \times 200 \times 39.9}{760 \times 82.0 \times 300} =$

e. $\dfrac{0.2182 \times 82.05 \times 414.3 \times 760}{45.00 \times 668.5} =$

f. $\dfrac{6.517 \times 82.05 \times 284.2 \times 760}{776.5 \times 135.1}$

A-5 Solve the following with logs:

a. 26.2^2 b. 2.713^3

c. 43.68^3 d. 0.1641^4

e. 0.2424^2 f. 81.4^5

g. 653.4^2 h. $\sqrt[3]{4.452}$

i. $\sqrt[3]{0.08610}$ j. $\sqrt{0.5432}$

A-6 Express the following numbers in exponential form:

a. 156,000   b. 8420

c. 0.000419   d. 0.00502

e. 63,080,000   f. 0.0005004

A-7 Compute the following:

    a. $6.40 \times 10^{-4} \times 2.00 \times 10^{8}$ =

    b. $3.00 \times 10^{-2} \times 480$ =

    c. $0.00286 \times 3,000$ =

    d. $0.00194 \times 2.00 \times 10^{-7}$ =

    e. $680,000 \times 0.00400$ =

A-8 Solve the following:

    a. $(2.00 \times 10^{-9})^{3}$ =   b. $(4.00 \times 10^{-8})^{\frac{1}{2}}$ =

    c. $(27.0 \times 10^{-15})^{1/3}$ =   d. $(25 \times 10^{4})^{2}$ =

    e. $(270 \times 10^{-7})^{1/3}$ =

A-9 What are the logs of

    a. 3.00×10^{5} b. 6.250×10^{-4}

    c. 87.61×10^{-3} d. 1.63×10^{-9}

    e. 4.07×10^{-11}

A-10 What are the numbers which correspond to the following logs?

    a. 3.7125 b. $\overline{4}.8274$

    c. $4.8725 - 10$ d. $\overline{3}.1486$

    e. $8.1942 - 10$

Appendix B

THE SLIDE RULE

The slide rule is a mechanical device for performing
mathematical manipulations such as multiplying, dividing, raising
to powers, and extracting roots. The slide rule is based on
logarithms. The distances marked on the C and D scales of the
slide rule are proportional to the logarithms of the numbers indicated
on the scales. The common 10-in. slide rule is useful for the
multiplication and division of numbers not exceeding three significant
figures.

MULTIPLICATION

To multiply two numbers, for example 2 x 4, place the number 1
on the left end of the C scale over one of the numbers to be
multiplied, say 2 on the D scale (Fig. B-1), and then move the slide

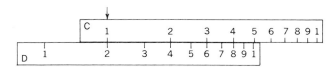

FIGURE B-1

to place the line over the other number, 4 in this case, on the C
scale (Fig. B-2). The product, 8, is under the line of the slide
on the D scale.

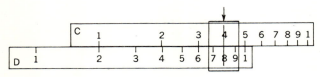

FIGURE B-2

If the number under the line on the C scale is beyond the limit
of the D scale, place the right end of the C scale over the larger
of the numbers to be multiplied and repeat the procedure described
but in the opposite direction. For example, to multiply 2 x 6,
place the 1 of the <u>right</u> end over the C scale over 6 on the D
scale; then place the line of the slide over 2 on the C scale. The
product, 12, is under 2 on the D scale (Fig. B-3).

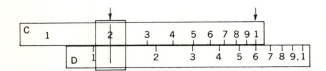

FIGURE B-3

In these operations, distances on the slide rule proportional
to the logarithms of the numbers to be multiplied have been added.

DIVISION

To divide two numbers, for example 9÷3, place the line on the
slide over the <u>numerator</u> on the D scale. Move the C scale until
the denominator is just under the line on the slide. The quotient,
3 in this case, is found under number 1 of the C scale on the D
scale (Fig. B-4).

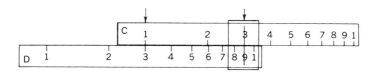

FIGURE B-4

Examples

    3 x 2   Place 1 on C over 3 on D and the slide over 2 on C; the
            product is under the slide on D.

    2.5 x 3.0   Place 1 on C over 2.5 and the slide over 3 on C; the
            product is under the slide on C.

    8 ÷ 4 Place the slide over 8 on D and 4 on C under the slide;
            the quotient is under 1 on D.

    12 ÷ 3 Place the slide over 12 on D and 3 on C under the slide;
            the quotient is under 1 on D.

MARKINGS

Since the slide rule scales are logarithmic, the distances
between the digits printed on the scales are not equal. Between
1 and 2, ten subdivisions also are numbered and further subdivided
into ten parts. Between 2 and 3 and 3 and 4 the ten subdivisions
are not marked, and they are further subdivided into only five parts.

Examples
    1.36, 13.6, 136, etc. (Fig. B-5).

FIGURE B-5

3.41, 0.341, 341, etc. (Fig. B-6).

FIGURE B-6

For the remainder of the scale the ten subdivisions are further subdivided into only two parts.

Examples

7.62, 762, 0.0762, etc. (Fig. B-7)

FIGURE B-7

DECIMAL POINTS

The position of the decimal point in the answer obtained by slide rule manipulation is often a perplexing problem to the beginner The best method is to carry out a simple multiplication or division with easily handled numbers having the same number of significant figures.

Example

52.3 x 624 = ?

The digits 326 are read from the D scale after the manipulation. Since 50 x 600 = 30,000, the answer is 32,600.

For many of the calculations in the problems of chemistry, the numbers are best expressed as powers of 10. When they are expressed

in this way, it is usually easy to estimate the approximate answer
and fix the position of the decimal point.

Example

$$\frac{0.252 \times 82.0 \times 298}{1.06 \times 282} = \text{?}$$

This may be written as

$$\frac{2.52 \times 10^{-1} \times 8.20 \times 10 \times 2.98 \times 10^2}{1.06 \times 2.82 \times 10^2} =$$

$$\frac{2.52 \times 8.20 \times 2.98 \times 10^2}{1.06 \times 2.82 \times 10^2} = 218$$

Then, using nearest whole numbers,

$$\frac{3 \times 8 \times 3 \times 10^2}{1 \times 3 \times 10^2} = 24, \text{ approx.}$$

Therefore the answer is 21.8

The rules required for simple multiplication and division have
been outlined here, and practice will lead to proficiency in these
simple operations. The best advice is to use the slide rule
on all of your assignments as soon as you know the simple operations.
The use of the slide rule for other operations, such as powers and
roots, is described in the instruction manuals provided with most
slide rules.

For an extended discussion of the use of the slide rule, the
student is referred to the book by C. C. Bishop, Slide Rule,
3d ed., Barnes & Noble, New York, 1955.

Appendix C

ANSWERS TO PROBLEMS

1-1 a. 3   b. 3   c. 6   d. 3
     e. 2   f. ?   g. 2   h. 4
     i. 2   j. 4

1-2 a. 98.5 g   b. 99.9%   c. 4.47 x 10^{-2} mg
     d. 77.8%   e. 14.33 g

1-3 a. 6.15 ml   b. 23.8 liters   c. 223.6 g
     d. 0.0159 g

1-4 a. 2.809 g   b. 2.615 liters

1-5 a. 1.85 x 10^5 b. 24.4   c. 24.47
     d. 24.47   e. 6.067 x 10^{23}

1-6 a. 2.732   b. 1.077   c. 3.28 x 10^{-4}
     d. 2.54   e. 3.20 x 10^{-4}

1-7 a. 7.4 x 10^{-2} nm   b. 7.4 x 10^{-5} μm
     c. 7.4 x 10^{-8} mm   d. 7.4 x 10^{-9} cm

1-8 a. 6.939 mg   0.006939 kg

1-9 1,000 γ   1-10 50 cc

1-11 2.580 g/liter   1-12 0.463 g

1-13 272 ml   1-14 $12.30 per 100 g

1-15 214 g per package   1-16 10 x 10^{-3} msec; 10 x 10^{-6} μsec

1-17 yes; 4 oz. = 113.5 g   1-18 9.52 x 10^{-2} mm

1-19   475 g   1-20   128 g

1-21   yes; 383 g = 0.845 lb   1-22   484 cc

1-23   348 g   1-24   a. 1776 g   b.   965 g

1-25   42.0°F   1-26   a. 98.6°F   b.   body
                                                                          temperature

1-27   6.65 cal   1-28   25.4°

1-29   36.1°C   1-30   a. 10°C   b. 1°C
                                                   c. 18°F; 1.8°F

1-31   a. 1.10 x 10^{-1} nm   b. 1.10 x 10^{-4} µm
       c. 1.10 x 10^{-7} mm   d. 1.10 x 10^{-8} cm

1-32   a. 1.32 x 10^{-8} cm   b. 1.32 x 10^{-7} mm
       c. 1.32 x 10^{-4} µm   d. 1.32 x 10^{-1} nm

1-33   200,590 mg   0.20059 kg   1-34   10 ml

1-35   250 ml   1-36   11,508 cc

1-37   5.155 g   1-38   1.51 lb

1-39   5.5 lb   1-40 $14.00 per liter

1-41   61.5 cc   1-42   188 ml

1-43   11.2 kg   1-44   205

1-45   0.003 sec   3,000 µsec   1-46   0.0545 g

1-47   113.5 g   1-48   880 g

1-49   3.77 ℓ   3,770 ml   1-50   472 ml   0.472 liters

1-51   22.7 g   431 g   1-52   66.4 g

1-53   35.6%   1-54   43.9°F

1-55   15°C   1-56   0.171 cal

1-57   21.2°

2-1   61.32% Cd   38.66% Cl   2-2   1.585

2-3   0.92188   2-4   47.970% Zn   52.028% Cl

2-5   1.008 g   2-6   16.03 g

2-7 3.82×10^{-23} g   2-8 1.97×10^{-22} g
      3.82×10^{-20} mg

2-9 4.77×10^{23} atoms 2-10   0.525 g atom

2-11 1.15×10^{23} atoms 2-12   12.84 g atoms 7.73×10^{24} atoms

2-13 5.33×10^{-11} g   2-14 3.48×10^{8} atoms

2-15 3.28×10^{24} atoms 2-16 1.46×10^{24} atoms

2-17   107.87   2-18   35.46

2-19   69.72   2-20   83.80

2-21   112.4   2-22   14.007

2-23   63.9   2-24   7.945 g

2-25   1.0025   2-26   63.90 g

2-27   63.80 g   2-28 3.01×10^{-19} mg

2-29   104 g   2-30   3.56 g atoms

2-31 2.14×10^{22} atoms/g   2-32 6.03×10^{10} atoms
       2.14×10^{19} atoms/mg
       2.14×10^{16} atoms/μg
       2.14×10^{16} atoms/γ

2-33 5.19×10^{20} atoms 2-34 2.69×10^{-6} g atom

2-35 1.30×10^{-19} g atom 2-36 4.42×10^{16} atoms

2-37   0.338 g atom   2-38   2.42 g atoms

2-39 1.14×10^{23} atoms 2-40   35.453

2-41   32.06   2-42   69.723

2-43   79.89   2-44   65.366

2-45 127.58   2-46 87.617

2-47 15.999

3-1   304 g   3-2   1.29 moles

3-3 1.03×10^{23} P atoms   3-4 3.39×10^{22} molecules
      4.12×10^{23} O atoms 6.78×10^{22} atoms of oxygen

3-5 1.34×10^{-22} g   3-6   a. 160.2 g b. 288 g
                                             c. 1.15×10^{-5} g
                                             d. 169 g   e. 1000 g

3-7 1.27×10^{24} Ca^{++} 3-8 3.55×10^{-2} g
      2.54×10^{24} Cl^{-}

3-9   1.10   3-10 0.644

3-11 32.8% As, 14.9% Si   3-12 39.59% Ga; 60.40% Cl

3-13 40.36% Me; 59.65% Cl

3-14 | % K | % M | % F |
      |------|------|------|
      | 25.9 | 36.4 | 37.7 |
      | 25.7 | 36.8 | 37.5 |
      | 25.2 | 38.0 | 36.8 |
      | 25.3 | 37.9 | 36.8 |
      | 24.5 | 39.8 | 35.7 |
      | 24.2 | 40.5 | 35.3 |

3-15 $MgAs_4$ 3-16 Sr_2IrO_4

3-17 BNH_4 3-18 Si_2NH_5

3-19 Tl_2TiCl_6 3-20 TiF_3

3-21 $CeCl_3 \cdot 7H_2O$ 3-22 $Co(NH_4)_2Cl_4 \cdot 2H_2O$

3-23 Mn_2O_3 3-24 $C_{11}H_{22}O_2$

3-25 $S_8N_6F_5H_3$ 3-26 B_6H_{12}

3-27 $B_3N_3H_3F_3$ 3-28 $C_{28}H_{24}WO$

3-29 BF_3NH_3 3-30 $P_2F_2Cl_2O_3$

3-31 71.0 g   3-32 0.0859 mole

3-33 1.03×10^{22} molecules   3-34 a. 257 g   b. 128 g
                                                          c. 64.2 g   d. 77.0 g
                                                          e. 2.07×10^{-2} g

3-35 a. 591 g   b. 86.6   3-36 7.60×10^{20} Mg^{++}
      c. 0.806 g d. 1540 g 1.52×10^{21} Br^-
      e. 61.3 g

3-37 1.12   3-38 0.6891

3-39 1.933   3-40 22.96% N; 37.70% Na; 39.34% O

3-41 41.1% V; 46.0% F   3-42 19.3% NH_3; 36.1% Cu, 43.2% F

3-43 BI_3 3-44 $ZrBr_4$

3-45 WCl_6 3-46 S_4N_3Cl

3-47 ThF_4 3-48 ReO_3F

3-49 K_2ReCl_6 3-50 PSI_2

3-51 ICl_3 3-52 TiF_4

3-53 $ZrBr_3$ 3-54 K_2TeI_6

3-55 BI_2 3-56 WBr_3

3-57 RhF_6 3-58 B_2O_2N BN

3-59 CdP_4 3-60 HgNCl

3-61 K_2AsO_3F 3-62 $LiBCH_6$

3-62 $(CH_3)_2AlH$ 3-64 $AgNH_2$

3-65 $Ag_2TeO_4 \cdot 2H_2O$ 3-66 $CaCl_2 \cdot 6H_2O$

3-67 Sc_2S_3 3-68 S_3Cl_2

3-69 Si_4OCl_{10} 3-70 $P_3N_3F_6$ $P_4N_4F_8$
      $Si_6O_2Cl_{14}$
      $Si_8O_3Cl_{18}$

3-71 CO_2F_2 3-72 $B_{12}Cl_{11}$

4-1 a. 3 4 1 4   b. 2 2 1 2   c. 1 2 1
     d. 1 2 1 2   e. 2 2 1   f. 2 2 1
     g. 2 10 6 1   h. 2 2 2 1   i. 8 4 7 2 8 8
     j. 3 1 2 1   k. 1 1 3 1   l. 1 3 1 2
     m. 1 12 4 3   n. 1 6 2 2 3   o. 1 3 2 3
     p. 2 5 2 2 4   q. 2 1 2 1   r. 1 3 1 1
     s. 1 2 1 1   t. 1 2 1 2   u. 2 3 2 2
     v. 2 1 2 1   w. 1 2 3 1   x. 1 1 2 2
     y. 1 1 2   z. 1 1 2

4-2 a. 2 9 1 6   b. 4 11 4 3   c. 6 2 6 6 1
     d. 1 12 3 4   e. 3 4 2 6 1   f. 6 3 2 8
     g. 1 6 2 4 1   h. 2 2 2 2 1

4-3 6 and 12 moles   4-4 24 moles

4-5 1.84×10^{-3} g   4-6 0.146 g

4-7 0.647 g   4-8 2.62 g

4-9 0.0458 g   4-10 14.1 g

4-11 1.81 mM   4-12 $WCl_4 + 2H_2 \rightarrow W + 4HCl$
                                    $HCl + NaOH \rightarrow NaCl + H_2O$

4-13 $TlVO_3$
      $Tl_2CO_3 + V_2O_5 \rightarrow 2TlVO_3 + CO_2$

4-14 $Bi + 2S \rightarrow BiS_2$
      $4BiS_2 + 11 O_2 \rightarrow 2Br_2O_3 + 8SO_2$
      $2BiS_2 \rightarrow Bi_2S_3 + S$

4-15 3.58 mM F_2 required   4-16 $2 O_2BF_4 \rightarrow 2BF_3 + 2O_2 + F_2$

4-17 Hg_3N_2 4-18 XeO_2F_2

4-19 $6CCl_4 + 9NH_4Cl \rightarrow C_6N_9H_3 + 33HCl$

4-20 58.8% $WOCl_4$, 41.2% WOF_4 based on W
      $2WO_2 + 2CCl_2F_2 \rightarrow WOF_4 + WOCl_4 + 2CO$

4-21 1.88 g $InAlCl_4$ 4-22 3.25×10^{-7} moles
                                       7.52×10^{-5} g

4-23 1.06×10^{-3} g   4-24 88.7%

4-25 39.55%   4-26 99.530%

4-27 43.82 g Br_2 in excess   4-28 18.89 g

4-29 No 56.8%   4-30 99.84%

4-31 14.00   4-32 40.09

4-33 50.94   4-34 22.99

4-35 a. 4 5 4 6   b. 3 1 2 1   c. 4 4 1 2
     d. 1 2 1 4   e. 1 3 5 3 5 1   f. 1 3 1 3
     g. 1 6 2 3   h. 3 1 3   i. 1 1 2 1
     j. 4 4 1 2   k. 1 1 1 1   l. 3 2 1 3
     m. 1 1 2   n. 6 1 2   o. 1 6 4
     p. 1 4 1 2 1   q. 1 3 1 3 1   r. 1 3 3 2
     s. 2 3 1   t. 2 3 3 3 1   u. 1 1 1 2
     v. 1 4 1 4   w. 2 1 1 1   x. 2 2 4 1
     y. 1 6 3 2   z. 1 10 2 10 4

4-36 20.4 mM   4-37 190 g

4-38 86.3%   4-39 $Zr + 2Cl_2 \rightarrow ZrCl_4$
                          $3ZrCl_4 + Zr \rightarrow 4ZrCl_3$

4-40 67.3 g   4-41 $3WCl_4 \rightarrow 2WCl_5 + WCl_2$

4-42 $SnS + 5F_2 \rightarrow SnF_4 + SF_6$ 4-43 $Gd_2O_3 + 3CCl_4 \rightarrow 3COCl_2 + 2GdCl_3$

4-44 99.970%   4-45 $B_3N_3H_{12}$

4-46 98.52%   4-47 $B_3N_3H_{12}(77\%) \rightarrow B_3N_3H_6 + 3H_2$

4-48 33.0%   4-49 71.2%

4-50 0.01979 g   4-51 18.15 g

4-52 11.1% (theoretical loss)   4-53 64.5 g

4-54 284 g   4-55 94%

4-56 0.48 mM HF   4-57 0.3128 g

4-58 0.635 g   4-59 1.81 g

4-60 76.68%   4-61 94.45%

4-62 41%   4-63 1.289

4-64 9.013   4-65 107.87

4-66 55.845   4-67 40.08

4-68 50.96

5-1 1.37   5-2 219 ml

5-3 542 ml   5-4 69.7 g/mole

5-5 3.17 g/liter   5-6 a. 39.92 g/mole
                              b. 3.998 g/mole
                              c. 2.013 g/mole
                              d. 83.06 g/mole
                              e. 20.16 g/mole
                              f. 28.01 g/mole
                              g. 32.01 g/mole
                              h. 218 g/mole
                              i. 131.2 g/mole

5-7 127.4 g/mole   5-8 8.80 atm

5-9 303.5 g/mole   5-10 133.5 g/mole $C_7H_{14}O_2$

5-11 B_4Cl_4 5-12 9.674×10^{-2} g

5-13 46.3 cc   5-14 17.2%

5-15 14.7 mm   5-16 VCl_4

5-17 871 ml   5-18 298 ml

5-19 39.3 g/mole   5-20 44.54 g/mole

5-21 540 ml   5-22 100%

5-23 B_2Cl_4 5-24 Ga_2Cl_6

5-25 22.1-22.6 liters 5-26 0.225 liter/min

5-27 1.08/1   5-28 20.167

5-29	39.941	5-30	0.683 atm
5-31	33.74 atm	5-32	0.00364
5-33	210 ml	5-34	67.3 ml
5-35	1.97 g/liter	5-36	a. 17.26 g/mole

5-36
  a. 17.26 g/mole
  b. 127.3 g/mole
  c. 39.92 g/mole
  d. 78.5 g/mole
  e. 67.2 g/mole
  f. 81.60 g/mole
  g. 36.74 g/mole
  h. 38.3 g/mole

5-37 176 g/mole $VOCl_3$

5-38 $B_3N_3Cl_3$

5-39 3.27 g/liter

5-40 $B_5H_9 + 15H_2O \rightarrow 5H_3BO_3 + 12H_2$

5-41 96.7% 92% 95.3%

5-42 49.25 vol% 57.8 wt.%

5-43 $SO_3 + H_2O$ molecules in vapor

5-44 No 43.2 g/mole

5-45 30.01 g/mole

5-46 28.00 g/mole

5-47 68.2 g/mole

5-48 B_2H_6

5-49 55.9%

5-50 $Ga_2Cl_6 + Ga \rightarrow 3GaCl_2$

5-51 4.95 g   1.53 g

5-52 7.72%

5-53 10/1

5-54 0.0987 g

5-55 0.335 g

5-56 1 min 43 sec

5-57 5.97×10^{-5} ml/sec

5-58 1.005/1

5-59 34.004 g/mole   30.980

5-60 15.9994

5-61 94.5 atm

6-1

      :Br: H
a. :Br: C :Br: *b.* H:Si:H   *c.* :O: F:
      :Br: H   : F:

d. H:Se:H   *e.* [H:N:H]$^+$:Cl:$^-$ *f.* H:O:Cl:
                H / H

g.   :O: *h.*   :O: *i.* H:O:Br:
    H:O:S:O: H:O:Cl:O:
      :O: :O:
      H

j. 2Na$^+$ [:O:S:O:]$^{--}$ *k.* :O: : C : :O:
            :O:

l. :Cl: O: *m.* :F:N:F: *n.* H: P:H
    :Cl: :F: H

o. Ca^{++}2 [N: :O: / :O:]$^-$

6-2   *a.* :Cl: Hg:Cl: *b.* Br. B .Br / :Br: *c.* :Cl. / :Cl. Te .Cl. / :Cl:

d. :Cl. Sb .Cl. / :Cl. .Cl. / :Cl: *e.* F / F. Se .F / F. .F / F

6-3   *a.* :O: :O:O: ⟷ :O:O: :O: *b.* :C: : :O:

        H
c. H:C: :O:

d. 2Li^+ $\left[\begin{array}{c} :\ddot{O}:C::\ddot{O}: \\ :\ddot{O}: \end{array}\right]^{--}$ ⟷ $\left[\begin{array}{c} :\ddot{O}: :C:\ddot{O}: \\ :\ddot{O}: \end{array}\right]^{--}$

$\updownarrow$

$\left[\begin{array}{c} :\ddot{O}:C:\ddot{O}: \\ \ddot{O}: \end{array}\right]^{-}$

e. $:N::N::\ddot{O}:$

f. Na^+ $\left[\begin{array}{c} :\ddot{O}:N::O: \\ :\ddot{O}: \end{array}\right]^{-}$ ⟷ $\left[\begin{array}{c} :\ddot{O}: :N:\ddot{O}: \\ :\ddot{O}: \end{array}\right]^{-}$ ⟷ $\left[\begin{array}{c} :\ddot{O}:N:\ddot{O}: \\ \ddot{O}: \end{array}\right]^{-}$

g. $:\ddot{O}:S::\ddot{O}:$ ⟷ $:\ddot{O}::S:\ddot{O}:$ h. $H:\ddot{N}::N::\ddot{N}:$

i. K^+ $\left[\begin{array}{c} :N:\ddot{O}: \\ \ddot{O}: \end{array}\right]^{-}$ ⟷ $\left[\begin{array}{c} :N::O: \\ :\ddot{O}: \end{array}\right]^{-}$

j. $:\ddot{O}:$ $\ddot{O}:$ $:\ddot{O}:$
   $S::\ddot{O}:$ ⟷ $S:\ddot{O}:$ ⟷ $S:\ddot{O}:$
   $:\ddot{O}:$ $:\ddot{O}:$ $\ddot{O}:$

6-4 a. $:\ddot{Cl}::\ddot{Cl}:$ b. $:N:::C:C:::N:$
        $:\ddot{Cl}:B:B:\ddot{Cl}:$

c. $:\ddot{F}:N:\ddot{O}:$ ⟷ $:\ddot{F}:N::O:$ d. $:\ddot{Cl}:$
        $\ddot{O}\cdot$ $:\ddot{O}:$ $:\ddot{Cl}:P:O:$
                                                              $:\ddot{Cl}:$

e. $2\text{Na}+$ $\left[\begin{array}{c} :\ddot{O}:S:S:\ddot{O}: \\ :\ddot{O}::O: \end{array}\right]^{--}$ f. $H:C::\ddot{O}:$
                                                                                    $:\ddot{F}:$

6-5
```
        F
        ··       ··
  F :O : S : O:
        ··   ··   ··
        : O:
        ··
```

6-6   *a.* H—Se
                 \
                  H

b.. F—O
           \
            F

c.   H   +
          |
          N
        ⁄ | \
      H H H

d. Cl—O
          \
           Cl

e. Cl—Hg—Cl

f. I   I
        \ ⁄
         B
         |
         I

g.   ··
          N
        ⁄ | \
      F | F
          F

h. Cl   Cl
        \ ⁄
   Cl —Se —:
        ⁄ \
      Cl   Cl

i. F F F
       \ | ⁄
        Se
       ⁄ | \
      F F F

j.   F F
        \ |
         As—F
        ⁄ |
      F   F

6-7 a. V shape   b. V shape   c. tetrahedron
     d. tetrahedron   e. trigonal pyramid   f. tetrahedron
     g. plane triangle   h. trigonal bipyramid   i. T shape
     j. octahedron

6-8 *a.* H—C ≡ N

b. Cl
        \
         C=O
        ⁄
     Cl

c. O—S
          \
           O

d.   O
         ‖
         P
       ⁄ | \
     Cl | Cl
         Cl

e.   O⁻
            |
      O—N
            \
             O

f.   ··
         S
       ⁄ | \
     F | O
         F

g. Cl—N
          \
           O

h.   O
            ‖
      F—N
            \
             O

6-9 $\ddot{S}e = C = N$ linear

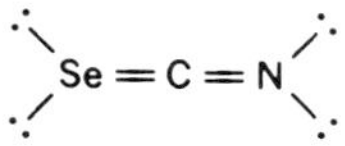

6-10   ClF5 structure (square pyramidal with F atoms)

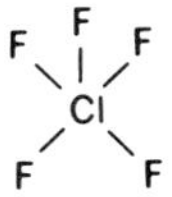

6-11   S2F10 structure (two SF5 groups connected)

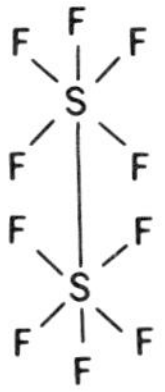

6-12   a. $B(SCN)_3$ b. BCl_3 slight excess

c. 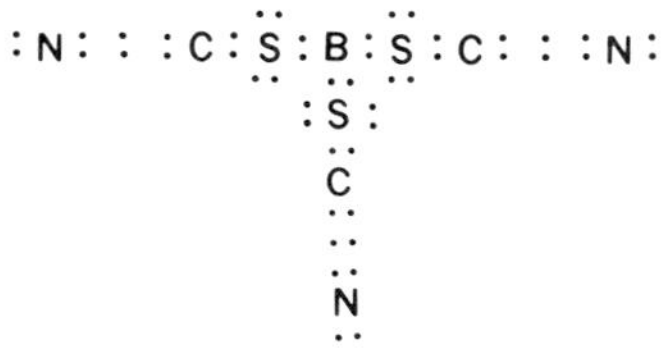   :N: : :C:$\ddot{S}$:B:$\ddot{S}$:C: : :N:

with :$\ddot{S}$: , C, N below the B

6-13   a. 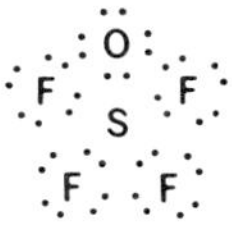   SOF4 structure

b. 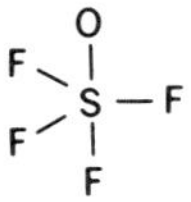 $F\diagdown \underset{\underset{F}{|}}{\overset{\overset{O}{|}}{S}} - F$

6-14

$\underset{F \diagup \underset{F}{|} \diagdown F}{\overset{\overset{N}{\overset{|||}{S}}}{}}$

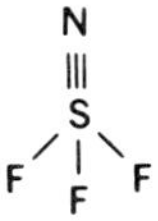

7-1   10 ppm   7-2   100 ppm

7-3   a.   50.4 g   b.   78.62 g   c.   7.5 g
      d.   7.24 g   e.   174 g   f.   84.8 g
      g.   825 g   h.   0.787 g   i. 2.43×10^{-2} g
      j.   19.5 g

7-4   a.   1.00 g   b.   150 g   c.   0.44 g
      d.   58.2 g   e. 8.75×10^{-4} g

7-5   a.   1.88 M   b.   2.87 M   c.   1.30 M

7-6   a.   24.5 ml   b.   100 ml   c.   0.781 ml
       d.   0.00830 ml   e.   0.0329 ml

7-7   6725 ml   7-8   12.4 M

7-9   16.09 M   7-10   12.1 M   16.0 M

7-11   7.576 M   7-12   15.6 m

7-13   0.0922 m   7-14   1.362 M   1.500 m

7-15   a.   7.059 m   b.   HF(aq)   c.   19.2 g

7-16   3.14 ml   7-17   0.977 solvent   0.023 solute

7-18   0.0158   7-19   5798 ppm

7-20   a.   66.46 g   b.   1.25 g   c.   6.044 g
       d.   0.605 g   e.   2.411 g   f.   157.6 g
       g.   39.8 g   h.   2.64 g   i. 3.87×10^{-3} g
       j.   0.401 g

7-21   a.   2.0 g   b.   1.7 g   c.   2.12 g
       d.   0.0602 g   e.   0.0917 g

7-22 5.54×10^{-3} g   7-23   a.   0.0603 M
                                                 b. 5.03×10^{-4} M
                                                 c.   0.0151 M

7-24   a.   0.5770 M   b.   0.212 M   c.   0.0194 M
       d.   0.00579 M   e.   0.00845 M

7-25   1481 ml   7-26   4094 ml

7-27   8.71 M   7-28   a.   25.0 ml b.   1.67 ml
                                                 c.   3.47 ml

7-29   23.0 M   7-30   11.7 M

7-31   3.99 m   7-32   0.495 m

7-33   4.3110 m   7-34   2.68 m

7-35   16.0 liters   7-36   85.2 g

8-1	17.35 mm	8-2	634.4 mm
8-3	35.5 mm	8-4	From Raoult's Law

8-4   106.5 mm   146.0 mm   250.4 mm
      346.7 mm   446.1 mm   552.9 mm

8-5	0.240°	8-6	40.6°
8-7	-6.42°C	8-8	0.362°
8-9	2.40 atm	8-10	182 g/mole
8-11	128.7 g/mole	8-12	463 g/mole
8-13	128 g/mole	8-14	258 g/mole S_8
8-15	P_4	8-16	179 g/mole
8-17	289 g/mole	8-18	28.2 g/mole
8-19	-7.30°	8-20	49 g/mole
8-21	1.75×10^6 g/mole	8-22	$S_4N_4F_4$
8-23	613.9 mm	8-24	0.156 mm
8-25	32.18 mm	8-26	From Raoult's Law

8-26   46.10 mm   176.9 mm   243.6 mm
       286.9 mm   309.4 mm   356.8 mm

8-27	71.8 mm	8-28	0.170°
8-29	46.5°C	8-30	-8.80°C
8-31	-0.0374°	8-32	-0.1485°
8-33	-4.10°	8-34	0.518 g
8-35	2.40×10^{-3} atm	8-36	129 g/mole
8-37	53.8 g/mole	8-38	258.6 g/mole
8-39	185 g/mole	8-40	758 g/mole
8-41	P_2I_4	8-42	S_8
8-43	P_4	8-44	341 g/mole

8-45   186 g/mole   8-46   258 g/mole

8-47   2.32 x 10^5 g/mole

9-1   10.3 cal   9-2   2595 cal

9-3   640 cal   9-4 B_2H_6

9-5   a. $B_2H_6 + 3F_2 \rightarrow BF_3 + 6HF$ -33,400 kcal/kg
      b. $CH_4 + 4F_2 \rightarrow CF_4 + 4HF$ -28,600 kcal/kg
      c. $N_2H_4 + 5F_2 \rightarrow 2NF_3 + 4HF$ -10,300 kcal/kg

9-6   -0.1996 kcal/g atom   9-7   -140.94 kcal/mole

9-8   -257.1 kcal/mole   9-9   -68.350 kcal/mole

9-10 + 131.5 kcal/mole   9-11 -118.4 kcal/mole
                                           -118.9 kcal/mole

9-12 -186.6 kcal/mole   9-13 -60.0 kcal/mole

9-14 -311.4 kcal/mole   9-15 -15.8 kcal/mole

9-16 -22.6 kcal/mole   9-17 -91.7 kcal/mole

9-18 -43.97 kcal/mole   9-19 -294.7 kcal/mole
                                           92.6 kcal/mole

9-20 -214.7 kcal/mole   9-21 -44.20 kcal/mole

9-22 -252.2 kcal/mole   9-23 460.8 kJ/mole   110.1 kcal/mc

9-24 3206.0 cal/°C   9-25 37.8 kcal/mole

9-26 76.9 kcal/mole   9-27 103.1 kcal/mole 1st bond
                                           93.4 kcal/mole   ave.

9-28 87.5 kcal/mole   9-29 28.935 kcal

9-30 -4.4 kcal/mole   9-31 -259.2 kcal/mole

9-32 -247.9 kcal/mole   9-33 -15.11 kcal/mole

9-34 -56.8 kcal/mole   9-35 -9.63 kcal/mole

9-36 -331.22 kcal/mole   9-37 -221.3 kcal/mole
         5.75 kcal/mole

9-38   -60.99 kcal/mole   9-39   -145.3 kcal/mole

9-40   -80.7 kcal/mole   9-41   -53.7 kcal/mole

9-42   215 kcal/mole   9-43   -217.72 kcal/mole

9-44   -64.92 kcal/mole   9-45   -410.7 kcal/mole

9-46   -21.34 kcal/mole   9-47   -31.89 kcal/mole

9-48   -110.3 kcal/mole   9-49   +33.55 kcal/mole

9-50   -221.5 kcal/mole   9-51   3212.6 g

9-52   -723.3 kcal/mole   9-53   103.0 kcal/mole

9-54   59.3 kcal/mole 1st bond   9-55   87.4 kcal/mole
       66.7 kcal/mole ave.

9-56   78.0 kcal/mole   9-57   12.03 kcal/mole

10-1   a.   7.16   b.   9.06   c. 2.92
       d.   1.95   e.   3.43

10-2   2.42   3.17   3.75   3.27   2.01

10-3   a. 1.12×10^{-10} mole/liter   b. 1.07×10^{-2} mole/liter
       c. 1.78×10^{-6} mole/liter   d. 8.32×10^{-11} mole/liter
       e. 3.80×10^{-3} mole/liter

10-4   a.   1.74   b. 1.94   c. 2.96

10-5   2.32   10-6   1.16

10-7 9.46×10^{-3} ml   10-8 8.46×10^{-2} ml

10-9 2.15×10^{-2} mole/liter   10-10 0.244 mole/liter

10-11 24.5 g   10-12 27.7 g

10-13 0.0130 g   10-14 0.226 mm

10-15 0.0810°   10-16 0.0165°

10-17 0.0934°   10-18 0.216° ions 0.108° no ions

10-19 From Raoult's Law   10-20 361
     17.41 mm   16.91 mm
     17.28 mm   16.79 mm
     17.16 mm   16.61 mm
     17.04 mm   16.32 mm
     16.97 mm

10-21 57.2   10-22 0.0188   0.1179
                                 0.0358   0.2164
                                 0.0613   0.4344

10-23 0.101 M   0.202 M   10-24 0.108 M

10-25 2.42

11-1 a. $$K_c = \frac{[N_2][O_2]}{[NO]^2}$$ b. $$K_c = \frac{[NH_3]^2}{[N_2][H_2]^3}$$

    c. $$K_c = \frac{[NH_3]^2[H_2O]^4}{[NO_2]^2[H_2]^7}$$ d. $$K_c = \frac{[CO]^2}{[O_2]}$$

    e. $$K_c = \frac{[H_2O]^4}{[H_2]^4}$$

11-2 a. $$K_p = \frac{P_{N_2} \cdot P_{O_2}}{P^2_{NO}}$$ b. $$K_p = \frac{P^2_{NH_3}}{P_{N_2} \cdot P^3_{H_2}}$$

    c. $$K_p = \frac{P^2_{NH_3} \cdot P^4_{H_2O}}{P^2_{NO_2} \cdot P^7_{H_2}}$$ d. $$K_p = \frac{P^2_{CO}}{P_{O_2}}$$

    d. $$K_p = \frac{P^4_{H_2O}}{P^4_{H_2}}$$

11-3 1.06×10^{-5}

11-4 9.64×10^{-4}

11-5 45.4

11-6 0.0218

11-7 8.16×10^{-3}

11-8 6.20

11-9 0.49

11-10 144 atm

11-11 2.95 atm

11-12 3.29×10^{-4}

11-13 3.13×10^{-2}

11-14 3.92

11-15 0.298 mole/liter

11-16 0.176

11-17 a. 0.992 atm
       b. 0.984 atm

11-18 158 mm

11-19 26.7

11-20 85.0 mm

11-21 2.76×10^{-4}

11-22 0.00478

11-23 2.90×10^{-4} 2.87×10^{-4} 3.05×10^{-4} 2.66×10^{-4}

$K_p(\text{ave}) = 2.87 \times 10^{-4}$

11-24 4.65×10^{-7} mole/liter

11-25 1.57

11-26 1.36

11-27 3.87×10^{-3} mole/liter

11-28 1.17×10^{-3} mole/liter

11-29 0.56 mole 0.44 mole

11-30 0.661 mole/liter

11-31 a.
$$K_c = \frac{[H_2O][Cl_2]}{[HCl]^2[O_2]^{\frac{1}{2}}}$$

b.
$$K_p = \frac{P_{NH_3}}{P_{N_2}^{\frac{1}{2}} P_{H_2}^{\frac{1}{2}}}$$

c.
$$K_p = \frac{P_{H_2O} \, P_{Cl_2}}{P_{HCl}^2 \, P_{O_2}^{\frac{1}{2}}}$$

d.
$$K_c = \frac{[NH_3]}{[N_2]^{\frac{1}{2}}[H_2]^{\frac{1}{2}}}$$

11-32 1.46×10^{-5}

11-33 1.46×10^{-5}

11-34 6.85×10^4

11-35 85.6 mm 0.113 atm

11-36 0.0375 mm^{-1}

11-37 4.32×10^{-3}

11-38 5.34×10^{-3}

11-39 2.20×10^{-5}

11-40 4.15×10^{11}

11-41 1.91 atm

11-42   31.6 mm   11-43   1.59

11-44   0.597 mm   11-45   0.787

11-46 1.71×10^{-11} 11-47 $K_c = 169$; $K_p = 5.82$

11-48 9.56×10^{-3} 11-49 4.48×10^{-10} mm^{-2}

11-50   5.376   11-51   27.8

11-52   94.44 mm   11-53   0.118 atm NO_2

11-54   insufficient data   11-55   0.165 atm NH_3
                                                 4.918 atm N_2
                                                 4.752 atm H_2

11-56   0.0770 atm   11-57   0.714

11-58   0.428 mole/liter   11-59   0.84%

11-60   90.8%   11-61   21.4 mole%

11-62 1.41×10^{-3} mole/liter   11-63 1.86×10^{-3} mole/liter

11-64 1.76×10^{-2} mole/liter
        for monomer

12-1 1.84×10^{-5} 12-2 1.51×10^{-5}

12-3 6.74×10^{-5} 12-4 6.5×10^{-4} mole/liter

12-5   2.73   12-6   0.925 M

12-7 2.97×10^{-3} mole/liter   12-8 7.75×10^{-3} mole/liter
                                                 6.78×10^{-3} mole/liter

                                                 5.66×10^{-3} mole/liter

12-9 2.94×10^{-6} mole/liter   12-10   10.39

12-11   3.09   12-12 1.51×10^{-6}

12-13 3.64×10^{-9} 12-14   2.21

12-15   2.60   12-16   11.50

12-17   1.56   12-18 1.05×10^{-5} mole/liter

12-19   3.88   12-20   0.455 g

12-21   11.3/1   12-22   4.54

12-23   0.075   12-24   a. 0.055/1   b.   17.8/1
                                            c.   2/1

12-25 1.88×10^{-5} 12-26   4.80

12-27 1.21×10^{-22} mole/liter   12-28 6.23×10^{-21} mole/liter

12-29   5.52   12-30 9.84×10^{-7} mole/liter
                                            pH = 7.99

12-31   12.7   12-32   5.98

12-33   4.21   12-34 1.86×10^{-5} 4.73

12-35   3.24   12-36   10.92

12-37   2.86   12-38   10.86

12-39   1.16%   3.66%   11.6%   12-40   2.86

12-41 6.26×10^{-5} mole/liter   12-42   4.30

12-43   2.26   12-44   1.98

12-45   2.47   12-46   2.2 g

12-47   3.46   12-48   0.077 M

12-49   1.40   12-50   3.97

12-51   1.16   12-52   5.07

12-53   2.21   12-54   2.60

12-55 1.48×10^{-3} mole/liter   12-56   0.109 M

12-57 3.50×10^{-6} mole/liter   12-58 1.12×10^{-3} mole/liter

12-59   0.0501 mole/liter   12-60   8.54

12-61   4.23   12-62   4.54

12-63   10.5   12-64   8.94

13-1 3.400×10^{-6} 13-2 7.02×10^{-7}

13-3 3.16×10^{-8} 13-4 5.67×10^{-8}

13-5 3.24×10^{-5} 13-6 3.48×10^{-5}

13-7 7.67×10^{-5} 13-8 $K_s = 2.13 \times 10^{-11}$

13-9 4.44×10^{-11} 13-10 1.09×10^{-8} mole/liter

13-11 3.74×10^{-4} mole/liter   13-12 1.09×10^{-3} mole/liter

13-13 4.31×10^{-4} mole/liter   13-14 1.59×10^{-2} mole/liter

13-15 $[Ag^+] \times 10^6$ 13-16 before 2.30×10^{-4} mole/liter
       a. 31.9 b. 12.1   after 4.9×10^{-9} mole/liter
       c. 2.91 d. 0.60
       e. 0.47 f. 0.07

13-17 8.0×10^{-7} mole/liter   13-18 1.52×10^{-7} mole/liter

13-19 0.0565 mole/liter   13-20 3.85×10^{-5} mole/liter

13-21 4.31×10^{19} 13-22 9.23

13-23 1.18×10^{-8} mole/liter   13-24 9.24

13-25 4.61   13-26 2.75×10^{-5}

13-27 9.44   13-28 8.73

13-29 5.21   13-30 8.90

13-31 1.69×10^{-11} 13-32 1.53×10^{-4}

13-33 2.67×10^{-12} 13-34 3.40×10^{-6}

13-35 3.08×10^{-11} 13-36 3.77×10^{-10}

13-37 6.22×10^{-5} 13-38 2.64×10^{-3} mole/liter

13-39 6.5×10^{-5} mole/liter   13-40 7.99×10^{-4} mole/liter

13-41 1.37×10^{-6} mole/liter   13-42 2.64×10^{-3} mole/liter

13-43 1.4×10^{-5} mole/liter   13-44 7.51×10^{-20} mole/liter

13-45 9.04×10^{-4} mole/liter   13-46 6.0×10^{-9} mole/liter

13-47 7.02×10^{-16} mole/liter   13-48 2×10^{-10} mole/liter

13-49 1.73×10^{-20} mole/liter   13-50 3×10^{-6} mole/liter

13-51 0.286 mole/liter   13-52 1.008 mole/liter

13-53 5.13 g/100 ml   13-54 2.48×10^{-5}

13-55 1.02%   13-56 8.41

13-57 5.36   13-58 8.71

14-1   Oxidized   Reduced   14-2   a. $NO_2 \rightarrow NO_3^-$ (ox)

a.   H   Fe $HClO \rightarrow Cl^-$ (red)

b.   Cl   Mn   b. $Mg \rightarrow Mg^{2+}$ (ox)

c.   S   Cr $NO_3^- \rightarrow N_2$ (red)

d.   C   P   c. $H_2S \rightarrow SO_2$ (ox)

e.   None $IO_3^- \rightarrow I_2$ (red)

f.   O   Cl   d. $HNO_2 \rightarrow NO_3^-$ (ox)

g.   None $CrO_4^{2-} \rightarrow Cr^{3+}$ (red)

h.   N   N   e. $Cl^- \rightarrow Cl_2$ (ox)

i.   None $MnO_2 \rightarrow Mn^{2+}$ (red)

j.   Br   Br

14-3 a. $I_2 + 2S_2O_3^{2-} \rightarrow 2I^- + S_4O_6^{2-}$

b. $6H^+ + 5Br^- + BrO_3^- \rightarrow 3Br_2 + 3H_2O$

c. $64H^+ + 16NO_3^- + 24CuS \rightarrow 24Cu^{2+} + 3S_8 + 16NO + 32H_2O$

d. $16H^+ + 2MnO_4^- + 10Cl^- \rightarrow 2Mn^{2+} + 5Cl_2 + 8H_2O$

e. $4Zn + 10H^+ + NO_3^- \rightarrow 4Zn^{2+} + NH_4^+ + 3H_2O$

f. $3Cu + 8H^+ + 2NO_3^- \rightarrow 3Cu^{2+} + 2NO + 4H_2O$

g. $Cu + 3H^+ + HSO_4^- \rightarrow Cu^{2+} + SO_2 + 2H_2O$

h. $64H^+ + 24H_2S + 8Cr_2O_7^{2-} \rightarrow 3S_8 + 16Cr^{3+} + 56H_2O$

i. $2Cu^{2+} + 4I^- \rightarrow 2CuI + I$

j. $H^+ + 2ClO_3^- + SO_2 \rightarrow 2ClO_2 + HSO_4^-$

k. $3H_2O + 5ICl \rightarrow 2I_2 + IO_3^- + 5Cl^- + 6H^+$

1. $3H_2SO_3 + 5H^+ + Cr_2O_7^{2-} \rightarrow 2Cr^{3+} + 3HSO_4^- + 4H_2O$

m. $2Mn^{2+} + 5S_2O_8^{2-} + 8H_2O \rightarrow 2MnO_4^- + 10HSO_4^- + 6H^+$

n. $6H^+ + 2MnO_4^- + 5H_2O_2 \rightarrow 2Mn^{2+} + 5O_2 + 8H_2O$

o. $2H^+ + Ag + NO_3^- \rightarrow Ag^+ + NO_2 + H_2O$

14-4 a. $2OH^- + 2H_2O + SbO_2^- + 2ClO_2 \rightarrow 2ClO_2^- + Sb(OH)_6^-$

b. $Cl_2 + 2OH^- + IO_3^- \rightarrow 2Cl^- + IO_4^- + H_2O$

c. $3I_2 + 6OH^- \rightarrow 5I^- + IO_3^- + 3H_2O$

d. $H_2O + 6Fe_3O_4 + 2MnO_4^- \rightarrow 9Fe_2O_3 + 2MnO_2 + 2OH^-$

e. $2OH^- + 2MnO_2 + 3H_2O_2 \rightarrow 2MnO_4^- + 4H_2O$

14-5 a. 3   2   8   3   2   2   11

b. 1   40   40   12   2   5

c. 2   27   64   6   2   54   32

d. 3   14   18   14   6   9

14-6 2.54 g   14-7 4372 ml

14-8 83.92 g 100.7 g   14-9 151 g

14-10 11.2 g   14-11 S excess 96.2%

14-12 307 g 25.8% 6340 ml   14-13 8.69%

14-14 19.0 g   14-15 81.1%

15-1 3671.8 C 4.8288 g   15-2 0.83046 g

15-3 3880 sec   15-4 4.99 g

15-5 0.680 g   15-6 1.6021×10^{-19} C

15-7 17.5 ml   15-8 Cu 94.1% Te 78.4%

15-9 96,490 C   15-10 4.1042 g

15-11 0.010447 mg/C   15-12 a. + 2.1223 volts
       0.082909 mg/C   + 0.6251 volt
       0.3043 mg/C   - 1.523 volts
                                               + 0.3588 volt
                                               - 1.1373 volts

15-12 b. $Zn + Cl_2 \rightarrow Zn^{2+} + 2Cl^-$

   $V + Cu^{2+} \rightarrow V^{2+} + Cu$

   $2Ag + Cl_2 \rightarrow 2AgCl$

   $Cd + 2AgCl \rightarrow Cd^{2+} + 2Ag + 2Cl^-$

   $Pb + 2H_3O^+ + SO_4^{2-} \rightarrow PbSO_4 + H_2 + 2H_2O$

15-13 0.958 volt   15-14 0.5738 volt

15-15 a. + 0.56 volt, yes   b. + 0.71 volt, yes

   c. + 0.82 volt, yes   d. - 0.486 volt, no

   e. - 0.935 volt, no

15-16 0.731 volt   15-17 - 0.414 volt

15-18 5.0×10^{71} 1.3×10^{21} 15-19 a. $\frac{1}{2}S_2(g, 1\ atm) + 2e^- \rightleftarrows S^{2-}$

   3.5×10^{-52} 1.3×10^{12} b. $2Ag \overset{\circ}{\rightleftarrows} 2Ag^+ + 2e^-$

   3.8×10^{-39} c. $2Ag(s) + \frac{1}{2}S_2(g, 1\ atm) \rightleftarrows$

                                                              Ag_2S

                                                       d. $E = E° - \frac{RT}{nF} \ln 1/p_{S_2}^{\frac{1}{2}}$

                                                       e. at 490°C slope = 0.0378

15-20 a. $2Ag^+ + O^{2-} + \frac{1}{2}O_2 + 2Ag$

   b. $E = E° - \frac{RT}{2F} \ln \frac{P_{O_2}^{\frac{1}{2}}}{[O^{2-}]}$

   c. yes slope = - 0.0289 V

   d. 0.6167 V

15-21 4.8274 g   15-22 0.02000 F

15-23 96,475 C   15-24 1990 sec

15-25 0.603 liters   15-26 2.27 g

15-27 3,6665 C   15-28 3,745.0 C

15-29 229 C   15-30 CuTe; 76.5%

15-31 0.0454 volt   15-32 0.115 volt

15-33   a.   no, $2CuF \rightleftarrows CuF_2 + Cu$,   + 0.368 volt

        b.   no, $2FeI_3 \rightleftarrows 2FeI_2 + I_2$,   + 0.235 volt

        c.   no, $TlI_3 \rightleftarrows TlI + I_2$,   + 0.71 volt

        d.   yes, $2AgF_2 \rightleftarrows 2AgF + F_2$,   -0.89 volt

15-34   -0.391 volt   15-35   0.2214 volt

15-36   0.269 volt

15-37   a. 5.7×10^{56} b. 9.2×10^{59} c. 1.3×10^{83}

        d. 3.8×10^{-17} e. 2.6×10^{-32}

15-38   a. $2Ag^+ + O^{2-} \rightleftarrows 2Ag + \tfrac{1}{2}O_2$

        b. $E = E° - \dfrac{RT}{2F} \ln \dfrac{P^{\frac{1}{2}}_{O_2}}{[O^{2-}]}$

        c.   yes, slope = 0.0583

        d.   0.6192 V

16-1   0.247 cal/deg-g   16-2   18.92 cal/deg-mole

16-3   77.8 cal/deg-mole   16-4   548.7 cal/mole

16-5   - 1.83 cal/deg   16-6   0.429 eu

16-7   0.52 eu   16-8   0.346 eu

16-9   1.38 eu/mole   16-10   0.81 eu/mole

16-11   1   16-12   a.   0.70 eu/mole C

                                           b.   4.74 eu/mole H_2

                                         c.   - 20.74 eu/mole CO

                                       d.   - 19.88 eu/mole S

                                       e.   - 22.7 eu/mole SO_2

                                         f.   - 10.6 eu/mole H_2

16-13   17.05 eu/mole

16-14   a. $\Delta G° = $ -283 kcal/mole B_2O_3 spontaneous

        b. $\Delta G° = $ -314.08 kcal/mole C_2H_4 spontaneous

        c. $\Delta G° = $ -318.18 kcal/mole C_2H_4 spontaneous

d. $\Delta G°$ = -227.83 kcal/mole C_2H_6 spontaneous
e. $\Delta G°$ = -136.72 kcal/mole CaO spontaneous
f. $\Delta G°$ = 51.98 kcal/mole HF nonspontaneous

16-15 - 261.3 kcal/mole

16-16 - 22.72 kcal/mole

16-17 - 9350 cals/mole

16-18 3452 cal/mole

16-19 - 20.76 kcal/mole

16-20 - 14.156 kcal/mole

16-21 1.032 V

16-22 0.306 eu/g

16-23 9.87 eu/mole

16-24 66.76 eu/mole

16-25 0.215 eu

16-26 58.56 eu/mole

16-27 8.49 eu

16-28 ΔS system = 8.49 eu
ΔS surr. = 0
ΔS universe = 8.49 eu

16-29 0

16-30 2

16-31 - 45.40 kcal/mole

16-32 7.65 kcal/mole

16-33 -10.22 kcal/mole

16-34 - 8.77 kcal

16-35 - 265.3 cal/mole

16-36 - 2.324 kcal/mole

16-37 - 22.59 kcal/mole

16-38 - 18.14 kcal/mole

16-39 - 58.71 kcal/mole

17-1 0.268 min^{-1}

17-2 0.271 min^{-1}

17-3 0.0152 min^{-1}

17-4 0.08145 min^{-1}

17-5 24.7 x 10^{-4} sec^{-1}

17-6 1.7 x 10^{-4} sec^{-1}(from graph)

17-7 2.16 x 10^3 sec

17-8 2.18 x 10^{-4} sec^{-1}
3.18 x 10^3 sec

17-9 first order
6.0 x 10^{-4} sec^{-1}
(from graph)

17-10 1165 sec 11.7%

17-11 1.32 x 10^{-3} mole/liter

17-12 1.50 x 10^{-2} mole/liter

17-13 0.10 sec^{-1} 17-14 $\dfrac{1}{1.8 \times 10^{-4}\ a}$

17-15 0.30 $M^{-1}\ min^{-1}$ 17-16 second order

17-17 $6.61 \times 10^{-2}\ M^{-1}\ sec^{-1}$ 17-18 3.06 $M^{-1}\ sec^{-1}$

17-19 a. $2.90 \times 10^{-3}\ min^{-1}$

      b. 0.140 $M^{-1}\ min^{-1}$ (Divide a. by 0.02 M to get b.)

17-20 a. second order   b.   second order   c.   fourth order

17-21 $5.69 \times 10^{-4}\ sec^{-1}$ 17-22 0.2688 min^{-1}

17-23 $1.44 \times 10^{-4}\ sec^{-1}$ 17-24 $3.30 \times 10^{-4}\ sec^{-1}$

17-25 $5.93 \times 10^{-4}\ sec^{-1}$ 17-26 $1.65 \times 10^{-4}\ sec^{-1}$

17-26 $6.93 \times 10^{-3}\ min^{-1}$ 17-28 10.5 sec^{-1}

      $6.22 \times 10^{-3}\ min^{-1}$

17-29 2.33×10^{-5} mole/g   17-30 3466 sec 78.7%

17-31 4.90 moles/liter   17-32 7.14×10^{-3} mole/liter

17-33 rate = $k[ClO_2^-][HSO_5^-]$ 17-34 0.0493 0.0481

17-35 11.7 $M^{-1}\ min^{-1}$ 17-36 1.24×10^{-3} mole/liter; I^-
      second order

17-37 8.25 sec

18-1   a.   tetraamminecopper(II) ion
      b.   potassium hexafluoronickelate(II)
      c.   tetrathiocyanatocolbaltate(II) ion
      d.   tetraisothiocyanatocolbaltate(II) ion
      e.   tetracarbonylnickel (0)
      f.   tris(ethylenediamine)cobalt(III) sulfate
      g.   potassium trichlorotrifluoroplatinate(IV)
      h.   triamminetrichlorocobalt(III)

18-2 a. $[Cd(NH_3)_4]$ $(NO_3)_2$ b. $[Ni(NH_3)_6]$ Cl_2

c. $Cr(CO)_6$ d. $[Co(NH_3)_4Cl_2]$ NO_3

e. Na_2 $[PtCl_2F_2]$ f. $[Rh(en)_3]Cl_3$

g. $[Cr(NH_3)_6]$ $[Cu(Cl)_5]$

18-3 a. trans-dibromodichloroplatinate(II) ion

b. trans-triamminetrichlorocobalt(III)

c. trans-diamminebis(ethylenediamine)cobalt(III) ion

18-4

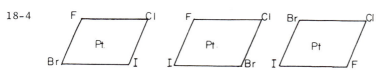

18-5 a. hexaamminechromium(III) pentachlorocuprate(II)

b. hexamminechromium(III) pentacyanonickelate(II)

c. hexaaminecobalt(III) hexachloroferrate(III)

18-6 potassium hexathiocyanatoniobate(V)

$K[Nb(SCN)_6]$

18-7 diamminecobalt(II) nitrate

$[Co(NH_3)_2]$ $(NO_3)_2$

18-8 90%   18-9   42.9%

18-10 $K[Nb(SCN)_6]$ potassium hexathiocyanatoniobate(V)

18-11 ⥮ ⥮ ↑ |__ ____| d^2sp^3 octahedron

18-12 a. low spin   b. low spin   c. low spin

d. low spin   e. high spin

18-13 a. ↑__ |__ ____| d^2sp^3 octahedron

b. e_g __

t_{2g} ↑__   = 20,300 cm^{-1}

c. λ (absorbed) = 4926 Å

18-14 $CoCl_2^{2-}$ > $CoBr_4^{2-}$ > CoI_4^{2-}

18-15   a.   tetraamminecadmium(II) ion

       b.   dicyanoargentate(I) ion

       c.   diamminetetrachlorocobaltate(III) ion

       d.   potassium octacyanomolybdate(IV)

       e.   pentacarbonyliron(0)

       f.   pentaammineaquocobalt(III) perrhenate

       g.   pentaamminechloroiridium(III) chloride

       h.   bis(ethylenediamine)dichlororuthenium(III) chloride

       i.   tris(ethylenediamine)chromium(III) pentacyanonickelate(II)

18-16   a. $[Co(NH_3)_4 ClNO_2]$ Cl

       b. $[Co(NH_3)_4 Cl_2]NO_2$

       c. $Co(CO)_3 NO$

       d. $K_2Zn(CN)_4$

       e. $[Ru(NH_3)_5Cl]Cl_2$

       f. $[Cr(en)_3]_2 (SO_4)_3$

       g. $K_4[Mo(CN)_8]$

18-17   a.   cis-dibromodichloroplatinate(IV) ion

       b.   cis-triamminetrichlorocobalt(III)

       c.   cis-diamminebis(ethylenediamine)cobalt(III) ion

18-18   a.   pentaammineaquocobalt(III) trioxalatocobaltate(III)

       b.   pentaammineaquocobalt(II) diamminetetrathiocyanato-
                                    chromate(II)

       c.   pentaammineaquocobalt(II) diamminetetranitrocobaltate(II)
       or
       pentaammineaquocobalt(II) diamminetetranitritocobaltate(I

18-19 $CoC_6H_{20}N_7O_7$

18-20 $K[Ta(SCN)_6]$ potassium hexathiocyanatotantalate(V)

18-21   a. $C_{12}H_{24}CoN_6S_4$

       b.   tetramethylammonium tetrathiocyanatocobaltate(II)
       $[(CH_3)_4N]_2Co(SCN)_4$

       c.   87.5%

18-22   70.4%   18-23   92.6

18-24 4 high spin 0 low spin

18-25 hexaaquoruthenium(II) ion
    tris(ethylenediamine)ruthenium(II) ion
    bis(propylenediamine)ruthenium(II) ion
    bis(diethylenediamine)ruthenium(II) ion
    hexaammineruthenium(II) ion
    $[Ru(dien)_2]^{2+} > [Ru(pn)_3]^{2+} > [Ru(en)_3]^{2+} > [Ru(NH_3)]^{2+} > [Ru(H_2O)_6]^{2+}$

19-1 1.0080   19-2 8.62×10^6 C

19-3 0.556 atm CO and H_2O 19-4 2.62
     0.444 atm CO_2 and H_2

19-5 9820 ml   19-6 76 kcal/mole

19-7 30.3439 g   19-8 111 kcal/mole   119 kcal/mole
                        102 kcal/mole

19-9 96.4   19-10 0.0364 M
                        1.44

19-11 165.5   19-12 a. +0.2223 volt b. 0.3936V

19-13 2.01   19-14 0.511 V

19-15 2.96×10^{-2} atm   19-16 15.9994

19-17 1.60×10^{-3} mole Cu 19-18 70.8%
     7.95×10^{-4} mole O_2

19-19 3.92×10^{-6} atm   19-20 $CaCr_2O_7 \cdot 3H_2O$
     1.96×10^{-6} atm

19-21 -46.96   19-22 45.9 kcal/mole

19-23 -54.6 kcal   19-24 0.458 volt

19-25 1.04 liters   19-26 0.75 atm

19-27 +0.82 volt   19-28 1.000 g

19-29 WO_3 19-30 0.2888 g N_2 0.3298 g O_2

19-31 6.939   19-32 $(C_2H_5Li)_6$

19-33 10.646 g   19-34 -13.60 kcal/mole

19-35   -21.34 kcal/mole   19-36   40.09

19-37   320 ml   19-38 1.3×10^{-9} mole/liter

19-39   -28 kcal   19-40   0.727 liters

19-41   10.80   19-42 BCl_3

19-43   250   19-44   28.1

19-45 $B_{18}H_{22}$ 19-46   BN

19-47   67.2   19-48   Plane triangle

19-49 $BF_3 \cdot NH_3$ 19-50 B_2Cl_4

19-51   13.6 cc H_2 19-52 $2Ga + 3I_2 \rightarrow 2GaI_3$
                                        $2GaI_3 + O_2 \rightarrow 2GaOI + 2I$
                                        $GaOI + 3F_2 \rightarrow GaOF + IF_5$

19-53   -117.9 kcal/mole AlF_3 19-54   0.936 atm due to dimer

19-55   Yes   19-56 SiS_2

19-57 SiI_4 $GeCl_4$ tetrahedra   19-58   a.   NaF   b.   81.4%

                                        c. $:\overset{..}{\underset{..}{F}}-C=\overset{..}{\underset{}{O}}:$

                                        d. $\underset{}{\overset{F}{\diagdown}}C=O$

19-59   O   C   C   C   O   19-60   Plane triangle

                                        $H-C=O$ Plane
                                        $\quad\quad|$ triangle
                                        $\quad\quad F$

19-61   62.5   19-62   114 ml

19-63   2.20 g   19-64   Plane triangle

                                        $\overset{F}{\diagdown}C=S$ Plane
                                        $\underset{F}{\diagup}$ triangle

19-65   28.3 g   19-66   0.45 kcal/g atom

19-67   0.473 volt   19-68   Tetrahedron, octahedron

19-69 118.69

19-70 1.78×10^{-4} M

19-71 14.007

19-72 124.1

19-73 78.4

19-74 Trigonal pyramid

19-75 $3NH_2OH \rightarrow NH_3 + N_2 + 3H_2O$

19-76 $N_3H_{10}Cl_3$

19-77 a. $2NF_3 + Cu \rightarrow$
                  $CuF_2 + N_2F_4$
       b. 10.2 g

19-78 Yes

19-79 NF_3

19-80 a. 71.3   b. NF_3

19-81

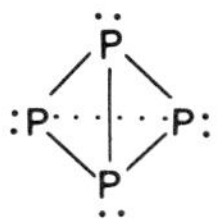

19-82 PIF_2 Trigonal pyramid

19-83 a. P_2F_4 b.

       c. two tetrahedra   d. $2PIF_2 + Hg \rightarrow HgI_2 + P_2F_4$

19-84 69%

19-85 75.5% CaF_2

19-86 Octahedron

19-87 Electron pair repulsion of basal electron pairs and apical electron pairs.

19-88 $AsOF_3$

19-89 126

19-90 14.053

19-91 420 mm

19-92 14.4 M 0.523

19-93 SF_2

19-94 V shape

19-95 a. SeF_5OF b. Octahedron

19-96 P_4S_3

19-97 32.07

19-98 1.21×10^{-22} M

19-99 Cu^{2+} 6.9×10^{-23} M
       Zn^{2+} 0.1 M

19-100 SO_3F_2

19-101 CF_3SF_5

19-102 S_4N_3Cl

19-103 $2P_4S_3 + 6S \rightarrow P_4S_5 + P_4S_7$

19-104 S_8 19-105 TeO_2

19-106   79.926   19-107   2.58 g

19-108 2.16×10^{-3} M   19-109   0.15 volt

19-110 0.19 volt   19-111 1240 mm

19-112 1.303×10^{-3} 19-113 9.33×10^{-2} M

19-114   T shape   19-115

$$:\overset{..}{O}::\overset{..}{O}:$$
$$:\overset{..}{O}:\overset{..}{I}::\overset{..}{I}:\overset{..}{O}:$$

19-116   Square-based pyramid   19-117   1600 mm
         Pentagonal bipyramid

19-118   94.7 mm   19-119   a. Square pyramid
                                                   b. Repulsion of
                                                       unshared pair

19-120   562.1 mm   19-121 IF_5

19-122   0.0547 M   19-123   V shape

19-124   -1126 cal   19-125   371 cal

19-126 XeF_4 19-127   Yes

19-128   Square plane   19-129   31.8 kcal/mole

19-130   39.964   19-131 7.52×10^{-4} g/ml

19-132 XeF_6 19-133 XeF_6

19-134   Yes   19-135   b. Distorted octahedron

19-136 Ba_3XeO_6 19-137   0.266 g

19-138 XeF_6 19-139   V shape

19-140   Trigonal pyramid   19-141 $XePtF_6$

19-142   0.2106 g   0.1924 g   19-143   245.4

19-144   30.6 kcal/mole   19-145   10.0 kcal/mole

19-146   5.40 g   19-147   0.7772 volt

19-148 166 ml

19-149 3.6×10^{-9} mole/liter

19-150 52 kcal/mole

19-151 $8HCl + K_2TeO_4 \rightarrow$
$\quad Cl_2 + TeCl_4 + 4H_2O + 2KCl$

19-152 0.073 volt

19-153 +0.490 volt

19-154 2.85×10^{-4} M

19-155 4.01 g

19-156 288

19-157 a. $HgCl_2$ molecule
b. Covalent
c. Linear

19-158 CdI_2 molecules

19-159 V shape

19-160 0.627

19-161 0.513 volts

19-162 -0.351 volt

19-163 200.60

19-164 1.70×10^{-16} M

19-165 0.756 g

19-166 $CrPO_4 \cdot 6H_2O$

19-167 $CrPO_4$

19-168 $FeCl_3$

19-169 $NiFe_2O_4$

19-170 NbF_5

19-171 8.3265 g

19-172 47.89

19-173 0.909 volt

19-174 O_2PtF_6

19-175 0.55 volt

19-176 -0.18

19-177 -1.182 volt

19-178 a. $14H^+ + Cr_2O_7^{2-} + 6I^- \rightarrow 2Cr^{3+} + 3I_2 + 7H_2O$
b. $4H^+ + 2CrO_4^{2-} + 3SO_2 \rightarrow 2Cr^{3+} + 3SO_4^{2-} + 2H_2O$
c. $2MnO_4^- + 3H_2S \rightarrow 2MnO_2 + 3S + 2H_2O + 2OH^-$
d. $6H^+ + 2MnO_4^- + 5H_2S \rightarrow 2Mn^{2+} + 5S + 8H_2O$
e. $4Fe^{2+} + 2H_2O + O_2 \rightarrow 4Fe^{3+} + 4OH^-$

19-179 1.31×10^{-4} M

19-180 3.06×10^3 M
3.7×10^{-17} M

19-181 50.942

19-182 $FeCl_3$

19-183 Fe_2Cl_6

19-184 -36.9 kcal/mole

19-185 -236.6 kcal/mole

19-186 11.54 F

19-187   0.423 g   19-188   0.1 mole $KMnO_4$

                                             0.1 mole $KHSO_4$

                                             0.5 mole H_2

19-189   5.0 ml   19-190   69.1 ml

19-191   a. $4Co^{2+} + 24\ NH_3 + O_2 + 2H_2O \rightarrow 4Co(NH_3)_6^{3+} + 4OH^-$

         b. $2[Fe(CN)_6]^{4-} + Cl_2 \rightarrow 2[Fe(CN)_6]^{3-} + 2Cl^-$

         c. $2Cr(OH)_3 + 3Na_2O_2 \rightarrow 2Na_2CrO_4 + 2H_2O + 2NaOH$

19-192 $YCl_3 \cdot 6H_2O$ 19-193 Cr_3C_2

19-194   71 hr., 24 min   19-195   5.23 liters

19-196 $3UO_2 + O_2 \rightarrow U_3O_8$ 19-197 $Pa_2O_5 + 5C + 4Br_2 \rightarrow$

                                                 $PaBr_5 + PaBr_3 + 5CO$

19-198   -24.972 kcal/mole   19-199 PuF_6

19-200   0.336 liters

Appendix D

TABLES

TABLE D-1A General Physical Constants*

Constant	Value
Avogadro's number	$6.022 \times 10^{23} \ mol^{-1}$
Atomic mass unit	$1.660 \times 10^{-27} \ kg$
Boltzmann constant	$1.38 \times 10^{23} \ J \ K^{-1}$
Charge on electron	$1.60 \times 10^{-19} \ C$
Faraday	$96,486.70 \ C \ mol^{-1}$
Gas constant	$8.314 \ J \ K^{-1} \ mol^{-1}$
Planck's constant	$6.626 \times 10^{-34} \ J \ S$
Speed of light in vacuum	$3.00 \times 10^{8} \ ms^{-1}$

*Most are taken from NBS reprint in J. Chem. Ed., 48, 569 (1971)

TABLE D-1B Vapor Pressure of Water

Temp., °C	Vapor pressure, torr
0	4.579
5	4.750
10	9.209
15	12.788
20	17.535
25	23.756
30	31.824

TABLE D-2 Four-place Common Logarithms

Natural numbers	0	1	2	3	4	5	6	7	8	9	Proportional parts								
											1	2	3	4	5	6	7	8	9
10	0000	0043	0086	0128	0170	0212	0253	0294	0334	0374	4	8	12	17	21	25	29	33	37
11	0414	0453	0492	0531	0569	0607	0645	0682	0719	0755	4	8	11	15	19	23	26	30	34
12	0792	0828	0864	0899	0934	0969	1004	1038	1072	1106	3	7	10	14	17	21	24	28	31
13	1139	1173	1206	1239	1271	1303	1335	1367	1399	1430	3	6	10	13	16	19	23	26	29
14	1461	1492	1523	1553	1584	1614	1644	1673	1703	1732	3	6	9	12	15	18	21	24	27
15	1761	1790	1818	1847	1875	1903	1931	1959	1987	2014	3	6	8	11	14	17	20	22	25
16	2041	2068	2095	2122	2148	2175	2201	2227	2253	2279	3	5	8	11	13	16	18	21	24
17	2304	2330	2355	2380	2405	2430	2455	2480	2504	2529	2	5	7	10	12	15	17	20	22
18	2553	2577	2601	2625	2648	2672	2695	2718	2742	2765	2	5	7	9	12	14	16	19	21
19	2788	2810	2833	2856	2878	2900	2923	2945	2967	2989	2	4	7	9	11	13	16	18	20
20	3010	3032	3054	3075	3096	3118	3139	3160	3181	3201	2	4	6	8	11	13	15	17	19
21	3222	3243	3263	3284	3304	3324	3345	3365	3385	3404	2	4	6	8	10	12	14	16	18
22	3424	3444	3464	3483	3502	3522	3541	3560	3579	3598	2	4	6	8	10	12	14	15	17
23	3617	3636	3655	3674	3692	3711	3729	3747	3766	3784	2	4	6	7	9	11	13	15	17
24	3802	3820	3838	3856	3874	3892	3909	3927	3945	3962	2	4	5	7	9	11	12	14	16
25	3979	3997	4014	4031	4048	4065	4082	4099	4116	4133	2	3	5	7	9	10	12	14	15
26	4150	4166	4183	4200	4216	4232	4249	4265	4281	4298	2	3	5	7	8	10	11	13	15
27	4314	4330	4346	4362	4378	4393	4409	4425	4440	4456	2	3	5	6	8	9	11	13	14
28	4472	4487	4502	4518	4533	4548	4564	4579	4594	4609	2	3	5	6	8	9	11	12	14
29	4624	4639	4654	4669	4683	4698	4713	4728	4742	4757	1	3	4	6	7	9	10	12	13
30	4771	4786	4800	4814	4829	4843	4857	4871	4886	4900	1	3	4	6	7	9	10	11	13
31	4914	4928	4942	4955	4969	4983	4997	5011	5024	5038	1	3	4	6	7	8	10	11	12
32	5051	5065	5079	5092	5105	5119	5132	5145	5159	5172	1	3	4	5	7	8	9	11	12
33	5185	5198	5211	5224	5237	5250	5263	5276	5289	5302	1	3	4	5	6	8	9	10	12
34	5315	5328	5340	5353	5366	5378	5391	5403	5416	5428	1	3	4	5	6	8	9	10	11
35	5441	5453	5465	5478	5490	5502	5514	5527	5539	5551	1	2	4	5	6	7	9	10	11
36	5563	5575	5587	5599	5611	5623	5635	5647	5658	5670	1	2	4	5	6	7	8	10	11
37	5682	5694	5705	5717	5729	5740	5752	5763	5775	5786	1	2	3	5	6	7	8	9	10
38	5798	5809	5821	5832	5843	5855	5866	5877	5888	5899	1	2	3	5	6	7	8	9	10
39	5911	5922	5933	5944	5955	5966	5977	5988	5999	6010	1	2	3	4	5	7	8	9	10
40	6021	6031	6042	6053	6064	6075	6085	6096	6107	6117	1	2	3	4	5	6	8	9	10
41	6128	6138	6149	6160	6170	6180	6191	6201	6212	6222	1	2	3	4	5	6	7	8	9
42	6232	6243	6253	6263	6274	6284	6294	6304	6314	6325	1	2	3	4	5	6	7	8	9
43	6335	6345	6355	6365	6375	6385	6395	6405	6415	6425	1	2	3	4	5	6	7	8	9
44	6435	6444	6454	6464	6474	6484	6493	6503	6513	6522	1	2	3	4	5	6	7	8	9
45	6532	6542	6551	6561	6571	6580	6590	6599	6609	6618	1	2	3	4	5	6	7	8	9
46	6628	6637	6646	6656	6665	6675	6684	6693	6702	6712	1	2	3	4	5	6	7	7	8
47	6721	6730	6739	6749	6758	6767	6776	6785	6794	6803	1	2	3	4	5	5	6	7	8
48	6812	6821	6830	6839	6848	6857	6866	6875	6884	6893	1	2	3	4	4	5	6	7	8
49	6902	6911	6920	6928	6937	6946	6955	6964	6972	6981	1	2	3	4	4	5	6	7	8
50	6990	6998	7007	7016	7024	7033	7042	7050	7059	7067	1	2	3	3	4	5	6	7	8
51	7076	7084	7093	7101	7110	7118	7126	7135	7143	7152	1	2	3	3	4	5	6	7	8
52	7160	7168	7177	7185	7193	7202	7210	7218	7226	7235	1	2	2	3	4	5	6	7	7
53	7243	7251	7259	7267	7275	7284	7292	7300	7308	7316	1	2	2	3	4	5	6	6	7
54	7324	7332	7340	7348	7356	7364	7372	7380	7388	7396	1	2	2	3	4	5	6	6	7

TABLE D-2 Four-place Common Logarithms

Natural numbers	0	1	2	3	4	5	6	7	8	9	1	2	3	4	5	6	7	8	9
											Proportional parts								
55	7404	7412	7419	7427	7435	7443	7451	7459	7466	7474	1	2	2	3	4	5	5	6	7
56	7482	7490	7497	7505	7513	7520	7528	7536	7543	7551	1	2	2	3	4	5	5	6	7
57	7559	7566	7574	7582	7589	7597	7604	7612	7619	7627	1	2	2	3	4	5	5	6	7
58	7634	7642	7649	7657	7664	7672	7679	7686	7694	7701	1	1	2	3	4	4	5	6	7
59	7709	7716	7723	7731	7738	7745	7752	7760	7767	7774	1	1	2	3	4	4	5	6	7
60	7782	7789	7796	7803	7810	7818	7825	7832	7839	7846	1	1	2	3	4	4	5	6	6
61	7853	7860	7868	7875	7882	7889	7896	7903	7910	7917	1	1	2	3	4	4	5	6	6
62	7924	7931	7938	7945	7952	7959	7966	7973	7980	7987	1	1	2	3	3	4	5	6	6
63	7993	8000	8007	8014	8021	8028	8035	8041	8048	8055	1	1	2	3	3	4	5	5	6
64	8062	8069	8075	8082	8089	8096	8102	8109	8116	8122	1	1	2	3	3	4	5	5	6
65	8129	8136	8142	8149	8156	8162	8169	8176	8182	8189	1	1	2	3	3	4	5	5	6
66	8195	8202	8209	8215	8222	8228	8235	8241	8248	8254	1	1	2	3	3	4	5	5	6
67	8261	8267	8274	8280	8287	8293	8299	8306	8312	8319	1	1	2	3	3	4	5	5	6
68	8325	8331	8338	8344	8351	8357	8363	8370	8376	8382	1	1	2	3	3	4	4	5	6
69	8388	8395	8401	8407	8414	8420	8426	8432	8439	8445	1	1	2	2	3	4	4	5	6
70	8451	8457	8463	8470	8476	8482	8488	8494	8500	8506	1	1	2	2	3	4	4	5	5
71	8513	8519	8525	8531	8537	8543	8549	8555	8561	8567	1	1	2	2	3	4	4	5	5
72	8573	8579	8585	8591	8597	8603	8609	8615	8621	8627	1	1	2	2	3	4	4	5	5
73	8633	8639	8645	8651	8657	8663	8669	8675	8681	8686	1	1	2	2	3	4	4	5	5
74	8692	8698	8704	8710	8716	8722	8727	8733	8739	8745	1	1	2	2	3	4	4	5	5
75	8751	8756	8762	8768	8774	8779	8785	8791	8797	8802	1	1	2	2	3	3	4	5	5
76	8808	8814	8820	8825	8831	8837	8842	8848	8854	8859	1	1	2	2	3	3	4	5	5
77	8865	8871	8876	8882	8887	8893	8899	8904	8910	8915	1	1	2	2	3	3	4	4	5
78	8921	8927	8932	8938	8943	8949	8954	8960	8965	8971	1	1	2	2	3	3	4	4	5
79	8976	8982	8987	8993	8998	9004	9009	9015	9020	9026	1	1	2	2	3	3	4	4	5
80	9031	9036	9042	9047	9053	9058	9063	9069	9074	9079	1	1	2	2	3	3	4	4	5
81	9085	9090	9096	9101	9106	9112	9117	9122	9128	9133	1	1	2	2	3	3	4	4	5
82	9138	9143	9149	9154	9159	9165	9170	9175	9180	9186	1	1	2	2	3	3	4	4	5
83	9191	9196	9201	9206	9212	9217	9222	9227	9232	9238	1	1	2	2	3	3	4	4	5
84	9243	9248	9253	9258	9263	9269	9274	9279	9284	9289	1	1	2	2	3	3	4	4	5
85	9294	9299	9304	9309	9315	9320	9325	9330	9335	9340	1	1	2	2	3	3	4	4	5
86	9345	9350	9355	9360	9365	9370	9375	9380	9385	9390	1	1	2	2	3	3	4	4	5
87	9395	9400	9405	9410	9415	9420	9425	9430	9435	9440	0	1	1	2	2	3	3	4	4
88	9445	9450	9455	9460	9465	9469	9474	9479	9484	9489	0	1	1	2	2	3	3	4	4
89	9494	9499	9504	9509	9513	9518	9523	9528	9533	9538	0	1	1	2	2	3	3	4	4
90	9542	9547	9552	9557	9562	9566	9571	9576	9581	9586	0	1	1	2	2	3	3	4	4
91	9590	9595	9600	9605	9609	9614	9619	9624	9628	9633	0	1	1	2	2	3	3	4	4
92	9638	9643	9647	9652	9657	9661	9666	9671	9675	9680	0	1	1	2	2	3	3	4	4
93	9685	9689	9694	9699	9703	9708	9713	9717	9722	9727	0	1	1	2	2	3	3	4	4
94	9731	9736	9741	9745	9750	9754	9759	9763	9768	9773	0	1	1	2	2	3	3	4	4
95	9777	9782	9786	9791	9795	9800	9805	9809	9814	9818	0	1	1	2	2	3	3	4	4
96	9823	9827	9832	9836	9841	9845	9850	9854	9859	9863	0	1	1	2	2	3	3	4	4
97	9868	9872	9877	9881	9886	9890	9894	9899	9903	9908	0	1	1	2	2	3	3	4	4
98	9912	9917	9921	9926	9930	9934	9939	9943	9948	9952	0	1	1	2	2	3	3	4	4
99	9956	9961	9965	9969	9974	9978	9983	9987	9991	9996	0	1	1	2	2	3	3	3	4

TABLE D-3 Table of Relative Atomic Weights, 1969
Based on the Atomic Mass of ^{12}C = 12

Name	Symbol	Atomic number	Atomic weights
Actinium	Ac	89	....
Aluminum	Al	13	26.9815
Americium	Am	95	....
Antimony	Sb	51	121.75
Argon	Ar	18	39.948
Arsenic	As	33	74.9216
Astatine	At	85	....
Barium	Ba	56	137.34
Berkelium	Bk	97	....
Beryllium	Be	4	9.0122
Bismuth	Bi	83	208.980
Boron	B	5	10.811
Bromine	Br	35	79.909
Cadmium	Cd	48	112.40
Calcium	Ca	20	40.08
Californium	Cf	98	....
Carbon	C	6	12.01115
Cerium	Ce	58	140.12
Cesium	Cs	55	132.905
Chlorine	Cl	17	35.453
Chromium	Cr	24	51.996
Cobalt	Co	27	58.9332
Copper	Cu	29	63.54
Curium	Cm	96	....
Dysprosium	Dy	66	162.50
Einsteinium	Es	99	....
Erbium	Er	68	167.26
Europium	Eu	63	151.96
Fermium	Fm	100	....
Fluorine	F	9	18.9984
Francium	Fr	87	....
Gadolinium	Gd	64	157.25
Gallium	Ga	31	69.72
Germanium	Ge	32	72.59
Gold	Au	79	196.967
Hafnium	Hf	72	178.49
Helium	He	2	4.0026
Holmium	Ho	67	164.930
Hydrogen	H	1	1.00797
Indium	In	49	114.82
Iodine	I	53	126.9044
Iridium	Ir	77	192.2
Iron	Fe	26	55.847
Krypton	Kr	36	83.80
Lanthanum	La	57	138.91
Lead	Pb	82	207.19
Lithium	Li	3	6.939
Lutetium	Lu	71	174.97
Magnesium	Mg	12	24.312

Name	Symbol	Atomic number	Atomic weights
Manganese	Mn	25	54.9380
Mendelevium	Md	101	....
Mercury	Hg	80	200.59
Molybdenum	Mo	42	95.94
Neodymium	Nd	60	144.24
Neon	Ne	10	20.183
Neptunium	Np	93	....
Nickel	Ni	28	58.71
Niobium	Nb	41	92.906
Nitrogen	N	7	14.0067
Nobelium	No	102	....
Osmium	Os	76	190.2
Oxygen	O	8	15.9994
Palladium	Pd	46	106.4
Phosphorus	P	15	30.9738
Platinum	Pt	78	195.09
Plutonium	Pu	94	....
Polonium	Po	84	....
Potassium	K	19	39.102
Praseodymium	Pr	59	140.907
Promethium	Pm	61	....
Protactinium	Pa	91	....
Radium	Ra	88	....
Radon	Rn	86	....
Rhenium	Re	75	186.2
Rhodium	Rh	45	102.905
Rubidium	Rb	37	85.47
Ruthenium	Ru	44	101.07
Samarium	Sm	62	150.35
Scandium	Sc	21	44.956
Selenium	Se	34	78.96
Silicon	Si	14	28.086
Silver	Ag	47	107.870
Sodium	Na	11	22.9898
Strontium	Sr	38	87.62
Sulfur	S	16	32.064
Tantalum	Ta	73	180.948
Technetium	Tc	43	....
Tellurium	Te	52	127.60
Terbium	Tb	65	158.924
Thallium	Tl	81	204.37
Thorium	Th	90	232.038
Thulium	Tm	69	168.934
Tin	Sn	50	118.69
Titanium	Ti	22	47.90
Tungsten	W	74	183.85
Uranium	U	92	238.03
Vanadium	V	23	50.942
Xenon	Xe	54	131.30
Ytterbium	Yb	70	173.04
Yttrium	Y	39	88.905
Zinc	Zn	30	65.37
Zirconium	Zr	40	91.22

TABLE D- 4 Periodic Table

IA	IIA	IIIB	IVB	VB	VIB	VIIB	VIII	VIII	VIII	IB	IIB	IIIA	IVA	VA	VIA	VIIA	
1 H																	2 He
3 Li	4 Be											5 B	6 C	7 N	8 O	9 F	10 Ne
11 Na	12 Mg											13 Al	14 Si	15 P	16 S	17 Cl	18 Ar
19 K	20 Ca	21 Sc	22 Ti	23 V	24 Cr	25 Mn	26 Fe	27 Co	28 Ni	29 Cu	30 Zn	31 Ga	32 Ge	33 As	34 Se	35 Br	36 Kr
37 Rb	38 Sr	39 Y	40 Zr	41 Nb	42 Mo	43 Tc	44 Ru	45 Rh	46 Pd	47 Ag	48 Cd	49 In	50 Sn	51 Sb	52 Te	53 I	54 Xe
55 Cs	56 Ba	57 La	72 Hf	73 Ta	74 W	75 Re	76 Os	77 Ir	78 Pt	79 Au	80 Hg	81 Tl	82 Pb	83 Bi	84 Po	85 At	86 Rn
87 Fr	88 Ra	89 Ac															

58 Ce	59 Pr	60 Nd	61 Pm	62 Sm	63 Eu	64 Gd	65 Tb	66 Dy	67 Ho	68 Er	69 Tm	70 Yb	71 Lu
90 Th	91 Pa	92 U	93 Np	94 Pu	95 Am	96 Cm	97 Bk	98 Cf	99 Es	100 Fm	101 Md	102 No	103 Lw

INDEX

483

39/405

$$\frac{53.1}{12} = \quad \wedge 4.5 \quad 5.5 \quad = \qquad 1P \quad - \quad C8$$

$$\dot{H}_7$$

$$\frac{6.2}{1} = \quad 6.2 \quad \frac{6.2}{0.85} \quad ⑦\frac{1}{2} \quad 15 \qquad N$$

$$O_2$$

$$\frac{12.4}{14_7} = \qquad \frac{62}{7} = \frac{31.6}{3.5.7} \quad \underline{0.85} \quad ① N \quad 2$$

$$10 \cdot \frac{28.3}{16.} = \underline{1.85} \qquad \frac{1.75}{0.85} = ② \quad o \quad 4$$

$$\frac{8}{4}$$

$$\frac{28.3}{16} = \frac{64.15}{8.}$$

$$\frac{7.08}{4} \qquad \frac{7}{4}$$

$$4 | \overline{17.8} | \overline{76}$$
$$\cdot 20$$
$$\cdot 8$$
$$20$$

$$62$$
$$8.5$$

$$6 \qquad\qquad 4.5$$

$$85 \qquad 51.0 \qquad 5.1 \qquad 0.85$$
$$51.0 \qquad 0.0 \qquad .85 \qquad 7$$
$$\overline{3} \qquad\qquad \overline{5.95} \qquad \overline{5.95}$$
$$.85 \qquad 5^3 \qquad 0.85$$
$$\overline{6.80} \qquad\qquad \times 5$$
$$1 1 \qquad\qquad \overline{425}$$
$$\cdot 40$$
$$\overline{4.65}$$

$$32$$
$$+22$$
$$7$$
$$+14.$$
$$\overline{125}$$
$$1 \qquad ⊘-12.$$